Fat Manifolds and
Linear Connections

**Alessandro De Paris**
University of Naples Federico II, Italy

**Alexandre Vinogradov**
University of Salerno, Italy

# Fat Manifolds and Linear Connections

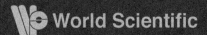

NEW JERSEY · LONDON · SINGAPORE · BEIJING · SHANGHAI · HONG KONG · TAIPEI · CHENNAI

*Published by*

World Scientific Publishing Co. Pte. Ltd.
5 Toh Tuck Link, Singapore 596224
*USA office:* 27 Warren Street, Suite 401-402, Hackensack, NJ 07601
*UK office:* 57 Shelton Street, Covent Garden, London WC2H 9HE

**British Library Cataloguing-in-Publication Data**
A catalogue record for this book is available from the British Library.

Picture of Sorrentine Peninsula taken by Vincenzo Moreno from mount Molare on June 25, 2008

**FAT MANIFOLDS AND LINEAR CONNECTIONS**

Copyright © 2009 by World Scientific Publishing Co. Pte. Ltd.

*All rights reserved. This book, or parts thereof, may not be reproduced in any form or by any means, electronic or mechanical, including photocopying, recording or any information storage and retrieval system now known or to be invented, without written permission from the Publisher.*

For photocopying of material in this volume, please pay a copying fee through the Copyright Clearance Center, Inc., 222 Rosewood Drive, Danvers, MA 01923, USA. In this case permission to photocopy is not required from the publisher.

ISBN-13 978-981-281-904-8
ISBN-10 981-281-904-5

Printed in Singapore.

# Preface

In the winter 1999-2000 the second author held a series of lectures on linear connections and gauge transformations. Notes of these lectures taken by G. Manno, F. Pugliese, L. Vitagliano and the first author constituted that raw material on the basis of which he wrote the first version of this text. This version was then substantially revised and new material added. Various parts of these notes were circulated among 'diffiety people' while we were working on them, and the feed back contributed to the final version. Luca Vitagliano and our friend and colleague G. Rotondaro, who tragically left us in October 2004, should be especially mentioned in this connection.

We highly acknowledge the support of this work by MIUR (Ministero dell'Università e della Ricerca), the Istituto Nazionale di Fisica Nucleare (Italy) and the Istituto Italiano per gli Studi Filosofici.

Scauri, S. Stefano del Sole,   *Alessandro De Paris*
October, 2008   *Alexandre M. Vinogradov*

# Foreword

The idea of parallel transport along a path in a Riemannian manifold gave birth to the concept of a linear connection on $M$ at the end of 19th century. Subsequently, it was extended to arbitrary vector bundles and much later, at the time of the Second War, to general bundles. According to the now standard approach, which is mainly due to Ch. Ehresmann, a connection in a fiber bundle is just a distribution of 'horizontal planes' on its total space. Duly specified to various types of fiber bundles this approach leads to connections of a particular interest, such as affine or linear. Geometrical clarity and apparent simplicity is an important advantage of Ehresmann's approach, which, unfortunately, is well balanced by a not negligible disadvantage. Namely, it gives almost no constructive indications on the operative machinery to work with. In particular, it does not allow an immediate natural extension of the theory to some recently emerged situations of a noteworthy importance such as supermanifolds (graded commutative algebras) or secondary calculus (see [Vinogradov (2001)]). Indeed, it would be hardly possible even to imagine what is a secondary ('quantized') connection in terms of a distribution of horizontal planes. Moreover, in field theory one deals directly with fields which may be, or not be interpreted as sections of a vector bundle but not with the bundle as such. So, in this context a connection *must* be defined as a construction which is pertinent to the fields 'in person'. This kind considerations and the fact that differential calculus is, in reality, an aspect of commutative algebra (see [Nestruev]) plainly indicate that a natural framework for the theory of linear connections is differential calculus in the category of modules over a (graded) commutative ground algebra. This point of view combines naturally with the idea to treat a vector bundle as a 'fat' manifold composed of 'fat' points that are its fibers. By using the term 'fat point' we refer to an object possessing an 'inner structure' whose constituents, nevertheless, cannot be directly observed, *i.e.*, something like an elementary particle. In the theory of gauge fields one deals, as a matter of fact, with fat points. In this context unobservability of the constituents is formalized by means of a suitable symmetry group that produce the necessary inseparable mixture.

These and other similar considerations leads to suppose existence of a 'fat' analogue of differential calculus on a fat manifold well adopted to treat various questions concerning a given vector bundle(s) and, in particular, connections in it. Such an analogue positively exists and the gauge freedom is an inherent feature of it. On the other hand, connections in the context of this 'fat' calculus play the role of a mechanism naturally effecting interrelations among fat points.

In these notes we present some basic elements of the fat calculus and then, on its basis, develop the theory of linear connections. In a sense this text may be viewed as a translation of the classical theory of linear connections in smooth vector bundles into its native language. An extension of the domain of the theory of linear connections much beyond its traditional differential geometry frames is one of results of this translation. For instance, this way one discovers that families of vector spaces over a smooth manifold different from vector bundles can also possess connections as well as vector bundles over manifolds with singularities. Another advantage of this new language is that it simplifies noteworthy working techniques and manipulations with connections by offering simple algebraic computations as a substitute for non infrequently ponderous geometrical constructions. In addition, it makes much easier to perceive more delicate aspects of the theory. An instance of that is the notion of compatibility of two connections along a morphism of vector bundles, introduced and studied in these notes for the first time.

These notes are structured along the following lines. The introductive zeroth chapter contains an algebraic interpretation of some basic facts of differential calculus on smooth manifolds that are brought to the form allowing a direct 'fat' generalization. Materials gathered in this chapter make the subsequent exposition self-contained and accessible for graduate students.

Fat manifolds and first elements of 'fat calculus' are introduced and discussed in the 1-st chapter. A fat manifold is simply a pair composed of a smooth manifold and a vector bundle on it. This notion, synonymous by itself to that of vector bundle, acquires, nevertheless, a new meaning in the context of fat calculus. This subtle but important difference is similar to that between 'just a particle' and a charged particle. A general algebraic counterpart of fat manifolds is a pair composed of a commutative algebra and a module over it. A good deal of fat calculus can be developed in this algebraic context and we do that as much as possible. In the 1-st chapter we discuss only simplest elements of fat calculus such as fat tangent vectors, fat vector fields, *etc.*, simultaneously, with their algebraic counterparts. Other fat notions are introduced as required in the course the exposition.

A fat manifold may be viewed as the result of a 'thickening' of the underlying ordinary manifold, say, $M$. A natural question is whether this thickening can be extended to other geometrical structures on $M$. In particular, the problem of a simultaneous thickening of vector fields on $M$ leads to discover the notion of a linear connection in the corresponding vector bundle. In chapter 2 the theory of linear connections is build on the basis of this idea. The main tools in doing that are fat differential calculus on $M$ and its algebraic counterpart. Among other things, here we construct some exotic examples of connections already mentioned above and describe basic operations of linear algebra with connections.

More fine elements of the theory of connections are developed in the 3-rd chapter. Covariant differential, duly interpreted, is the conceptual center of our exposition here. In particular, we show that a connection can be understood as a *cd-module structure* in the graded algebra of *thickened differential forms*. This fact makes possible to introduce the concept of *compatibility* of two connections and the concept of a *connection along a fat map*. From one side, this enriches the standard theory of connections with morphisms and relative objects and, from the other side, allows to develop a more satisfactory theory of the covariant Lie derivation.

The covariant differential of a flat connection transforms the algebra of thickened form into a complex. This kind of cohomology is studied at the beginning of the concluding 4-th chapter. The main result here is the fat homotopy formula, which is surprisingly valid even for cd-modules. As a curiosity we show that the parallel translation along a curve is described naturally by the 'fat Newton-Leibniz formula'.

A cd-module associated with a connection is not, generally, a complex. Nevertheless, there are naturally related with it differential complexes furnishing connections with cohomological invariants. We interpret Maxwell's equations as dynamics of gauge equivalence classes of connections over the fat Minkowski space-time in order to illustrate importance of this aspect in the theory of connections. The theory of characteristic classes of gauge structures is the final accord of these notes. Indeed, many elements of the previously developed theory are here shown in a common action.

Linear connections appear naturally in many areas of mathematics by starting from abstract algebra and up to mathematical physics. A finite separable extension of an algebraic field is supplied canonically with a flat connection. This elementary fact is easily seen from the point of view presented in these notes. On the other hand, the cohomology of the associated with this connection de Rham like complex is an invariant of the extension and a natural question is what are these and, in particular, how to compute them. Some hints about one can extract from differential geometry where flat connection cohomology appears as the de Rham cohomology with 'twisted coefficients'. Moreover, this kind cohomology appears in some situations in (physical) field theory, *etc.* This simple example illustrates why a unified point of view on connections could be of interest and our hopes are that these notes would be useful as a reference point to the subject.

# Contents

*Preface*     v

*Foreword*     vii

0. Elements of Differential Calculus over Commutative Algebras     1
   - 0.1 Algebraic Tools . . . . . . . . . . . . . . . . . . . . . . . 1
   - 0.2 Smooth Manifolds . . . . . . . . . . . . . . . . . . . . . . 7
   - 0.3 Vector Bundles . . . . . . . . . . . . . . . . . . . . . . . 23
   - 0.4 Vector Fields . . . . . . . . . . . . . . . . . . . . . . . . 41
   - 0.5 Differential Forms . . . . . . . . . . . . . . . . . . . . . 52
   - 0.6 Lie Derivative . . . . . . . . . . . . . . . . . . . . . . . 73

1. Basic Differential Calculus on Fat Manifolds     91
   - 1.1 Basic Definitions . . . . . . . . . . . . . . . . . . . . . . 92
   - 1.2 The Lie Algebra of Der-operators . . . . . . . . . . . . . 98
   - 1.3 Fat Vector Fields . . . . . . . . . . . . . . . . . . . . . . 104
   - 1.4 Fat Fields and Vector Fields on the Total Space . . . . . 110
   - 1.5 Induced Der-operators . . . . . . . . . . . . . . . . . . . 114
   - 1.6 Fat Trajectories . . . . . . . . . . . . . . . . . . . . . . . 117
   - 1.7 Inner Structures . . . . . . . . . . . . . . . . . . . . . . 130

2. Linear Connections     141
   - 2.1 Basic Definitions and Examples . . . . . . . . . . . . . . 141
   - 2.2 Parallel Translation . . . . . . . . . . . . . . . . . . . . . 147
   - 2.3 Curvature . . . . . . . . . . . . . . . . . . . . . . . . . . 156
   - 2.4 Operations with Linear Connections . . . . . . . . . . . 159
   - 2.5 Linear Connections and Inner Structures . . . . . . . . . 164

3. Covariant Differential 171

    3.1  Fat de Rham Complexes . . . . . . . . . . . . . . . . . . . 171
    3.2  Covariant Differential . . . . . . . . . . . . . . . . . . . . 178
    3.3  Compatible Linear Connections . . . . . . . . . . . . . . 187
    3.4  Linear Connections Along Fat Maps . . . . . . . . . . . 200
    3.5  Covariant Lie Derivative . . . . . . . . . . . . . . . . . . 208
    3.6  Gauge/Fat Structures and Linear Connections . . . . . . 224

4. Cohomological Aspects of Linear Connections 233

    4.1  An Introductory Example . . . . . . . . . . . . . . . . . 233
    4.2  Cohomology of Flat Linear Connections . . . . . . . . . 238
    4.3  Maxwell's Equations . . . . . . . . . . . . . . . . . . . . 247
    4.4  Homotopy Formula for Linear Connections . . . . . . . . 253
    4.5  Characteristic Classes . . . . . . . . . . . . . . . . . . . 257

*Bibliography* 281

*List of Symbols* 283

*Index* 289

Chapter 0

# Elements of Differential Calculus over Commutative Algebras

In this chapter all necessary notions and facts forming the starting point of the further exposition are collected. First of all this is done in order to make this book self-contained modulo 'undergraduate' mathematics. The suggested reference textbooks are [Singer and Thorpe (1976)] and [Mac Lane and Birkhoff (1967)]. On the other hand, we present some standard elementary topics in a different perspective which better fits our goals. The book [Nestruev (2003)] is highly recommended to the reader who is interested in better understanding the origin and motivation of the algebraic approach to Differential Calculus we follow in this book. Basically, terms that are not explicitly defined here are tacitly assumed to be borrowed from the aforementioned books (in reverse order of priority).

## 0.1 Algebraic Tools

In this section the needed algebraic terminology is set up. The degree of generality is tuned in view of applications in the subsequent exposition.

### 0.1.1 General Conventions

All *rings* are assumed to posses the identity element 1 (but not all rings will be commutative); all ring homomorphisms are assumed to preserve the identity element. A $k$-*algebra* is not necessarily commutative, but the base ring $k$ is always assumed to be commutative. Nevertheless, most of the algebras considered here will be commutative. In particular, in most cases there will be no distinctions between left and right modules. When the distinction takes place, 'module' stands for 'left module'. The dual module $\text{Hom}(P, A)$ of an $A$-module $P$ will be denoted by $P^\vee$. We say that a projective and finitely generated $A$-module $P$ has *constant rank* $r$ if for all maximal ideals $\mathfrak{m}$ of $A$ the dimension of the $A/\mathfrak{m}$–vector space $P/\mathfrak{m}P$ is $r$.

There will be generally no a priori choices for universal constructions. For instance, 'direct sum' and 'coproduct' in thus book are synonymous. When a direct sum, tensor product, extension of scalars, *etc.*, is invoked, the reader may fix any object that satisfies the appropriate universal property, unless a particular choice is explicitly indicated.

As usual, *graded algebras* and *graded modules* will be 'internally graded', that is, direct sums of their components (*cf.* [Mac Lane and Birkhoff (1967), Chap. XVI, Appendix to Sect. 3 (p. 546)]). The index set will always be $\mathbb{N}_0 = \{0, 1, 2, \ldots\}$, homogeneous components will be denoted by subscripts and a component with a negative subscript will be zero by convention. If $\mathcal{A}$ is a graded $k$-algebra and $\mathcal{P}$ is a graded $k$-module equipped with a '$k$-compatible' $\mathcal{A}$-module structure, then $\mathcal{P}$ will be called a *graded $\mathcal{A}$-module*, provided that ([1])

$$a_r p_s \in \mathcal{P}_{r+s}, \quad a_r \in \mathcal{A}_r, p_s \in \mathcal{P}_s .$$

A graded $k$-algebra $\mathcal{A}$ will be called *commutative* if it is commutative as a ring; it will be called *graded commutative* ([2]) if

$$a_r a'_s = (-1)^{rs} a'_s a_r, \quad a_r \in \mathcal{A}_r, a'_s \in \mathcal{A}_s$$

A homomorphism

$$\varphi : \mathcal{P} \to \mathcal{Q}$$

of graded $k$-modules will be called a *graded homomorphism of $n$-th degree* ($n \in \mathbb{Z}$) if for all $s$,

$$p_s \in \mathcal{P}_s \implies \varphi(p_s) \in \mathcal{Q}_{s+n} .$$

When $\mathcal{P}$ and $\mathcal{Q}$ are graded $\mathcal{A}$-modules, with $\mathcal{A}$ being a graded commutative $k$-algebra, $\varphi$ is a *graded homomorphism of $\mathcal{A}$-modules* (of $n$-th degree), if, in addition,

$$\varphi(a_r p_s) = (-1)^{rn} a_r \varphi(p_s), \quad a_r \in \mathcal{A}_r, p_s \in \mathcal{P}_s .$$

If $\mathcal{P}$ and $\mathcal{Q}$ are itself graded $k$-algebras, $\varphi$ will be a *graded algebra homomorphism* if it is both a ring homomorphism and a zeroth degree homomorphism of graded $k$-modules.

---

[1]As usual, the components of the direct sums (=coproducts) $\mathcal{A}$ and $\mathcal{P}$ are identified here with their images through the natural monomorphisms $\mathcal{A}_r \hookrightarrow \mathcal{A}$ and $\mathcal{P}_s \hookrightarrow \mathcal{P}$.

[2]The definition of a commutative graded algebra given in [Mac Lane and Birkhoff (1967), Chap. XVI, Sect. 4 (p. 551)] corresponds to the present definition of a *graded commutative* algebra.

In this book, a *cochain complex* (or, for short, *complex*) is a graded module $\mathcal{P}$ together with a first degree homomorphism $d : \mathcal{P} \to \mathcal{P}$ such that $d \circ d = 0$, which is called *differential*. If $(\mathcal{P}, d)$ and $(\mathcal{P}', d')$ are complexes, a zeroth degree homomorphism $\varphi : \mathcal{P} \to \mathcal{P}'$ will be called a *homomorphism of complexes* or a *cochain homomorphism*, if it commutes with the differentials, i.e., $\varphi \circ d = d' \circ \varphi$.

*Commutators* will always be understood in the sense of ring theory, i.e., $[a, b] = ab - ba$. If $\mathcal{A}$ is a graded $k$-algebra then the *graded commutator of* $a_r \in \mathcal{A}_r$ and $a'_s \in \mathcal{A}_s$ will be

$$[a_r, a'_s]^{(\mathrm{gr})} \stackrel{\mathrm{def}}{=} a_r a'_s - (-1)^{rs} a'_s a_r .$$

Similarly, if $\mathcal{P}$ is a graded $k$-module and $\varphi_r$, $\psi_s$ are graded endomorphisms of $\mathcal{P}$ of $r$-th and $s$-th degree, respectively, then the graded commutator of $\varphi_r$ and $\psi_s$ is the graded endomorphism $[\varphi_r, \psi_s]^{(\mathrm{gr})} = \varphi_r \circ \psi_s - (-1)^{rs} \psi_s \circ \varphi_r$.

## 0.1.2 Differential Operators

Let $A$ be a commutative $k$-algebra, with $k$ being a field, and $P$, $Q$ modules over $A$. If $a \in A$ and $\Delta : P \to Q$ is a $k$-homomorphism, the commutator

$$[\Delta, a] : P \to Q$$

makes sense provided that $a$ is identified with the multiplication by $a$ operators in $P$ and $Q$, respectively. Define inductively

$$\mathrm{Diff}_0(P, Q) \stackrel{\mathrm{def}}{=} \mathrm{Hom}_A(P, Q) = \{\Delta : [\Delta, a] = 0 \; \forall a \in A\},$$
$$\mathrm{Diff}_n(P, Q) \stackrel{\mathrm{def}}{=} \{\Delta : [\Delta, a] \in \mathrm{Diff}_{n-1}(P, Q) \; \forall a \in A\},$$
$$\mathrm{Diff}(P, Q) \stackrel{\mathrm{def}}{=} \bigcup_n \mathrm{Diff}_n(P, Q) .$$

Equivalently, $\Delta \in \mathrm{Diff}_n(P, Q)$ if and only if

$$[\ldots [[\Delta, a_0], a_1], \ldots, a_n] = 0, \qquad \forall a_0, a_1, \ldots, a_n \in A .$$

These sets admit two natural $A$-module structures

$$a\Delta \stackrel{\mathrm{def}}{=} a \circ \Delta, \quad a^+\Delta \stackrel{\mathrm{def}}{=} \Delta \circ a .$$

The notation $\mathrm{Diff}(P, Q)$ usually refers to the first one, while $\mathrm{Diff}^+(P, Q)$ is used for the second, and $\mathrm{Diff}^{(+)}(P, Q)$ is used to denote the bimodule. Elements of these modules are called *linear differential operators from $P$ to $Q$*. The interested reader is referred to [Nestruev (2003), 9.66, 9.67] for more details.

## 0.1.3 Derivations

Let $A$ be a commutative $k$-algebra and $P$ an $A$-module. A *derivation of $A$ into $P$* is a linear over $k$ function

$$\Delta : A \to P$$

that fulfills the *Leibnitz rule*

$$\Delta(ab) = a\Delta(b) + b\Delta(a), \quad a, b \in A \ .$$

Such a function is sometimes also called *$k$-derivation* (this may be useful when more than one algebra structure on the same ring are under consideration). The set of all derivations of $A$ into $P$, equipped with the natural $A$-module structure

$$(a\Delta)(a') \stackrel{\text{def}}{=} a(\Delta(a')), \quad a, a' \in A \ ,$$

will be denoted by $\mathrm{D}(P)$, or sometimes by $\mathrm{D}_k(P)$. In particular, $\mathrm{D}(A)$ is the $A$-module of all derivations of $A$ into itself (often shortly called 'derivations of $A$'). Take notice that $\mathrm{D}(A)$ is not, generally, a subring of $\mathrm{End}_k(A)$ (with the operation of function composition). However, it is easily checked that the commutator of elements of $\mathrm{D}(A)$ lies again in $\mathrm{D}(A)$ (see, *e.g.*, [Nestruev (2003), 9.53]).

If $\varphi : A \to B$ is a homomorphism of commutative $k$-algebras, a *derivation along $\varphi$* will be a derivation $A \to B$ with $B$ considered as an $A$-module via $\varphi$. The set of all derivations along $\varphi$, equipped with the natural $B$-module structure

$$(b\Delta)(a) \stackrel{\text{def}}{=} b(\Delta(a)), \quad a \in A, b \in B \ ,$$

will be denoted by $\mathrm{D}(A)_\varphi$.

Let $\mathcal{A}$ be a graded commutative algebra. An $n$-th degree graded module endomorphism $\Delta : \mathcal{A} \to \mathcal{A}$ is called a *graded derivation of $\mathcal{A}$* (into itself) if it fulfills the following *graded Leibnitz rule*:

$$\Delta(a_s a') = \Delta(a_s)a' + (-1)^{ns} a_s \Delta(a'), \quad a_s \in \mathcal{A}_s, a' \in \mathcal{A} \ ,$$

for all $s$.

## 0.1.4 Additive Functions on Tensor Products

Let $A$ be a commutative ring. As usual, an $A$-module homomorphism

$$P \otimes Q \to R$$

will often be determined by means of an assignment such as
$$p \otimes q \mapsto b(p,q),$$
provided that the expression $b(p,q)$ is $A$-bilinear. Occasionally in this book, there will be needed functions $P \otimes Q \to R$ that are not $A$-module homomorphisms. To recognize if an assignment
$$p \otimes q \mapsto f(p,q),$$
gives a well-defined additive function, it suffices to check that $f : P \times Q \to R$ is biadditive and satisfies
$$f(ap,q) = f(p,aq), \quad a \in A, p \in P, q \in Q$$
(see, *e.g.*, [Hilton and Stammbach (1971), Chap. III, Theorem 7.2]).

### 0.1.5 Some Basic Facts

Let $\varphi : A \to B$ be a homomorphism of commutative rings, $P, P_1, \ldots, P_n$ modules over $A$, $Q$ a module over $B$, $Q_A$ the $A$-module obtained from $Q$ by restriction of scalars, $P_B, P_{1B}, \ldots, P_{nB}$ the $B$-modules obtained from $P, P_1, \ldots, P_n$ by extension of scalars, and $\nu : P \to P_B$, $\nu_1 : P_1 \to P_{1B}$, ..., $\nu_n : P_n \to P_{nB}$ the universal homomorphisms. In the sequel the following simple facts are supposed to be known.

(1) If $P$ is projective, then $P_B$ is projective (see, *e.g.*, [Nestruev (2003), 11.52]).
(2) For every multilinear function of $A$-modules
$$b : P_1 \times \cdots \times P_n \to Q_A,$$
there is exactly one multilinear function of $B$-modules
$$\bar{b} : P_{1B} \times \cdots \times P_{nB} \to Q$$
such that
$$b = \bar{b} \circ (\nu_1 \times \cdots \times \nu_n).$$
(3) In the above situation, if $P_1 = \cdots = P_n$ and $b$ is alternating or symmetric, then $\bar{b}$ is, respectively, alternating or symmetric.
(4) There exists exactly one graded $A$-homomorphism between (fixed) exterior algebras
$$\bigwedge\nolimits^{\bullet} Q_A \to \bigwedge\nolimits^{\bullet} Q$$
such that the first degree component is the identity map of $Q$ ([3]).

---

[3] As usual, the first degree components of tensor, symmetric and exterior algebras of a module are supposed to be identified with the module itself. We use the symbol $\bigwedge^{\bullet}$ for exterior algebras.

(5) The graded algebra obtained from $\bigwedge^\bullet P$ by extension of scalars is an exterior algebra of $P_B$:
$$B \otimes_A \left(\bigwedge\nolimits^\bullet P\right) = \bigwedge\nolimits^\bullet (B \otimes_A P)$$
(it follows form (3)).

(6) If $P$ is projective and finitely generated then the natural homomorphism $P \to P^{\vee\vee}$ is an isomorphism ([4]).

(7) If either $P$ or $P_1$ is projective and finitely generated then the natural homomorphism
$$P^\vee \otimes P_1 \to \mathrm{Hom}\,(P, P_1)$$
is an isomorphism.

(8) If $P$ is projective and finitely generated, then $P_B{}^\vee$ is a module obtained from $P^\vee$ by extension of scalars via $\varphi$, where the universal homomorphism
$$\mu : P^\vee \to P_B{}^\vee$$
is determined by
$$\mu(\alpha)(\nu(p)) = \varphi(\alpha(p)), \qquad p \in P, \alpha \in P^\vee$$
(it follows from (7)).

(9) More generally, if $P$ is projective and finitely generated, then $\mathrm{Hom}_B\,(P_B, P_{1\,B})$ is a module obtained from $\mathrm{Hom}\,(P, P_1)$ by extension of scalars via $\varphi$.

(10) If $P$ is projective, finitely generated and of constant rank 1 then all its endomorphisms are multiplication by scalars operators ([5]).

(11) There exists a natural decomposition
$$\bigwedge\nolimits^\bullet (P \oplus P_1) = \bigwedge\nolimits^\bullet P \otimes \bigwedge\nolimits^\bullet P_1$$
(see [Bourbaki (1989), Chap. III, Sect. 7.7]).

(12) If $P$ is projective and finitely generated, then $\bigwedge^n P$ is projective and finitely generated for all $n \in \mathbb{N}_0$ (it follows from (11)).

---

[4] This result and the following (7) easily follow from the fact that the natural homomorphisms involved are compatible with finite direct sums.

[5] It follows from (9) and Nakayama's Lemma (see, e.g., [Atiyah and Macdonald (1969), Proposition 2.6]; take also into account [Atiyah and Macdonald (1969), Chap. 2, Exercises, n. 10 (p. 32) and Proposition 3.9]).

## 0.1.6 Equivalence of Categories

A functor $\mathcal{E} : \mathfrak{A} \to \mathfrak{C}$ is called an *equivalence of categories* if there exists a functor $\mathcal{F} : \mathfrak{C} \to \mathfrak{A}$ and natural isomorphisms $\eta : I_{\mathfrak{C}} \xrightarrow{\sim} \mathcal{E} \circ \mathcal{F}$ and $\varepsilon : \mathcal{F} \circ \mathcal{E} \xrightarrow{\sim} I_{\mathfrak{A}}$, where $I_{\mathfrak{A}}$, $I_{\mathfrak{C}}$ denote the identity functors (see [Mac Lane (1971), Chap. 4, Sect. 4 (p. 91)]).

Suppose that, in addition, the following *triangular identities* are fulfilled for all objects $C$ of $\mathfrak{C}$ and $A$ of $\mathfrak{A}$:

$$\mathcal{E}(\varepsilon_A) \circ \eta_{\mathcal{E}(A)} = \mathrm{id}_{\mathcal{E}(A)}, \quad \varepsilon_{\mathcal{F}(C)} \circ \mathcal{F}(\eta_C) = \mathrm{id}_{\mathcal{F}(C)} . \tag{0.1}$$

Then $\eta$ and $\varepsilon$ determine an adjunction $\varphi$ ([6]): see [Mac Lane (1971), Chap. IV, Sect. 1, Theorem 2, (v) (p. 81)]; *cf.* also [Mac Lane and Birkhoff (1967), Chap. XV, Sect. 8, Exercise 12 (p. 535)]. The transformation $\eta$ is called the *unit* and $\varepsilon$ the *counit* of the adjunction. In this case the triple $(\mathcal{F}, \mathcal{E}, \varphi)$ is called an *adjoint equivalence*: see [Mac Lane (1971), Chap. IV, Sect. 4 (p. 91)] ([7]).

A functor is said to be *full* if, for all pairs of objects, the map on morphisms are surjective. The notion of a *faithful* functor is obtained by replacing 'surjective' with 'injective'. Every equivalence $\mathcal{E} : \mathfrak{A} \to \mathfrak{C}$ is a full and faithful functor with the property that every object of $\mathfrak{C}$ is isomorphic to $\mathcal{E}(A)$ for some object $A$ of $\mathfrak{A}$: see [Mac Lane (1971), Chap. IV, Sect. 4, Theorem 1 (p. 91)]. By the same theorem, if a full and faithful functor $\mathfrak{A} \to \mathfrak{C}$ is such that every object of $\mathfrak{C}$ is isomorphic to the correspondent of some object of $\mathfrak{A}$, then it is part of an *adjoint* equivalence. In particular, if $\eta : I_{\mathfrak{C}} \xrightarrow{\sim} \mathcal{E} \circ \mathcal{F}$ and $\varepsilon : \mathcal{F} \circ \mathcal{E} \xrightarrow{\sim} I_{\mathfrak{A}}$ are natural isomorphisms, then $\mathcal{E}$ is part of an adjoint equivalence. However, this does *not* imply, generally, that $\eta$ and $\varepsilon$ satisfy the triangular identities (0.1), because the unit and counit of the so-obtained adjoint equivalence do not necessarily coincide with $\eta$ and $\varepsilon$.

## 0.2 Smooth Manifolds

In this section, we recall some basic facts concerning the algebraic interpretation of the theory of smooth manifolds. For additional information, see [Nestruev (2003)].

---

[6] Be aware that, when $\varphi$ is an adjunction of $\mathcal{F} : \mathfrak{C} \to \mathfrak{A}$ to $\mathcal{E} : \mathfrak{A} \to \mathfrak{C}$ in the sense of our reference book [Mac Lane and Birkhoff (1967)], then the triple $(\mathcal{F}, \mathcal{E}, \varphi)$ is called an adjunction from $\mathfrak{C}$ to $\mathfrak{A}$ in [Mac Lane (1971)].

[7] In [Mac Lane (1971)], when an adjunction of $\mathcal{F}$ to $\mathcal{E}$ is determined by $\eta$ and $\varepsilon$, it is also denoted by $\langle \mathcal{F}, \mathcal{E}, \eta, \varepsilon \rangle$: see [Mac Lane (1971), Chap. IV, Sect. 1 (p. 81)].

## 0.2.1 Dual Space

Let $k$ be a field and $A$ a commutative $k$-algebra. A *$k$-point of $A$* is a $k$-algebra homomorphism

$$A \to k \; ;$$

the *dual space* $|A|$ of $A$ is the set of all $k$-points of $A$ ([8]).

## 0.2.2 Geometric Algebras

Each element $a$ in the commutative $k$-algebra $A$ gives rise to the real function

$$\widetilde{a} : |A| \to k$$

defined by the formula

$$\widetilde{a}(m) \stackrel{\text{def}}{=} m(a) \; .$$

Plainly, the set $\widetilde{A} = \{\widetilde{a} : a \in A\}$ is a subalgebra of the $k$-algebra of all $k$-valued functions defined on $|A|$ and

$$\tau : A \to \widetilde{A}, \quad a \mapsto \widetilde{a}$$

is a surjective $k$-algebra homomorphism (*cf.* [Nestruev (2003), 3.4]). When $\tau$ is an isomorphism (*i.e.*, it is also injective), each $a$ will often be identified with $\widetilde{a}$, and therefore $A$ with $\widetilde{A}$. This way, elements $a \in A$ will be viewed as functions $|A| \to k$ by means of the equality

$$a(m) = m(a) \quad m \in M, a \in A$$

(where the left-hand side is 'abusive', while the right-hand side is formally correct; *cf.* [Nestruev (2003), 3.8]).

**Definition.** When $k = \mathbb{R}$ and $\tau$ is an isomorphism, the commutative $\mathbb{R}$-algebra $A$ is said to be *geometric* ([9]).

---

[8] This definition is taken from [Nestruev (2003), 3.4] with $\mathbb{R}$ replaced by an arbitrary field $k$ (*cf.* the footnote of [Nestruev (2003), Preface]). In [Nestruev (2003), Definition 8.4], one may find an extension of this concept, for a $K$-point of $A$ is introduced, with $K \supseteq k$ being a ring without zero divisors. When $K = k$, this notion reduces to the former, up to an obvious identification of a $k$-point with its singleton. Readers acquainted with *schemes* will easily recognize that $K$-points of $A$ correspond to points of the $k$-scheme $\operatorname{Spec} A$ with a residue field isomorphic over $k$ to the quotient field of $K$ (*cf.* [Hartshorne (1977), Chap. II, Exercise 2.7]). It follows that $|A|$ may be identified with the set of all points of $\operatorname{Spec} A$ that are rational over $k$.

[9] Although this notion would make sense over an arbitrary field, a geometric algebra will be always understood over the field $\mathbb{R}$.

### 0.2.3 Natural Topology

Let $A$ be a geometric algebra. Among all topologies on the dual space $|A|$ such that all functions $a \in A$ are continuous (with respect to the usual topology of $\mathbb{R}$), there obviously exists a weakest one. This is called the *natural topology on* $|A|$. The dual space $|A|$ will always be understood as a topological space by means of its natural topology ([10]).

A basis for the natural topology of $|A|$ is

$$\mathcal{B} = \{a^{-1}(U) : a \in A \text{ and } U \text{ is open in the usual topology of } \mathbb{R}\} \; ;$$

*cf.* [Nestruev (2003), 3.12] ([11]).

### 0.2.4 Dual Map

If $\varphi : A \to B$ is a homomorphism of geometric algebras, the map

$$|\varphi| : |A| \to |B|, \quad h \mapsto h \circ \varphi$$

will be called the *dual map of* $\varphi$.

Note that, for all $a \in A$,

$$\varphi(a) = a \circ |\varphi|$$

(up to the identification introduced in n. 0.2.2). Thus, $\varphi$ may be recovered from $|\varphi|$ ([12]).

It is easy to show that the dual map is continuous (see [Nestruev (2003), 3.19]).

---

[10] For algebras over an arbitrary field $k$, the topology which is customarily considered is the Zariski topology, that is, induced on $|A|$ from the usual Zariski topology of Spec $A$ (see [Nestruev (2003), 8.8–8.10]). It generally differs from the natural one when $k = \mathbb{R}$; however (a bit surprisingly) they coincide in the cases of our main interest. For instance, with the help of [Nestruev (2003), 2.4, 3.16], it is not a difficult exercise to show this fact in the case when $A = C^\infty(U)$, the algebra of infinitely differentiable real-valued functions on an open set $U \subseteq \mathbb{R}^n$.

[11] It follows from

$$a^{-1}(\,]a(m) - \varepsilon,\, a(m) + \varepsilon\,[\,) \cap a'^{-1}(\,]a'(m) - \varepsilon,\, a'(m) + \varepsilon\,[\,) \supseteq b^{-1}(\,]-1,\, \varepsilon^2\,[\,)$$

with $a, a' \in A$, $m \in M$, $\varepsilon > 0$ and

$$b = (a - a(m))^2 + (a' - a'(m))^2 \; .$$

[12] The definition of the dual map could be given for non-geometric algebras, but with this general setting, the assertion would be false (even over $\mathbb{R}$).

## 0.2.5

Let $A$, $B$ be geometric algebras and
$$f : |B| \to |A|$$
a map such that
$$a \in A \Rightarrow a \circ f \in B \, .$$
The correspondence
$$a \mapsto a \circ f$$
clearly preserves the algebra operations and therefore, it defines an $\mathbb{R}$-algebra homomorphism
$$\varphi : A \to B \, .$$
It is easily seen that $f = |\varphi|$.

## 0.2.6  Restriction Algebra

Let $A$ be a geometric algebra and $N$ a subset of $|A|$. Consider the set $A|_N$ of all functions
$$N \to \mathbb{R}$$
that are locally restrictions of elements of $A$. This means that $f \in A|_N$ if and only if every $n \in N$ admits a neighborhood $U$ in the subspace $N \subseteq |A|$ such that the restriction $f|_U$ coincides with $a|_U$ for some $a \in A$.

Plainly, $A|_N$ is a subalgebra of the $\mathbb{R}$-algebra of all real-valued functions on $N$. Following [Nestruev (2003), 3.23], the $\mathbb{R}$-algebra $A|_N$ will be called the *restriction of $A$ to $N$*.

For each $n \in N$, define an evaluation homomorphism as
$$e_n : A|_N \to \mathbb{R}, \quad f \mapsto f(n)$$
and, therefore, a map
$$\mu : N \to |A|_N| \quad n \mapsto e_n \, .$$

Suppose that $f \in A|_N$ is such that $f(h) = 0$ for all $h \in |A|_N|$. In particular, $f(e_n) = 0$ for all $n \in N$. But $f(e_n) = e_n(f) = f(n)$, hence $f(n) = 0$ for all $n \in N$, that is, $f = 0$ in $A|N$. This shows that $A|_N$ is geometric.

The *restriction homomorphism* is defined as the map
$$\rho : A \to A|_N, \quad f \mapsto f|_N \, .$$

It is immediate to see that the composition

$$N \xrightarrow{\mu} |A|_N \xrightarrow{|\rho|} |A|$$

is nothing but the inclusion map of $N \subseteq |A|$; moreover, it is not difficult to check that $\mu$ is a homeomorphism onto its image (see [Nestruev (2003), 3.29]).

Finally, following [Nestruev (2003), 3.28], a geometric algebra $A$ will be said to be a *complete algebra* if the restriction homomorphism $A \to A|_{|A|}$ is surjective (pay attention to $N = |A|$).

### 0.2.7 Smooth Algebras

A *smooth algebra* is defined below as in [Nestruev (2003), 4.1]. The dimension $n$ is allowed to be zero. With this respect, the reasonable convention that all real-valued functions on the single-point Euclidean space $\mathbb{R}^0$ are smooth is assumed, so that $C^\infty(\mathbb{R}^0) \cong \mathbb{R}$.

**Definition.** A *smooth algebra of dimension $n$* is a complete (geometric) algebra $A$ that admits a finite or countable open covering $\{U_i\}_{i \in \mathcal{I}}$ of $|A|$ such that $A|_{U_i}$ is isomorphic to $C^\infty(\mathbb{R}^n)$ for all $i$.

According to the above definition, $\mathbb{R}$ is a zero-dimensional smooth algebra. The zero $\mathbb{R}$-algebra may be considered as a smooth algebra of indeterminate dimension.

A *smooth algebra with boundary* (of dimension $n$) is defined in the same way, with the only change being that now the algebras $A|_{U_i}$ are allowed to be isomorphic, either to $C^\infty(\mathbb{R}^n)$ or to $C^\infty(\mathbb{R}^n_H)$, with

$$\mathbb{R}^n_H = \{(r_1, \ldots, r_n) \in \mathbb{R}^n : r_1 \geq 0\}$$

being the (upper) half-space and the smooth functions on it being defined as restrictions of smooth functions on $\mathbb{R}^n$ (see [Nestruev (2003), 4.2]).

### 0.2.8 $C^\infty$-closed Algebras

According to [Nestruev (2003), Definition 3.32], a geometric algebra $A$ is said to be $C^\infty$-*closed* if, for all $f_1, \ldots f_k \in A$ and $g \in C^\infty(\mathbb{R}^k)$, the function

$$f : |A| \to \mathbb{R}, \quad h \mapsto g(f_1(h), \ldots, f_k(h))$$

belongs to $A$.

**Proposition.** Smooth algebras and (more generally) smooth algebras with boundary are $C^\infty$–closed.

**Proof.** See [Nestruev (2003), 4.4]. □

### 0.2.9 Smooth Manifolds

The Nestruev's book presents two definitions of the term 'smooth manifold' and discusses their interplay. Below, the algebraic definition (from [Nestruev (2003), 4.1]) is assumed.

**Definition.** A *(smooth) manifold* is a pair $(M, A)$, such that $A$ is a smooth algebra and $M = |A|$ is the dual space of $A$.

Although $M$ is determined by $A$, to make a concession to geometric intuition, it is generally said that

$$M \text{ is a smooth manifold}$$

and $A$ is often implicitly assumed as given. Since smooth algebras are geometric by definition, the convention of n. 0.2.2 and 0.2.3 apply. Accordingly, $M$ will be considered as a topological space and the elements of $A$ will be identified with real-valued functions by means of the equality

$$a(m) = m(a) \quad m \in M, a \in A \,.$$

They will be called *smooth functions on $M$*. The *algebra of smooth functions $A$ on $M$* is generally denoted by $C^\infty(M)$.

The definition of a *(smooth) manifold with boundary* is plainly obtained from Definition 0.2.9 by replacing 'smooth algebra' with 'smooth algebra with boundary'.

### 0.2.10 Smooth Maps

**Definition.** Let $M$ and $N$ be smooth manifolds, possibly with boundary. A *smooth map of $N$ into $M$* is the dual map $|\varphi|$ of some $\mathbb{R}$-algebra homomorphism

$$\varphi : C^\infty(M) \to C^\infty(N) \,.$$

A *diffeomorphism* is a smooth map that admits a smooth inverse.

## 0.2.11 Associated Homomorphism

If $f : N \to M$ is a smooth map, according to n. 0.2.4, an $\mathbb{R}$-algebra homomorphism $\varphi : C^\infty(M) \to C^\infty(N)$ such that $f = |\varphi|$ is uniquely determined.

**Definition.** The homomorphism $\varphi$ will be said to be *associated with* $f$ and denoted by $f^*$.

## 0.2.12 Smoothness Condition

**Proposition.** A map $f : N \to M$ is smooth if and only if
$$a \in C^\infty(M) \Rightarrow a \circ f \in C^\infty(N)$$
and, in this case, $f^*$ is given by
$$a \mapsto a \circ f.$$

*Proof.* It trivially follows from nn. 0.2.4 and 0.2.5. □

As an immediate consequence we have that a composition $g \circ f$ of smooth maps $f : N \to M$ and $g : V \to N$ is smooth and
$$(g \circ f)^* = f^* \circ g^* ;$$
besides, the identity $\mathrm{id}_M$ is smooth and $\mathrm{id}_M^* = \mathrm{id}_{C^\infty(M)}$ (*cf.* [Nestruev (2003), 6.6]).

Obviously, smooth manifolds, possibly with boundary, and smooth maps constitute a category. Smooth manifolds without boundary constitute a full subcategory which will be denoted by $\mathfrak{SM}_a$ (*cf.* [Nestruev (2003), 6.6]; the subscript is to remind one that the algebraic definition is assumed).

## 0.2.13 Classical Definition of Manifolds

An extensive discussion about the consistency of the algebraic setting about smooth manifolds with the classical one is the matter of [Nestruev (2003), Chap. 7]. The main equivalence theorems will be reported below, after a little preparation.

To fix the definition of an *atlas*, the reader is referred to [Nestruev (2003), 5.5]. See [Nestruev (2003), 5.17] for the notion of a *smooth function with respect to an atlas*. When referring to [Nestruev (2003), Chap. 5], take into account that the term 'smooth manifold' is used there with its standard 'coordinate meaning' (see [Nestruev (2003), 5.8]). If $\mathcal{A}$, $\mathcal{B}$ are atlases on the sets $M$, $N$, respectively, a map $N \to M$ will be said *smooth with respect to $\mathcal{B}$ and $\mathcal{A}$* if it is smooth in the sense of [Nestruev (2003), 6.14].

**Proposition.** In the above notation, a map $f : N \to M$ is smooth with respect to $\mathcal{B}$ and $\mathcal{A}$ if and only if, for every function $a : M \to \mathbb{R}$ that is smooth with respect to $\mathcal{A}$, the function $a \circ f$ is smooth with respect to $\mathcal{B}$.

**Proof.** See [Nestruev (2003), 7.16]. □

If $\mathcal{A}$ is a maximal atlas on a set $M$, that satisfies countability and Hausdorff conditions (see [Nestruev (2003), 5.7]), then $(M, \mathcal{A})$ is what is called a smooth manifold in the common usage. Maps are said to be smooth when they are smooth with respect to the structure atlases. The so-obtained category will be denoted by $\mathfrak{SM}_c$ (where the subscript is to remind the 'coordinate' definition).

### 0.2.14 *Associated Atlas*

**Theorem.** Let $M$ be a smooth manifold (without boundary). Then $M$ is a Hausdorff topological space with a countable basis, and there is a unique maximal atlas $\mathcal{A}$ on $M$ such that the $\mathbb{R}$-algebra of functions that are smooth with respect to $\mathcal{A}$ equals $C^\infty(M)$.

**Proof.** The space $M$ is Hausdorff by [Nestruev (2003), 3.13]. The fact that $M$ has a countable basis immediately follows from [Nestruev (2003), 7.8] and the fact that $\mathbb{R}^n$ has a countable basis. By [Nestruev (2003), 7.7], there exists an atlas $\mathcal{A}'$ such that the algebra of functions that are smooth with respect to $\mathcal{A}'$ is $C^\infty(M)$. It is a simple consequence of the definition of the compatibility of atlases that an atlas $\mathcal{A}''$ determines the same smooth functions as $\mathcal{A}'$ if and only if it is compatible with $\mathcal{A}'$. Hence, the required atlas $\mathcal{A}$ is just the unique maximal atlas containing $\mathcal{A}'$ (see [Nestruev (2003), 5.5]). □

**Definition.** The maximal atlas $\mathcal{A}$ will be said to be *associated with $M$*.

Let $f : N \to M$ be a map and $\mathcal{A}, \mathcal{B}$ be the atlases associated with $M$, $N$, respectively. From Propositions 0.2.12 and 0.2.13, it follows that $f$ is smooth if and only if it is smooth with respect to $\mathcal{B}$ and $\mathcal{A}$.

### 0.2.15

**Theorem.** Let $(M, \mathcal{A})$ be an object of $\mathfrak{SM}_c$ and $A$ the $\mathbb{R}$-algebra of functions that are smooth with respect to the atlas $\mathcal{A}$. Then

(1) $A$ is a smooth $\mathbb{R}$-algebra (without boundary);

(2) the map
$$\theta_M : M \to |A|, \quad m \mapsto h_m ,$$
where
$$h_m : A \to \mathbb{R}, \quad f \mapsto f(m) ,$$
is a homeomorphism.

**Proof.** See [Nestruev (2003), Theorem 7.2]. □

In the above notation, if $a \in A$, then $a \circ \theta_M^{-1}$ coincides with $a$ as a function on $|A|$. Therefore, Proposition 0.2.13 easily implies that $\theta_M$ and $\theta_M^{-1}$ are smooth with respect to the $\mathcal{A}$ and the atlas $\mathcal{A}_{|A|}$ associated with the smooth manifold $|A|$.

### 0.2.16 The Equivalence $\mathcal{E} : \mathfrak{SM}_a \to \mathfrak{SM}_c$

From n. 0.2.14 it immediately follows that the assignments
$$\mathcal{E}(M, \mathrm{C}^\infty(M)) \stackrel{\text{def}}{=} (M, \mathcal{A}), \quad \mathcal{E}(f) \stackrel{\text{def}}{=} f ,$$
where $\mathcal{A}$ denotes the atlas associated with the smooth manifold $M$, define a full and faithful functor
$$\mathcal{E} : \mathfrak{SM}_a \to \mathfrak{SM}_c .$$
According to n. 0.2.15, every object in $\mathfrak{SM}_c$ is diffeomorphic to the correspondent through $\mathcal{E}$ of some object in $\mathfrak{SM}_a$. Therefore, $\mathcal{E}$ is an equivalence of categories (see n. 0.1.6).

More explicitly, denote by $\mathrm{I}_a$ and $\mathrm{I}_c$ the identity functors of $\mathfrak{SM}_a$ and $\mathfrak{SM}_c$, respectively, and, in notation of Theorem 0.2.15, let $\mathcal{F} : \mathfrak{SM}_c \to \mathfrak{SM}_a$ be the functor
$$(M, \mathcal{A}) \mapsto (|A|, A), \quad f \mapsto |\varphi|$$
with $f : N \to M$ and $\varphi$ sending $a \in A$ to $a \circ f$. Then
$$M \mapsto \theta_M$$
defines a natural isomorphism
$$\theta : \mathrm{I}_c \to \mathcal{E} \circ \mathcal{F} ,$$
and
$$(M, \mathrm{C}^\infty(M)) \mapsto |\tau_M| ,$$
with $\tau_M$ being the identification isomorphism of $\mathrm{C}^\infty(M)$ (see n. 0.2.2), defines a natural isomorphism
$$\varepsilon : \mathcal{F} \circ \mathcal{E} \to \mathrm{I}_a .$$

## 0.2.17

It is easy to see that, in addition, $\mathcal{E}$ and $\mathcal{F}$ are part of an *adjoint* equivalence with unit $\theta$ and counit $\varepsilon$, that is, the triangular identities (0.1), p. 7, are satisfied (with $\theta$ in place of $\eta$).

## 0.2.18

Some basic results about smooth manifolds that are well-know in the coordinate approach will sometimes be used: on the basis of the equivalence theorem, recognizing their validity in the algebraic setting is a matter of straightforward details. The same remark holds for the extension of these results to manifolds with boundary (*cf.* [Nestruev (2003), 5.14, 7.12]). By these reasons, references to elementary results about 'coordinate smooth manifolds' will often be applied to 'algebraic smooth manifolds with boundary' with no more explanations, except when in the presence of some nontrivial details.

## 0.2.19

By the equivalence theorem, each manifold $(M, \mathcal{A})$ in the classical sense could be identified with the corresponding manifold $(|A|, A)$, with $A$ being the algebra of smooth functions in the classical sense (*i.e.*, smooth with respect to $\mathcal{A}$; *cf.* [Nestruev (2003), 7.19]). In the formal setting of this book, it will not be necessary to make use of this identification, with the following exception. An open subset $U$ of a finite-dimensional vector space $E$ over $\mathbb{R}$, which is canonically an object of $\mathfrak{SM}_c$ (through whatever vector space isomorphism $E \xrightarrow{\sim} \mathbb{R}^n$), will be identified with a smooth manifold through the map

$$\theta_U : U \xrightarrow{\sim} |A|$$

defined in the statement of Theorem 0.2.15. This identification will particularly be used for the whole space $E$ and for open subsets of $\mathbb{R}^n$.

Note that each function on $|A|$ is identified with the function $f \circ \theta$ on $U$. This way, real-valued functions on $|A|$ that are smooth according to n. 0.2.9 are identified with functions on $U \subseteq \mathbb{R}^n$ that are smooth in the ordinary sense (*i.e.*, infinitely differentiable). This identification is the same as that in n. 0.2.2.

### 0.2.20  *Submanifolds*

Let $M$ be a smooth manifold, possibly with boundary, $N$ be a subset of $M$ and set $A = \mathrm{C}^\infty(M)$. Recall that if $\mu$ and $\rho$ are as in n. 0.2.6, then the composition

$$N \xrightarrow{\mu} |A|_N| \xrightarrow{|\rho|} M$$

is the inclusion map of $N \subseteq M$ and $\mu$ maps $N$ homeomorphically onto its image.

Suppose that $h \in |A|_N|$ and set $n = |\rho|(h)$. For all $a \in A|_N$ there exists a function $b \in A$ that coincides with $a$ locally around $n$ because of the definition of $A|_N$. Exploiting [Nestruev (2003), 4.17, (ii)] ([13]), one finds $f \in A$ such that $f(n) = 1$ and $a\rho(f) = \rho(bf)$. It follows that

$$h(a) = h(a)f(n) = h(a)h(\rho(f)) = h(a\rho(f))$$
$$= h(\rho(bf)) = b(n)f(n) = b(n) = a(n) \, .$$

Hence $h = \mu(n)$. This shows that $\mu$ is surjective and, hence, a homeomorphism. (See also [Nestruev (2003), 3.32–3.33] for a more general discussion.)

**Definition.** If $A|_N$ is smooth then the manifold $|A|_N|$ will be called a *(smooth) submanifold of* $M$ and the smooth map $|\rho| : |A|_N| \to M$ the *embedding of* $|A|_N|$.

The definition of a *submanifold with boundary* (and of its embedding) is the same, but with the weakened requirement that $A|_N$ be smooth with boundary.

Since the embedding $|\rho| : |A|_N| \to M$ corresponds, through the homeomorphism $\mu : N \to |A|_N|$, to the inclusion of $N$, it maps $|A|_N|$ homeomorphically onto $N$. Thus, the term 'embedding' is correct from a topological viewpoint.

Points of $|A|_N|$ will be identified with their images through the embedding (hence through $\mu^{-1}$). Thus, $|A|_N|$ may be identified with $N$. Accordingly, a submanifold $N \subseteq M$ may be introduced by shortly saying

*let $N$ be a submanifold of $M$.*

Note that each function $|A|_N| \to \mathbb{R}$ is identified with the function $a \circ \mu : N \to \mathbb{R}$. By the definition of $\mu$, smooth functions on $|A|_N|$ are so identified with elements of $A|_N$, so as to be consistent with n. 0.2.2. Therefore, $A|_N$ may be denoted by $\mathrm{C}^\infty(N)$.

---
[13] Plainly, this result holds for a manifold with boundary as well.

If $N$ is an open subset of $M$ then $A|_N$ is always a smooth algebra, possibly with boundary. Therefore, an *open submanifold of $M$* is nothing but an open subset of $M$, considered as a manifold according to the above introduced identification.

Now suppose that $C$ is a closed subset of $M$. Taking into account the definition of the restriction algebra and using a suitable partition of unity, one easily proves that the restriction homomorphism

$$\rho: C^\infty(M) \to C^\infty(M)|_C$$

is surjective. On the other hand, it is not difficult to show that, when a restriction homomorphism $C^\infty(M) \to C^\infty(M)|_N$ is surjective, $N$ is closed (no matter whether $C^\infty(M)|_N$ is smooth or not; cf. [Nestruev (2003), 4.12, (i)]). In conclusion, if the restriction algebra $C^\infty(M)|_C$ to a closed subset $C \subseteq M$ is smooth, then $C$ may be called a *closed submanifold of $M$*, consistently with [Nestruev (2003), 4.11]. Similarly, if $C^\infty(M)|_C$ is smooth with boundary then $C$ is a closed submanifold with boundary.

A closed subset $C$ together with

$$C^\infty(C) = C^\infty(M)|_C$$

(see [Nestruev (2003), 7.13]) constitute a *smooth set in $M$*, no matter whether $C^\infty(M)|_N$ is smooth or not.

### 0.2.21  *Restrictions*

Let $f: M \to M'$ be a smooth map between manifolds, possibly with boundary, and $N$ a submanifold of $M$, possibly with boundary. The set-theoretic restriction $f|_N : N \to M'$ is identified with a map $|A|_N| \to M'$ between manifolds, which is smooth because it coincides with the composition

$$|A|_N| \hookrightarrow M \xrightarrow{f} M',$$

where the first map is the embedding.

Let $N'$ be a submanifold of $M'$. If $f(M) \subseteq N'$, then, taking into account the definition of $C^\infty(M')|_{N'}$, one easily deduces from Proposition 0.2.12 that the restriction $g: M \to N'$ of $f$ on the codomain is smooth.

It immediately follows that, in general, if $f(N) \subseteq N'$ then the set-theoretic restriction

$$f|_{N,N'} : N \to N'$$

of $f$ is smooth.

### 0.2.22 Smooth Envelope

According to [Nestruev (2003), Definition 3.36], a *smooth envelope* of a geometric algebra $A$ is a pair $(\overline{A}, i)$ such that $\overline{A}$ is a $C^\infty$-closed geometric algebra and $i : A \to \overline{A}$ is an $\mathbb{R}$-algebra homomorphism that satisfies the following universal property: *for every homomorphism $\alpha : A \to A'$ of $A$ into a $C^\infty$-closed (geometric) algebra $A'$, there is exactly one homomorphism $\overline{\alpha} : \overline{A} \to A'$ such that $\alpha = \overline{\alpha} \circ i$.*

A smooth envelope exists and is unique up to an isomorphism preserving the envelope homomorphism (see [Nestruev (2003), 3.37]); accordingly, it will be generally introduced by simply saying

*let $\overline{A}$ be the smooth envelope of $A$.*

The (often understood) universal homomorphism $i : A \to \overline{A}$ will be called the *envelope homomorphism*.

### 0.2.23

Let $T = A \otimes_\mathbb{R} B$ be a tensor product of geometric algebras. According to [Nestruev (2003), Exercise 4.28], $T$ is geometric ([14]).

Now let $\overline{T}$ be a smooth envelope of $T$, with the envelope homomorphism $i : T \to \overline{T}$, denote by $\iota_A : A \to T$, $\iota_B : B \to T$ the universal homomorphisms and set

$$\overline{\iota}_A = i \circ \iota_A \quad \text{and} \quad \overline{\iota}_B = i \circ \iota_B \, .$$

**Proposition.** For every $C^\infty$-closed algebra $C$ and for every pair of algebra homomorphisms

$$\varphi_A : A \to C \quad \text{and} \quad \varphi_B : B \to C \, ,$$

there exists exactly one algebra homomorphism

$$\overline{\varphi} : \overline{T} \to C$$

such that

$$\varphi_A = \overline{\varphi} \circ \overline{\iota}_A \quad \text{and} \quad \varphi_B = \overline{\varphi} \circ \overline{\iota}_B \, .$$

**Proof.** It easily comes from the characteristic universal properties of tensor product algebras and smooth envelopes. □

---

[14] To solve the exercise, note that an element $t \in T$ may be written as $a_1 \otimes b_1 + \cdots + a_n \otimes b_n$ with $a_1, \ldots, a_n$ linearly independent over $\mathbb{R}$. For all $h \in |A|$ and $k \in |B|$, assuming $\mathbb{R} \otimes \mathbb{R} = \mathbb{R}$ one gets $h \otimes k \in |T|$. Suppose that $(h \otimes k)(t) = 0$ for all $h, k$. Keeping $k$ fixed, from the fact that $A$ is geometric, one deduces that $b_1(k)a_1 + \cdots + b_n(k)a_n = 0 \in A$. Then $b_i(k) = 0$ for all $i \in \{1, \ldots, n\}$ and $k \in |B|$, because of the linear independence of $a_1, \ldots, a_n$. Hence $b_i = 0 \in B$ because $B$ is geometric.

### 0.2.24 Smooth Tensor Product

Suppose now that $A$ and $B$ are smooth with boundary. According to [Nestruev (2003), 4.31], if at least one of $A$ and $B$ is smooth without boundary, then $\overline{T}$ is smooth, possibly with boundary; and moreover, $\overline{T}$ is without boundary if and only if $A$ and $B$ are both without boundary. Propositions 0.2.23 and 0.2.8 imply the following universal property. For every smooth algebra $C$, possibly with boundary, and for every pair of algebra homomorphisms

$$\varphi_A : A \to C \quad \text{and} \quad \varphi_B : B \to C \, ,$$

there exists exactly one algebra homomorphism

$$\overline{\varphi} : \overline{T} \to C$$

such that

$$\varphi_A = \overline{\varphi} \circ \overline{\iota}_A \quad \text{and} \quad \varphi_B = \overline{\varphi} \circ \overline{\iota}_B \, .$$

Conversely, a smooth algebra with boundary that matches the above property turns out to be a smooth envelope of $A \otimes B$.

In this situation, it is natural to say that $\overline{T}$ is *a smooth tensor product*. A smooth tensor product of $A$ and $B$ will be generally denoted by $A \overline{\otimes} B$, and the (usually understood) homomorphisms $\overline{\iota}_A$ and $\overline{\iota}_B$ will be called *natural homomorphisms into $A \overline{\otimes} B$*.

When both $A$ and $B$ are smooth with nonempty boundaries, although the smooth envelope $\overline{T}$ of $A \otimes B$ satisfies the property, it is not difficult to prove that it is not a smooth algebra with boundary (*cf.* [Nestruev (2003), Exercise 4.31, (ii)]): basically, it depends on the fact that a product of half-spaces is diffeomorphic to no half or whole Euclidean spaces. Therefore, a smooth tensor product does not exist in this case.

### 0.2.25 Cartesian Product

Let $M$, $N$ be smooth manifolds, possibly with boundary. Clearly, a manifold $V$ is a product of $M$ and $N$ in the category of manifolds with boundary if and only if $C^\infty(V)$ is a smooth tensor product of $C^\infty(M)$ and $C^\infty(N)$. It immediately follows that a pair of smooth manifolds with boundary admits a product when at least one is a smooth manifold (without boundary), and that a product of smooth manifolds is again a smooth manifold (hence it is also a product in the category $\mathfrak{SM}_a$). Consistent with the policy about algebraic constructions, a *product manifold of $M$ and $N$* in this book will be, by definition, a product in the category of manifold with boundary. Plainly, a product manifold $V$ will be denoted by $M \times N$ and the universal smooth functions

$$\pi_M : M \times N \to M \quad \text{and} \quad \pi_N : M \times N \to N$$

will be simply called the *projections maps of $M \times N$*. If $f : S \to M$ and $g : S \to N$ are smooth maps, the unique smooth map

$$f : S \to M \times N$$

such that

$$f = \pi_M \circ f \quad \text{and} \quad g = \pi_N \circ f$$

will be said to be *induced by $f$ and $g$*, and it will be sometimes identified with the pair

$$(f, g) \ .$$

(This map generally differs from $f \times g : S \times S \to M \times N$, of course.)

The natural identification of the set $M \times N$ with the set-theoretic product will always be assumed.

### 0.2.26 *Embeddings of Factors of a Product*

Let $M$ and $N$ be smooth manifolds, possibly with boundary, that admit a product $M \times N$.

For each $n_0 \in N$ the smooth map

$$i_{n_0} : M \to M \times N, \quad m \mapsto (m, n_0)$$

will be called the *embedding at $n_0$ into $M \times N$*. Similarly, the smooth map

$$j_{m_0} : N \to M \times N, \quad n \mapsto (m_0, n) \ ,$$

$m_0 \in M$, will be called the *embedding at $m_0$ into $M \times N$*.

### 0.2.27 *Tangent Vectors*

Let $M$ be a manifold, possibly with boundary, and $m \in M$ a point. Taking into account the notion of a derivation along an algebra homomorphism (see n. 0.1.3), one may define a tangent vector in the following way.

**Definition.** A *tangent vector to $M$ at $m$* is a derivation along

$$m : C^\infty(M) \to \mathbb{R} \ .$$

The set
$$T_m M \stackrel{\text{def}}{=} \mathrm{D}\left(\mathrm{C}^\infty(M)\right)_m$$
of all tangent to $M$ at $m$ vectors gets an obvious structure of real vector space. This vector space is called the *tangent space to $M$ at $m$*.

The usual coordinate description of a tangent vector is called the Tangent Vector Theorem in [Nestruev (2003), 9.6]; together with [Nestruev (2003), 9.12], it guarantees consistency with other classical definitions of a tangent vector (see, *e.g.*, [Berger and Gostiaux (1988), 2.5.9]). Tangent vectors are *local operators*, that is, if two functions coincides in a neighborhood of $m$, then each tangent vector at $m$ takes the same value on them (see [Nestruev (2003), 9.8]).

### 0.2.28 Differential of a Smooth Map

**Definition.** The *differential of a smooth map* $f: N \to M$ *at* $n \in N$ is the ($\mathbb{R}$-linear) map
$$\mathrm{d}_n f : T_n N \to T_{f(n)} M, \quad \eta \mapsto \eta \circ f^* .$$

As usual, $f$ is an *immersion* or a *submersion at $n$*, according to whether $\mathrm{d}_n f$ is injective or surjective; it is an *embedding* if it is an immersion at every point and maps $N$ homeomorphically onto its image. The embedding of a submanifold is actually an embedding, because of the definition of the restriction algebra and the fact that tangent vectors are local operators.

### 0.2.29 Local Description of Submanifolds

In the coordinate framework (without boundary), a submanifold of a manifold $M$ may be defined as a subset that locally looks like a coordinate subspace of $\mathbb{R}^n$; *cf.* [Berger and Gostiaux (1988), Definition 2.6.1]. This description holds, of course, in the algebraic setting too. More precisely, let $M$ and $N$ be smooth manifolds (without boundary), of respective dimensions $d \geq e$, and consider the map
$$\nu : \mathbb{R}^e \hookrightarrow \mathbb{R}^d, \quad (x_1, \ldots, x_e) \mapsto (x_1, \ldots, x_e, 0, \ldots, 0) .$$

If $N$ is a submanifold of $M$ then, for each $n \in N$, there exists a neighborhood $U$ of $n$ in $M$ such that the restriction $U \cap N \to U$ of the embedding of $N$ corresponds to $\nu$ through diffeomorphisms $U \cap N \xrightarrow{\sim} \mathbb{R}^e$, $U \xrightarrow{\sim} \mathbb{R}^d$ (it may be deduced, *e.g.*, from [Berger and Gostiaux (1988), Corollary 2.6.11], which substantially relies on a well-known basic result from Calculus: see [Berger and Gostiaux (1988), 0.2.24]).

An analogous description in the case when $N$ or $M$ are with boundary is left to the reader.

### 0.2.30 *Closed Embeddings*

The cited result [Berger and Gostiaux (1988), Corollary 2.6.11] also implies that if $f : N \to M$ is a closed embedding (that is, an embedding of manifolds that is topologically a closed map), then $f^*$ is surjective, because $f$ induces a diffeomorphism onto the closed submanifold $f(N) \subseteq M$.

## 0.3 Vector Bundles

The algebraic treatment of vector bundles is based on projective modules over smooth algebras (possibly with boundary). We recall below the construction of pseudobundles given in [Nestruev (2003), 11.11], which emphasizes the similarity with the smooth manifolds setting.

### 0.3.1 *Pseudobundles*

Let $k$ be a field, $A$ a commutative $k$-algebra, and $P$ an $A$-module. The *pseudobundle*

$$\pi_P : |P| \to |A|$$

is determined by $P$ in the following way. For each $h \in |A|$, the fiber of $\pi_P$ at $h$ is the $k$-module

$$P_h \stackrel{\text{def}}{=} \frac{P}{\mu_h P} = k_h \otimes_A P \,,$$

where $\mu_h = \operatorname{Ker} h$ and $k_h$ denotes $k$ with the $A$-algebra structure given by $h : A \to k$. Thus, the set $|P|$ is the (disjoint) union of all $P_h$, and the projection $\pi_P : |P| \to |A|$ sends every $p \in P_h$ into $h$.

With each $p \in P$, a *regular section* of $\pi_P$ is associated; that is, the map

$$s_p : |A| \to |P|$$

(satisfying $\pi_P \circ s_p = \operatorname{id}_M$) defined by setting for all $h \in |A|$

$$s_p(h) \stackrel{\text{def}}{=} p + \mu_h P = 1 \otimes p \in P_h \,.$$

The set $\{s_p\}_{p\in P}$ of all regular sections is denoted by $\Gamma(P)$. For all $p, p' \in P$ and $\widetilde{a} \in \widetilde{A}$ (see n. 0.2.2) define $s_p + s'_p$ and $\widetilde{a}s_p$ by setting, for all $h \in |A|$,
$$(s_p + s_{p'})(h) \stackrel{\text{def}}{=} s_p(h) + s_{p'}(h),$$
$$(\widetilde{a}s_p)(h) \stackrel{\text{def}}{=} \widetilde{a}(h)s_p(h)$$
(here the operations appearing in the right-hand sides are those in the module $P_h$). It is readily seen that
$$s_p + s_{p'} = s_{p+p'} \in \Gamma(P)$$
and
$$\widetilde{a}s_p = s_{ap} \in \Gamma(P).$$
This shows that $\Gamma(P)$ is an $\widetilde{A}$-module in a natural way. Of course, $\Gamma(P)$ may be also considered as an $A$-module by restriction of scalars via the natural homomorphism $A \to \widetilde{A}$.

A pseudobundle $\pi_P$ is said to be *equidimensional* if all fibers $P_h = \pi_P^{-1}(h)$ ($h \in |A|$) have the same dimension as vector spaces over $k$.

### 0.3.2  *Geometric Modules*

The $\widetilde{A}$-module $\Gamma(P)$ of all regular sections of the pseudobundle $\pi_P : |P| \to |A|$ determined by $P$ will be called the *geometrization of $P$*. The $A$-module homomorphism
$$P \to \Gamma(P), \quad p \mapsto s_p,$$
will be called the *geometrization homomorphism* ([15]). If the geometrization homomorphism is an isomorphism, then $P$ is said to be *geometric*. In this case, $P$ will be identified with $\Gamma(P)$ through this isomorphism.

By these conventions, when $A$ and $P$ are both geometric, there will usually be no distinction between the $A$-module $P$ and the $\widetilde{A}$-module $\Gamma(P)$.

It is not difficult to show that a projective module over a geometric algebra is geometric.

### 0.3.3

The following fact will be useful later. Let $\varphi : A \to B$ a homomorphism of geometric algebras and $P$ a geometric $B$-module. Then the $A$-module $P_A$ obtained from $P$ by restriction of scalars via $\varphi$ is also geometric. The proof is an easy exercise left to the reader.

---

[15] In [Nestruev (2003), 11.11] one may also find an alternative definition of the geometrization of $P$ as a quotient module of $P$. Consistency is obvious because the geometrization homomorphism is an epimorphism.

### 0.3.4 Vector Bundles

The 'algebraic definition' of a vector bundle is assumed below ([16]).

**Definition.** A *(smooth) vector bundle* over a smooth manifold $M$, possibly with boundary, is a pair
$$(\pi : E_\pi \to M, P)$$
such that:

(1) $P$ is a projective, finitely generated $C^\infty(M)$–module;
(2) $\pi : E_\pi \to M$ is the pseudobundle $\pi_P : |P| \to M$ determined by $P$;
(3) the pseudobundle $\pi$ is equidimensional ([17]).

The map $\pi = \pi_P$ will be called the *projection map*, $E_\pi = |P|$ the *total space*, and $M$ the *base* of the vector bundle $(\pi, P)$. The *fiber at a point* $m \in M$ is $\pi^{-1}(m) = P_m$. A *general fiber* is a vector space $E$ that is isomorphic to some, hence to all, fibers of $\pi$.

Although $\pi$ (and henceforth $E_\pi$) is determined by $P$, a vector bundle is generally identified by $\pi$ or by $E_\pi$, and the module $P$ is denoted by $\Gamma(\pi)$, or even by $\Gamma(E_\pi)$. Accordingly, a vector bundle is generally introduced by saying:

$$\text{let } \pi : E_\pi \to M \text{ be a vector bundle}$$

or, simply,

$$\text{let } E_\pi \text{ be a vector bundle}.$$

A *smooth section of* $\pi$ is an element of $\Gamma(P)$, i.e., a regular section of the pseudobundle $\pi = \pi_P$. In view of n. 0.3.2, $\Gamma(P)$ is identified with $P = \Gamma(\pi)$. Accordingly, $P$ will be often called the *module of smooth sections of* $\pi$.

### 0.3.5 Morphisms of Vector Bundles over the Same Base

Let
$$\pi : E_\pi \to M, \qquad \pi' : E'_{\pi'} \to M$$

---

[16] This approach is deduced from [Nestruev (2003), Chap. 11], where the reader is gradually led from the classical notion of a vector bundle to the algebraic one. Be aware that in [Nestruev (2003)], because of that pedagogical line, the term 'vector bundle' has to be interpreted according to its 'geometric definition' (see [Nestruev (2003), 11.2]). Here, the interplay between the 'algebraic' and the 'geometric' approaches will be briefly recalled later.

[17] We shall prove soon that Condition (3) is automatically verified when $M$ is connected.

be vector bundles over the same manifold $M$ (possibly with boundary) and set
$$P = \Gamma(\pi), \quad P' = \Gamma(\pi'), \quad A = C^\infty(M).$$
If
$$\varphi : P \to P'$$
is an $A$-module homomorphism then, for each $m \in M$, a homomorphism
$$\varphi_m \stackrel{\text{def}}{=} \text{id}_{\mathbb{R}_m} \otimes \varphi : P_m \to P'_m$$
is defined. More explicitly:
$$\varphi_m(p + \mu_m P) = \varphi(p) + \mu_m P',$$
with $\mu_m = \text{Ker}\, m$. Taking into account that
$$E_\pi = |P| = \bigcup_{m \in M} P_m, \quad E'_{\pi'} = |P'| = \bigcup_{m \in M} P'_m,$$
one may define a map
$$|\varphi| : E_\pi \to E'_{\pi'}, \quad \mathbf{e} \mapsto \varphi_{\pi(\mathbf{e})}(\mathbf{e}),$$
which is compatible with the projection maps, *i.e.*,
$$\pi = \pi' \circ |\varphi|.$$
Note that $|\varphi|$ is fiber-wise linear by definition, *i.e.*, each restriction $\varphi_m : \pi^{-1}(m) \to \pi'^{-1}(m)$ of $|\varphi|$ is linear over $\mathbb{R}$.

It is immediately seen that, for all $p \in P$,
$$\varphi(p) = |\varphi| \circ p$$
(according to the identification of $p$ with a smooth section). This formula shows that $\varphi$ may be recovered from $|\varphi|$.

**Definition.** The map $|\varphi|$ is, by definition, *a morphism over $M$ from $\pi$ to $\pi'$*.

A morphism between $\pi$ and $\pi'$, will be often introduced by directly saying
$$\text{let } \overline{f} : E_\pi \to E'_{\pi'} \text{ be a morphism.}$$
The restriction $\varphi_m$ of $\overline{f}$ to the fibers over $m \in M$ will be generally denoted by $\overline{f}_m$.

### 0.3.6 *Induced Bundle*

Let $\pi : E_\pi \to M$ be a vector bundle, $f : N \to M$ a smooth map between manifolds, possibly with boundary, and set

$$A = C^\infty(M), \quad B = C^\infty(N), \quad P = \Gamma(\pi) \ .$$

Denote by

$$f^*(\pi) : E_{f^*(\pi)} \to N$$

the pseudobundle determined by the $B$-module $P_B = B \otimes_A P$ obtained from $P$ by extension of scalars via $f^* : A \to B$. The universal homomorphism

$$\nu : P \to P_B$$

determines a map

$$\overline{f} : E_{f^*(\pi)} \to E_\pi$$

in the following way. Let

$$\mathbf{e} \in E_{f^*(\pi)} \ ,$$

set

$$n = f^*(\pi)(\mathbf{e}), \quad m = f(n)$$

and denote by $\mathbb{R}_n$, $\mathbb{R}_m$ the algebras determined on $\mathbb{R}$ by the homomorphism $n : B \to \mathbb{R}$, $m : A \to \mathbb{R}$, respectively. By definition,

$$\mathbf{e} \in f^*(\pi)^{-1}(n) = \mathbb{R}_n \otimes_B P_B = \frac{P_B}{(\operatorname{Ker} n) P_B}$$

and

$$\pi^{-1}(m) = \mathbb{R}_m \otimes_A P = \frac{P}{(\operatorname{Ker} m) P} \ .$$

Since $m = f(n)$ (*i.e.*, $m = n \circ f^*$), there exists a natural isomorphism

$$\iota_n : \mathbb{R}_m \otimes_A P \xrightarrow{\sim} \mathbb{R}_n \otimes_B P_B, \quad \lambda \otimes p \mapsto \lambda \otimes \nu(p) \ .$$

It may be also described by

$$[p]_m \mapsto [\nu(p)]_n \ ,$$

where square brackets denote cosets in the respective quotient modules. Let us define

$$\overline{f} : E_{f^*(\pi)} \to E_\pi, \quad \mathbf{e} \mapsto \iota_n^{-1}(\mathbf{e}) \ .$$

Since $\overline{f}$ sends each fiber $f^*(\pi)^{-1}(n)$ into the fiber $\pi^{-1}(m)$, the diagram

$$\begin{array}{ccc} E_{f^*(\pi)} & \xrightarrow{\overline{f}} & E_\pi \\ {\scriptstyle f^*(\pi)}\downarrow & & \downarrow{\scriptstyle \pi} \\ N & \xrightarrow{f} & M \end{array}$$

is commutative and the restriction

$$\overline{f}_n : f^*(\pi)^{-1}(n) \to \pi^{-1}(m)$$

of $\overline{f}$ is an isomorphism of vector spaces (it is just $\iota_n^{-1}$). It follows that $f^*(\pi)$ is equidimensional, because $\pi$ is equidimensional. Moreover, by n. 0.1.5, (1), $P_B$ is a projective $B$-module, which clearly is also finitely generated. This shows that

$$f^*(\pi) : E_{f^*(\pi)} \to N$$

is a vector bundle.

Since

$$[\nu(p)]_n = \iota_n([p]_m) = \overline{f}_n^{-1}\left([p]_{f(n)}\right), \quad p \in P, n \in N,$$

the map $\nu$ may be recovered from $\overline{f}$ according to the formula

$$\nu(p)(n) = \overline{f}_n^{-1}(p(f(n))), \quad p \in P, n \in N.$$

**Definition.** The vector bundle

$$f^*(\pi) : E_{f^*(\pi)} \to N,$$

together with $\overline{f}$, will be said to be *induced by $f$ from $\pi$*, or a *pull-back of $\pi$ by $f$*. The map $\overline{f}$ will be called *canonical morphism*, or *map induced by $f$*, or simply *induced map*.

An induced bundle, by definition, must always carry on a determined (though sometimes understood) induced map ([18]). Moreover, note that an induced bundle $f^*(\pi) : E_{f^*(\pi)} \to N$ is determined only up to an isomorphism that is compatible with the induced map, because $P_B$ is determined up to an isomorphism that is compatible with the universal homomorphisms $P \to P_B$.

---

[18] Note that, indeed, the same module $P$ could be a scalar extension of $P$ in different ways, by means of different natural functions $P \to P_B$. Consequently, the same bundle may be considered as an induced bundle in different ways, because of different induced maps.

### 0.3.7 Pull-back by a Composition of Maps

Let $\pi : E_\pi \to M$ be a vector bundle and
$$f : N \to M, \quad g : V \to N$$
be smooth maps between manifolds, possibly with boundary. Consider the induced by $f$ bundle $f^*(\pi) : E_{f^*(\pi)} \to N$, with induced map
$$\overline{f} : E_{f^*(\pi)} \to E_\pi ,$$
and then the induced by $g$ from $f^*(\pi)$ bundle $g^*(f^*(\pi)) : E_{g^*(f^*(\pi))} \to V$, with induced map
$$\overline{g} : E_{g^*(f^*(\pi))} \to E_{f^*(\pi)} .$$
From the construction in n. 0.3.6 and elementary properties of scalar extension, it easily follows that $g^*(f^*(\pi))$, together with $\overline{g} \circ \overline{f}$, is an induced by $g \circ f$ from $E_\pi$ bundle.

### 0.3.8 Morphisms of Vector Bundles

Let $\xi : E_\xi \to N$ and $\pi : E_\pi \to M$ be vector bundles and $f : N \to M$ a smooth map. Denote as usual by $f^*(\pi) : E_{f^*(\pi)} \to N$ the induced by $f$ from $E_\pi$ bundle and by $\overline{f}$ the induced map.

**Definition.** A map
$$\overline{f}' : E_\xi \to E_\pi$$
is a *morphism of vector bundles with the base smooth map $f$*, or simply a *morphism over $f$*, if there exists a morphism
$$\overline{g} : E_\xi \to E_{f^*(\pi)}$$
over $N$ (see Definition 0.3.5) such that
$$\overline{f}' = \overline{f} \circ \overline{g} .$$
If $\overline{g}$ is an isomorphism, then $\overline{f}'$ will be said to be *regular*.

A morphism is compatible with the projection maps $\xi$, $\pi$ and fiber-wise linear, because of similar properties of $\overline{f}$ and $\overline{g}$. For each $n \in N$,
$$\overline{f}'_n : \xi^{-1}(n) \to \pi^{-1}(f(n))$$
will denote the restriction of $\overline{f}'$ on the fibers at $n$ and at $f(n)$.

Since the induced bundle $f^*(\pi)$ is defined up to an isomorphism of bundles over $N$, the question of whether a map
$$\overline{f}' : E_\xi \to E_\pi$$
is a morphism over $f$ or not does not depend on the choice of $f^*(\pi)$. The same remark holds for the notion of regularity.

**0.3.9**

If a choice of the induced bundle $f^*(\pi)$ is fixed, then the morphism
$$\overline{g}: E_\xi \to E_{f^*(\pi)}$$
such that
$$\overline{f}' = \overline{f} \circ \overline{g}$$
is uniquely determined by $\overline{f}'$, because the induced map $\overline{f}$ induces isomorphisms on the fibers. This morphism, in turn, leads to the $C^\infty(N)$–module homomorphism
$$\varphi: \Gamma(\xi) \to \Gamma(f^*(\pi)) = C^\infty(N) \otimes_{C^\infty(M)} \Gamma(\pi),$$
that is uniquely determined by the condition
$$\overline{f}' = \overline{f} \circ |\varphi|.$$

Conversely, given $\varphi: \Gamma(\xi) \to C^\infty(N) \otimes_{C^\infty(M)} \Gamma(\pi)$, the above equality gives, by definition, a morphism $\overline{f}': E_\xi \to E_\pi$.

**0.3.10 Composition of Morphisms**

Let $\eta: E_\eta \to V, \xi: E_\xi \to N$, and $\pi: E_\pi \to M$ be vector bundles, and consider morphisms
$$\overline{f}: E_\xi \to E_\pi \quad \text{and} \quad \overline{g}: E_\eta \to E_\xi,$$
respectively over
$$f: N \to M \quad \text{and} \quad g: V \to N.$$

If $\varphi: \Gamma(\xi) \to C^\infty(N) \otimes_{C^\infty(M)} \Gamma(\pi)$ and $\gamma: \Gamma(\eta) \to C^\infty(V) \otimes_{C^\infty(N)} \Gamma(\xi)$ are the homomorphisms respectively corresponding to $\overline{f}$ and $\overline{g}$ (see n. 0.3.9), then the fact that $\overline{f} \circ \overline{g}$ is a morphism over $f \circ g$ may be easily deduced from the following commutative diagram:

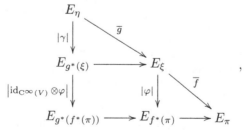

where the horizontal arrows are the induced maps.

## 0.3.11

In conclusion, vector bundles form a category, which will be denoted by $\mathfrak{VB}$. If $M$ is a fixed manifold, possibly with boundary, vector bundles and morphisms over $M$ form a subcategory $\mathfrak{VB}_M$ of $\mathfrak{VB}$. Note that, generally, it is *not* a full subcategory, because morphisms over $M$ are morphisms *over the identity map* of $M$.

The main subject of this book will be another important subcategory of $\mathfrak{VB}$: vector bundles with a fixed general fiber $F$ together with *regular* morphisms. They will be called *fat manifolds* and will be carefully studied starting from Chap. 1.

## 0.3.12

Let $\pi : E_\pi \to M$ be a vector bundle, $f : N \to M$ a smooth map, and $f^*(\pi) : E_{f^*(\pi)} \to N$ the induced bundle. The induced map $\overline{f} : E_{f^*(\pi)} \to E_\pi$ is obviously a regular morphism over $f$.

Conversely, if $\xi : E_\xi \to N$ is a vector bundle and
$$\overline{f}' : E_\xi \to E_\pi$$
is a regular morphism, then, by the definition of a regular morphism, $\xi$ is isomorphic to the induced bundle. But the induced bundle is defined up to an isomorphism compatible with the induced map. Therefore $\xi$, together with $\overline{f}'$, is an induced bundle.

In conclusion, assigning an induced by $f$ bundle is equivalent to assigning a regular morphism over $f$.

## 0.3.13  *Associated Homomorphism with a Regular Morphism of Vector Bundles*

Let $\xi : E_\xi \to N$ and $\pi : E_\pi \to M$ be vector bundles and
$$\overline{f} : E_\xi \to E_\pi$$
a regular morphism over $f : N \to M$. According to n. 0.3.12 and taking into account the construction in n. 0.3.6, one may assume $\Gamma(\xi) = C^\infty(N) \otimes_{C^\infty(M)} \Gamma(\pi)$ with universal homomorphism
$$\overline{f}^* : \Gamma(\pi) \to \Gamma(\xi)$$
given by
$$\overline{f}^*(p)(n) = \overline{f}_n^{-1}(p(f(n))), \quad p \in P, n \in N.$$

**Definition.** The homomorphism $\overline{f}^*$ will be called *associated with* $\overline{f}$.

Note that if $N = M$ and $f$ is the identity map, then $\overline{f} = \left| \overline{f}^{*-1} \right|$.

### 0.3.14 Universal Property of the Induced Bundle

Let $\pi : E_\pi \to M$ be a vector bundle and $f : N \to M$ a smooth map. As a trivial consequence of the definition of bundle morphisms, the induced bundle

$$f^*(\pi) : E_{f^*(\pi)} \to N,$$

together with the induced map

$$\overline{f} : E_{f^*(\pi)} \to E_\pi,$$

satisfies the following universal property. *For every vector bundle $\xi : E_\xi \to N$ and every bundle morphism $\overline{f}' : E_\xi \to E_\pi$ over $f$, there exists exactly one bundle morphism $\overline{g} : E_\xi \to E_{f^*(\pi)}$ over $N$ such that $\overline{f}' = \overline{f} \circ \overline{g}$* ([19]). In this book the following more general result will be needed.

**Proposition.** *(Universal property of the induced bundle.)* For every vector bundle

$$\xi : E_\xi \to V$$

and every morphism

$$\overline{g} : E_\xi \to E_\pi$$

of vector bundles such that the base map

$$g : V \to M$$

factors through $f$ and a smooth map

$$h : V \to N$$

(*i.e.*, $g = f \circ h$), there is exactly one morphism of vector bundles

$$\overline{h} : E_\xi \to E_{f^*(\pi)}$$

over $h$ such that $\overline{g}$ factors through $\overline{f}$ and $\overline{h}$.

Moreover, $\overline{h}$ is regular if and only if $\overline{g}$ is regular.

The proof is straightforward and left to the reader.

---

[19] A generalization of this property to *fiber* bundles may be found in [Nestruev (2003), 10.18].

### 0.3.15 Standard Trivial Bundles

Let $E$ be a finite-dimensional vector space over $\mathbb{R}$ and $\varepsilon_E : E_{\varepsilon_E} \to |\mathbb{R}|$ the vector bundle determined by $E$. The smooth manifold $|\mathbb{R}|$ contains only one point $\mathrm{id}_{\mathbb{R}}$, denoted below by a dot '$\bullet$', and the unique fiber of $\varepsilon_E$ coincides with the whole total space

$$E_{\varepsilon_E} = \frac{E}{0E} = \frac{E}{\{0\}} \;.$$

It may be assumed that

$$E_{\varepsilon_E} = E$$

up the usual identification of each $\mathbf{e} \in E$ with its coset $\mathbf{e} + \{0\} = \{\mathbf{e}\}$ ([20]).

Now, let $M$ be a smooth manifold, possibly with boundary.

**Definition.** A *standard trivial bundle over $M$ with standard fiber $E$* will be a vector bundle $\pi : E_\pi \to M$ induced from $\varepsilon_E$ by the constant map

$$M \to \{\bullet\} \;.$$

The induced map $E_\pi \to E$ will be also called the *trivializing morphism*. An arbitrary bundle $\pi : E_\pi \to M$ with general fiber $E$ will be said to be a *trivial vector bundle*, if there exists a regular morphism of bundles $E_\pi \to E$ over the constant map $M \to \{\bullet\}$.

In other words, to give a standard trivial bundle is the same as to give a trivial bundle $\pi : E_\pi \to M$ together with a fixed choice of a regular morphism $E_\pi \to E$ over the constant map (*i.e.*, the trivializing morphism). By the same reason, it may be said that a bundle is trivial if *it may be realized* as a bundle on $M$ induced from $E$ by the constant map.

---

[20] Note that, on the strictly formal side, the identification $\Gamma(E_{\varepsilon_E}) = \Gamma(E)$ admits now two interpretations. The first is given, as usual, by the geometrization isomorphism, because $E$ is the $\mathbb{R}$-module defining the bundle $\varepsilon_E$. The second interpretation is given by the identification $E_{\varepsilon_E} = E$ introduced above. According to these identifications, a vector $\mathbf{e} \in E$ is also interpreted as the coset

$$\mathbf{e} + \{0\} = \{\mathbf{e}\} \in E_{\varepsilon_E}$$

and as the section

$$s_{\mathbf{e}} \in \Gamma(E), \quad \bullet \mapsto \{\mathbf{e}\} \equiv \mathbf{e} \;.$$

### 0.3.16 Identification Isomorphisms

A canonical identification of the fibers of a standard trivial bundle $\pi : E_\pi \to M$ with its standard fiber $E$ is obtained by means of the trivializing morphism $E_\pi \to E$ as follows.

**Definition.** For all $m \in M$ the restriction

$$\tau_m : \pi^{-1}(m) \xrightarrow{\sim} E$$

of the trivializing morphism $\tau : E_\pi \to E$ will be called the *identification isomorphism at m*. Vectors in the fiber $\pi^{-1}(m)$ will often be identified with their correspondents in $E$.

Although the terms 'standard fiber' and 'general fiber' are usually synonymous, in this book there is a distinction between them: 'standard fiber' is reserved for standard trivial bundles, and refers to the fixed vector space which the fibers are identified with.

According to the identification isomorphisms, the notion of a *constant section* makes sense, that is, a section that associates a fixed $\mathbf{e} \in E$ with all $m \in M$. Such a section is smooth because it is identified with $1 \otimes \mathbf{e} \in \Gamma(E_\pi) = C^\infty(M) \otimes_\mathbb{R} E$.

### 0.3.17 Uniform Morphisms

Let $\xi : E_\xi \to N$ and $\pi : E'_\pi \to M$ be standard trivial bundles with standard fiber $E$ and $E'$, respectively. Consider a vector space homomorphism $\varphi : E \to E'$ and a smooth map $f : N \to M$. Because of the universal property of the induced bundle, there exists exactly one morphism

$$\overline{f}_\varphi : E_\xi \to E'_\pi$$

with base $f$ such that the diagram

$$\begin{array}{ccc} E_\xi & \xrightarrow{\overline{f}_\varphi} & E'_\pi \\ {\scriptstyle \tau}\downarrow & & \downarrow{\scriptstyle \tau'} \\ E & \xrightarrow{\varphi} & E' \end{array},$$

is commutative (here $\tau$ and $\tau'$ denote the trivializing morphisms). Up to the identification isomorphisms, $\overline{f}_\varphi$ may be characterized by

$$\left(\overline{f}_\varphi\right)_n = \varphi \quad \forall n \in N .$$

If $\varphi$ is an isomorphism, then $\overline{f}_\varphi$ is regular, and the associated homomorphism is

$$\overline{f}_\varphi{}^* = f^* \otimes_\mathbb{R} \varphi^{-1}.$$

**Definition.** The morphism $\overline{f}_\varphi$ will be called the *uniform morphism with base $f$ and fiber $\varphi$*.

### 0.3.18 Geometric Definition of Vector Bundles

Let $E$ be a finite-dimensional real vector space and consider it also as a smooth manifold (see n. 0.2.19). Let

$$\pi : E_\pi \to M$$

be a smooth map between manifolds, possibly with boundary, and suppose that, for each $m \in M$, it is given an $\mathbb{R}$-vector space structure on the set $\pi^{-1}(m)$.

**Definition.** The smooth map $\pi$, together with the assigned family of vector space structures, is said to satisfy the *vector property of local triviality* (with respect to $E$) if, for each $m \in M$, there is an open neighborhood $U_m \subseteq M$ and a map $\tau_m : \pi^{-1}(U_m) \to E$ such that:

(1) the manifold $\pi^{-1}(U_m)$, together with

$$\pi|_{\pi^{-1}(U_m), U_m} : \pi^{-1}(U_m) \to U_m$$

and

$$\tau_m : \pi^{-1}(U_m) \to E$$

is a product $U_m \times E$;
(2) for all $u \in U_m$, the restriction $\tau_m|_{\pi^{-1}(u)} : \pi^{-1}(u) \to E$ is linear (hence a vector space isomorphism).

A smooth map $\pi : E_\pi \to M$ that satisfies the vector property of local triviality ([21]) is what it is called a *vector bundle* according to (a version of) the classical 'geometric' definition: see [Nestruev (2003), 11.2].

---

[21] In the common usage, the family of vector space structures is generally understood.

## 0.3.19 Vector Bundle Morphisms from a Geometric Viewpoint

Let $\pi : E \to M$ and $\pi' : E' \to M'$ be smooth maps satisfying the vector property of local triviality, and $f : M \to M'$ a smooth map.

**Definition.** A map $\overline{f} : E \to E'$ will be called a *morphism with the base map* $f$, or simply a *morphism over* $f$, if it satisfies the following conditions:

(1) it is smooth;
(2) the diagram

$$\begin{array}{ccc} E & \xrightarrow{\overline{f}} & E' \\ \pi \downarrow & & \downarrow \pi' \\ M & \xrightarrow{f} & M' \end{array}$$

is commutative;
(3) for each $m \in M$, the restriction $\overline{f}_m : \pi^{-1}(m) \to \pi'^{-1}(f(m))$ is linear.

A morphism $\overline{f}$ is said to be *regular* if $\overline{f}_m$ is an isomorphism for all $m$.

It is immediately checked that, in this way, one gets a category $\mathfrak{VB}_g$.

## 0.3.20 Module of Smooth Sections of a Geometrically Defined Vector Bundle

A *section* of an object $\pi : E_\pi \to M$ of $\mathfrak{VB}_g$ is a map

$$s : M \to E_\pi$$

such that

$$\pi \circ s = \mathrm{id}_M \, ;$$

a *smooth section of* $\pi$ is a section that is smooth as a map between manifolds. Because of the vector spaces structures on the fibers of $\pi$, a natural $C^\infty(M)$-module structure on the set of all sections of $\pi$ is defined (*cf.* the operations on $\{s_p\}_{p\in P}$ defined in n. 0.3.1). Using *adapted coordinates* [22], it is also easily seen that the module operations preserve smoothness; thus, smooth sections form a submodule. The so-obtained

---

[22] See [Nestruev (2003), 11.3]; details about manifolds with boundary are left to the reader.

module of smooth sections of $\pi$ is denoted by $\Gamma(\pi)$ (cf. [Nestruev (2003), 11.7]), or even by $\Gamma(E_\pi)$.

A simple but important result about $\Gamma(E_\pi)$ is the following. Fix $m \in M$ and consider the homomorphism

$$\overline{h}_m : \Gamma(E_\pi) \to \pi^{-1}(m), \quad s \mapsto s(m)$$

of vector spaces over $\mathbb{R}$.

**Proposition.** The homomorphism $\overline{h}_m$ is surjective and its kernel is the module

$$\mu_m \Gamma(E_\pi) ,$$

where $\mu_m = \operatorname{Ker} m$.

***Proof.*** See [Nestruev (2003), 11.8, 11.9]. □

### 0.3.21

**Theorem.** Let $M$ be a manifold, possibly with boundary. A $C^\infty(M)$–module $P$ is isomorphic to $\Gamma(\pi)$ for some object $\pi : E_\pi \to M$ of $\mathfrak{VB}_g$ if and only if it is finitely generated, projective and determines an equidimensional pseudobundle, the last condition being superfluous when $M$ is connected.

***Proof.*** In the case when $M$ is connected, the statement with the equidimensionality condition left out is proved in [Nestruev (2003), Theorem 11.32] ([23]). The connectedness of $M$ is used only in the 'if' implication, to show that that the fibers $P_m$, $m \in M$, of $P$ are of the same dimension. On the other hand, if $P$ is isomorphic to $\Gamma(\pi)$ for some object $\pi : E_\pi \to M$ of $\mathfrak{VB}_g$, equidimensionality is an immediate consequence of Proposition 0.3.20. This explains why, for non-connected manifolds, it suffices to add the equidimensionality condition. □

In other words, the above result says that a $C^\infty(M)$–module is the module of smooth section of some object of $\mathfrak{VB}$ if and only if it is isomorphic to the module of smooth sections of some object of $\mathfrak{VB}_g$.

---

[23] Plainly, all the preceding Nestruev's results needed in the proof hold for manifolds with boundary (to check details about the Whitney's immersion theorem that is involved in [Nestruev (2003), Theorem 11.27], the interested reader may refer, e.g., to the proof given in [Lee (2003), Chap. 12]). To simplify the job, one may also avoid the invocation of the geometric construction of the induced bundle: in [Nestruev (2003), Corollary 11.28], it is sufficient to know that there exists a regular morphism of $E_\pi$ into the tautological bundle over an appropriate Grassmann manifold.

## 0.3.22

A projective and finitely generated $C^\infty(M)$–module $P$ determines a vector bundle (that is, the pseudobundle $|P| \to M$ is equidimensional) with $r$-dimensional fibers if and only if $P$ has constant rank $r$. Since we do not strictly need this algebraic fact in the sequel, we leave the details to the interested readers ([24]).

## 0.3.23 The Total Space as a Smooth Manifold

Let $\pi_P : |P| \to M$ be a vector bundle with module of smooth sections $P$. According to Theorem 0.3.21, it is possible to fix an object $\pi : E_\pi \to M$ of $\mathfrak{VB}_g$ and a $C^\infty(M)$–module isomorphism

$$\iota : P \xrightarrow{\sim} \Gamma(\pi) \ .$$

**Proposition.** There is a unique bijective map

$$\theta : |P| \to E_\pi$$

such that

$$\theta \circ p = \iota(p), \quad p \in P \ .$$

Moreover,

(1) the diagram

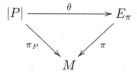

is commutative, i.e., for each $m \in M$, $\theta$ sends the fiber of $\pi_P$ over $m$ into the fiber of $\pi$ over $m$,

(2) for each $m \in M$ the restriction

$$\theta_m : \pi_P^{-1}(m) \to \pi^{-1}(m)$$

is an isomorphism of vector spaces.

The proof is straightforward and left to the reader.

---

[24] One may use Nakayama's Lemma (see, *e.g.*, [Atiyah and Macdonald (1969), Proposition 2.6]) and the fact that a prime spectrum Spec $A$ is connected if and only if the ring $A$ contains no nontrivial idempotents (see [Atiyah and Macdonald (1969), Chap. 1, Exercises, n. 22 (p. 14)]); *cf.* also [Hartshorne (1977), Chap. III, Example 12.7.2].

**Corollary.** In the notation of the proposition,

$s: M \to |P|$ is a smooth section of $\pi_P \iff \theta \circ s$ is a smooth section of $\pi$.

**Proof.** Since $\theta$ is bijective,
$$s \mapsto \theta \circ s$$
gives a bijection between the set of all sections of $\pi_P$, i.e., all the maps $s: M \to |P|$ such that $\pi_P \circ s = \mathrm{id}_M$, to the set of all sections of $\pi$. By the proposition, this bijection restricts to the subsets $\Gamma(P)$ and $\Gamma(\pi)$: indeed, the restriction coincides with $\iota$ up to the identification $P = \Gamma(P)$. This means exactly that $s$ is a smooth section of $\pi_P$ if and only if $\theta \circ s$ is a smooth section of $\pi$. □

**Definition.** The map $\theta$ will be said to be *associated with* $\iota$.

When necessary, the total space $|P|$ of a vector bundle will be identified with the smooth manifold $E_\pi$ (possibly with boundary) through the map $\theta$ associated with a fixed (understood) isomorphism $\iota: P \to \Gamma(\pi)$.

## 0.3.24

Let $P$ be the module of smooth sections of a vector bundle $\pi: E_\pi \to M$ and suppose that a map $a: E_\pi \to \mathbb{R}$ is linear on the fibers of $\pi$ and smooth when $E_\pi$ is considered as a smooth manifold according to n. 0.3.23. By virtue of Corollary 0.3.23,
$$p \in P \Rightarrow a \circ p \in C^\infty(M)$$
and by fiber-wise linearity of $a$, the map
$$\alpha: P \to C^\infty(M), \quad p \mapsto a \circ p$$
belongs to the dual module $P^\vee$.

It is easy to show that
$$a \mapsto \alpha$$
gives a bijective map between the set of smooth, fiber-wise linear functions on $E_\pi$ and the dual $P^\vee$.

Smooth, fiber-wise linear functions on $E_\pi$ will be generally identified with linear forms on $P$ according to the above bijection. Thus,
$$\alpha(p(m)) = \alpha(p)(m), \quad \alpha \in P^\vee, p \in P, m \in M.$$

Moreover, in this context, the identification of smooth functions on $M$ with their correspondent through $\pi^*$ (which are fiber-wise constant maps) will be assumed:

$$f(p(m)) = f(m) \qquad f \in C^\infty(M), p \in P, m \in M$$

([25]).

### 0.3.25 Geometric Criterion for Vector Bundle Morphisms

Let $f : N \to M$ be a smooth map between manifolds, possibly with boundary, and $\pi : E_\pi \to M$, $\xi : E_\xi \to N$ vector bundles with modules of smooth sections $P$ and $Q$, respectively.

Suppose that $\overline{f} : E_\xi \to E_\pi$ is a morphism when $\xi$ and $\pi$ are considered as elements of $\mathfrak{VB}_g$ according to n. 0.3.23. If $\alpha \in P^\vee$ is considered as a smooth function on $E_\pi$ according to n. 0.3.24, then $\alpha \circ \overline{f}$ is smooth and obviously fiber-wise linear. This way, one gets a homomorphism

$$P^\vee \to Q^\vee, \quad \alpha \mapsto \alpha \circ \overline{f}$$

and, therefore, a $C^\infty(N)$-module homomorphism

$$C^\infty(N) \otimes_{C^\infty(M)} P^\vee \to Q^\vee .$$

This homomorphism must be of the form $\varphi^\vee$ for a uniquely determined

$$\varphi : Q \to C^\infty(N) \otimes_{C^\infty(M)} P$$

because $P$ and $Q$ are finitely generated and projective (see n. 0.1.5, (6) and (8)). It is a good exercise to show that the morphism of vector bundles determined by $\varphi$ according to n. 0.3.9 coincides with $\overline{f}$.

In conclusion, if a fiber-wise linear map $\overline{f} : E_\xi \to E_\pi$ over $f$ is smooth as a map between manifolds, then it is a vector bundle morphism (even in the algebraic sense).

### 0.3.26 Equivalence Between $\mathfrak{VB}_g$ and $\mathfrak{VB}$

Theorem 0.3.21 and n. 0.3.25 allow us to define a faithful functor $\mathfrak{VB}_g \to \mathfrak{VB}$ that associates, with each $\pi$, the vector bundle determined by $\Gamma(\pi)$ and, with each morphism, the corresponding morphism obtained by means

---

[25]Note that the identification homomorphisms $C^\infty(M) \hookrightarrow C^\infty(E_\pi)$, $P^\vee \hookrightarrow C^\infty(E_\pi)$ induce a $C^\infty(M)$-algebra homomorphism $\iota : S(P^\vee) \to C^\infty(E_\pi)$, where $\iota : S(P^\vee)$ denotes the symmetric algebra of $P^\vee$ ($S^r$ will denote an $r$-th symmetric power). It may be proved that $\iota$ is injective, that $S(P^\vee)$ is geometric, and that $C^\infty(E_\pi)$, together with $\iota$, is a smooth envelope (cf. [Nestruev (2003), 11.58]).

of the identification maps associated with the identity homomorphisms, according to n. 0.3.23. It may be easily proved, using adapted coordinates, that this functor is also full (*cf.* [Nestruev (2003), 11.29 (b)]). Finally, again by Theorem 0.3.21, every vector bundle is isomorphic to the correspondent of some object of $\mathfrak{VB}_g$. This suffices to prove that $\mathfrak{VB}_g$ and $\mathfrak{VB}$ are equivalent categories (see n. 0.1.6).

## 0.4 Vector Fields

Basic properties of vector fields, with some extensions about vector fields *along maps*, are reviewed in this section.

### 0.4.1 Smooth Vector Fields

Let $M$ be a manifold, possibly with boundary. A *(smooth) vector field on $M$* is a derivation of the $\mathbb{R}$-algebra $C^\infty(M)$. The $C^\infty(M)$–module $D(C^\infty(M))$ of all vector fields will be also shortly denoted by $D(M)$.

For each point $m \in M$, a vector field $X$ gives, in a natural way, a tangent vector $X_m$ at $m$:

$$X_m = m \circ X \, .$$

The vector field $X$ may be recovered from the family $\{X_m\}_{m \in M}$, but not all families of tangent vectors give rise to a smooth vector field. With this description, it is easy to see that vector fields are local operators, because tangent vectors are as well: see [Nestruev (2003), 9.40]. The description of a vector field in local coordinates $U \to \mathbb{R}^n$ is given by

$$\sum_{i=1}^{n} \alpha_i \frac{\partial}{\partial x_i}, \quad \alpha_i \in C^\infty(U)$$

(see [Nestruev (2003), 9.41]).

### 0.4.2 Image of a Vector Field

**Definition.** Let $f : N \to M$ be a diffeomorphism and $Y$ a vector field on $N$. The *image of $Y$ through $f$* will be the vector field

$$f_*(Y) \stackrel{\text{def}}{=} f^{*-1} \circ Y \circ f^*$$

on $M$.

Note that
$$f_*(Y)_{f(n)} = (\mathrm{d}_n f)(Y_n), \quad n \in N.$$
In the more general context of vector fields along maps, which will be examined later, another natural notion of 'image' arises (*cf.* also [Nestruev (2003), Example 9.47]).

### 0.4.3 *Compatibility*

Let $f : N \to M$ be a smooth map, and $X$ and $Y$ be vector fields on $M$ and $N$, respectively.

**Definition.** The fields $X$ and $Y$ are said to be *compatible with respect to* $f$ (or, for short, $f$-*compatible*) if
$$Y \circ f^* = f^* \circ X.$$

A geometric characterization of compatibility is
$$(\mathrm{d}_n f)(Y_n) = X_{f(n)}, \quad n \in N.$$

In the case when $f$ is a diffeomorphism, note that the image $f_*(Y)$ is the unique vector field on $M$ that is $f$-compatible with $Y$.

### 0.4.4 *Commutator of Vector Fields*

According to n. 0.1.3, the commutator
$$[X, Y] = X \circ Y - Y \circ X$$
of vector fields is again a vector field. It follows that vector fields on $M$ constitute a Lie algebra.

Now, let $f : N \to M$ be a smooth map and suppose that $X_1, X_2 \in \mathrm{D}(M)$ are respectively $f$-compatible with $Y_1, Y_2 \in \mathrm{D}(N)$. Then it is easily checked that $[X_1, X_2]$ is $f$-compatible with $[Y_1, Y_2]$.

### 0.4.5 *Restriction on Open Submanifolds*

**Proposition.** Let $N$ be an open submanifold of a manifold $M$ (both possibly with boundary) and $X$ a vector field on $M$. Then, there is a unique vector field $X|_N$ on $N$ compatible with $X$ with respect to the embedding $i : N \hookrightarrow M$.

**Proof.** Let $A = C^\infty(M)$ and $f \in C^\infty(N) = A|_N$. By the definition of the restriction algebra, for each point $n \in N$, an open neighborhood $U_{f,n} \subseteq N$ and a function $f_n \in A$ such that

$$f|_{U_{f,n}} = f_n|_{U_{f,n}}$$

may be chosen. Hence, setting

$$X|_N (f)(n) \stackrel{\text{def}}{=} X(f_n)(n)$$

one gets a function $X|_N (f) : N \to \mathbb{R}$. Taking into account the fact that vector fields are local operators, it is easily checked that the function

$$X|_N : A|_N \to A|_N, \quad f \mapsto X|_N(f)$$

is well defined, that it is a vector field, and that it is the unique one $i$-compatible with $X$. ([26]). □

**Definition.** The vector field $X|_N$ will be called the *restriction of $X$ to $N$*.

### 0.4.6

Since $i^*$ is the restriction homomorphism $A \to A|_N$, the compatibility condition that characterizes the restriction $X|_N$ may be written

$$X|_N (f|_N) = X(f)|_N, \quad f \in A.$$

### 0.4.7

Let $M$ be a smooth manifold, possibly with boundary, $\{U_i\}_{i \in I}$ an open covering of $M$ and, for each $i \in I$, let $X_i$ be a vector field on the submanifold $U_i$.

**Proposition.** If the vector fields $X_i$ agree on the intersections, *i.e.*,

$$X_i|_{U_i \cap U_j} = X_j|_{U_i \cap U_j}, \quad i,j \in I,$$

then there exists a unique vector field $X$ on $M$ such that

$$X|_{U_i} = X_i \quad \forall i \in I.$$

---

[26] The above result could be proved in a even more algebraic fashion, once it is known that $A|_N$ may be realized as an algebraic localization of $A$ (see [Nestruev (2003), 10.7]). In fact, the restriction of $X$ could be defined, even in the general context of arbitrary commutative algebras, by the rule

$$X|_N \left(\frac{f}{g}\right) = \frac{X(f)g - fX(g)}{g^2}.$$

***Proof.*** Let $A = C^\infty(M)$, so that $C^\infty(U_i) = A|_{U_i}$ for all $i$. For all $f \in A$ and $i, j \in I$,

$$X_i(f|_{U_i})|_{U_i \cap U_j} = X_i|_{U_i \cap U_j}(f|_{U_i \cap U_j}) = X_j|_{U_i \cap U_j}(f|_{U_i \cap U_j})$$
$$= X_j(f|_{U_j})|_{U_i \cap U_j}.$$

Thus, the functions

$$X_i(f|_{U_i}) \in A|_{U_i}$$

agree on the intersections. Since $A$ is complete (being smooth with boundary), there exists a unique function $f' \in A$ such that

$$f'|_{U_i} = X_i(f|_{U_i}) \quad \forall i \in I.$$

It is straightforward to check that

$$X : A \to A, \quad f \mapsto f'$$

is the required vector field. □

***Definition.*** The vector field $X$ is said to be obtained by *gluing* the vector fields $X_i$.

### 0.4.8 Extension from Closed Submanifolds

***Proposition.*** Let $N$ be a closed submanifold of a manifold $M$ (both possibly with boundary) and $X$ a vector field on $N$. Then $X$ can be extended to $M$, i.e., there exists a (not necessarily unique) vector field $X^M$ on $M$ that is compatible with $X$ with respect to the embedding $N \hookrightarrow M$.

***Proof.*** According to the local description of closed submanifolds (see n. 0.2.29) and the coordinate description of vector fields, the result is true around each point of $N$. Thus the global result is easily deduced with the help of a partition of unity and a gluing procedure. □

### 0.4.9 Tangent Bundle

Let $M$ be a manifold, possibly with boundary. According to [Nestruev (2003), 11.6 (II)] and [Nestruev (2003), Exercise 9.40], there exists an object $\pi : TM \to M$ of $\mathfrak{VB}_\mathfrak{g}$ such that $\Gamma(\pi)$ is isomorphic to the module $D(M)$ of all vector fields on $M$. By the equivalence between $\mathfrak{VB}_\mathfrak{g}$ and $\mathfrak{VB}$, the pseudobundle $\pi_{D(M)} : |D(M)| \to M$ determined by $D(M)$ is a vector bundle. In this book, the *tangent bundle of* $M$ will be, by definition, $\pi_{D(M)}$.

## 0.4.10

By the Nestruev's construction of the tangent bundle in $\mathfrak{VB}_\mathfrak{g}$, the fiber at $m \in M$ of the tangent bundle in $\mathfrak{VB}$ turns out to be naturally isomorphic to the tangent space $T_m M$, and the value at $m$ of a vector field $X$, considered as a section of the tangent bundle, corresponds to the tangent vector $X_m$. More precisely, if $\mu_m = \operatorname{Ker} m$ is the ideal of functions vanishing at $m$, then there exists a natural isomorphism

$$T_m M \cong \frac{\mathrm{D}(M)}{\mu_m \mathrm{D}(M)}.$$

In other terms, $T_m M$ is an $\mathbb{R}$-vector space

$$\mathbb{R}_m \otimes_{\mathrm{C}^\infty(M)} \mathrm{D}(M)$$

obtained from $\mathrm{D}(M)$ by extension of scalars via $m : \mathrm{C}^\infty(M) \to \mathbb{R}$, with universal homomorphism

$$\mathrm{D}(M) \to T_m M, \quad X \mapsto X_m.$$

## 0.4.11

If $E$ is a vector space regarded as a smooth manifold (see n. 0.2.19), from the local description of vector fields it easily follows that for each vector $e \in E$ there exists exactly one vector field $X_e$ on $E$ such that

$$X_e(\varphi) = \varphi(e), \quad \varphi \in E^\vee,$$

and $\mathrm{D}(E)$ can be regarded as obtained from $E$ by extension of scalars via $\mathbb{R} \to \mathrm{C}^\infty(E)$, with the universal homomorphism given by $e \mapsto X_e$:

$$\mathrm{D}(E) = \mathrm{C}^\infty(E) \otimes_\mathbb{R} E.$$

Therefore the tangent bundle $TE$ is in a canonical way a standard trivial bundle with standard fiber $E$. We shall often tacitly identify tangent vectors to $E$ with vectors in $E$ through the identification isomorphisms and the isomorphisms of n. 0.4.10.

## 0.4.12 Vector Fields Along Maps

Let $f : N \to M$ be a smooth map between manifolds, possibly with boundary.

**Definition.** A *(smooth) vector field along $f$* is a derivation along $f^*$ (see n. 0.1.3).

The $C^\infty(N)$–module $D(C^\infty(M))_{f_*}$ of all vector fields along $f$ will be generally denoted by $D(M)_f$.

For each point $n \in N$, a vector field $X : C^\infty(M) \to C^\infty(N)$ along $f$ gives, in a natural way, a tangent vector $X_n$ to $M$ at $f(n)$:

$$X_n = n \circ X,$$

and $X$ is determined by the family $\{X_n\}_{n \in N}$ (but not every such family gives rise to a smooth vector field along $f$).

### 0.4.13

Ordinary vector fields may be understood as vector fields along the identity map. Note also that if $X$ is a vector field along $f : N \to M$ and

$$g : V \to N$$

is a smooth map, then the composition

$$g^* \circ X$$

is a vector field along $f \circ g$. Similarly, for every smooth map

$$g' : M \to V'$$

the composition

$$X \circ g'^*$$

is a vector field along $g' \circ f$.

In particular, every (ordinary) vector field $Y$ on $N$ gives rise to the vector field

$$Y_f \stackrel{\text{def}}{=} Y \circ f^*$$

along $f$, and every (ordinary) vector field $Z$ on $M$ gives rise to the vector field

$$Z^f \stackrel{\text{def}}{=} f^* \circ Z$$

along $f$.

With the above notation, the compatibility condition between $Y$ and $Z$ may also be written

$$Y_f = Z^f.$$

## 0.4.14

The vector field $Y_f$ along $f$ may be considered as the 'image' of $Y$ since, for all $n \in N$,
$$(Y_f)_n = (d_n f)(Y_n) \ .$$
One has to be careful when $f$ is a diffeomorphism: in this case, the image $Y_f$ along $f$ is different from the ordinary image $f_*(Y)$. Hence, instead of 'image', the following terminology will be used.

**Definition.** Let $f : N \to M$ be a smooth map, $X$ be a vector field along $f$, $Y$ be a vector field on $N$ and $Z$ be a vector field on $M$. Then

$$Y \text{ projects onto } X \text{ through } f \stackrel{\text{def}}{\iff} Y_f = X \ ;$$

$$X \text{ projects into } Z \text{ through } f \stackrel{\text{def}}{\iff} Z^f = X \ ;$$

$$Y \text{ projects into } Z \text{ through } f \stackrel{\text{def}}{\iff} Y \text{ and } Z \text{ are } f\text{-compatible}.$$

Moreover, $Y$ will be said to be *projectable through* $f$ if and only if it is $f$-compatible with some vector field on $M$.

## 0.4.15

Let $X$ be a vector field along $f : N \to M$. The following results are easy extensions of basic properties of ordinary vector fields.

(1) $X$ is a local operator, that is, if $g, g' \in C^\infty(M)$ coincide on an open $U \subseteq M$, then $X(g)$ and $X(g')$ coincide on $f^{-1}(U)$.

(2) If $U \subseteq M$ and $V \subseteq N$ are open submanifolds such that
$$f(V) \subseteq U \ ,$$
then there is a unique vector field $X|_{V,U}$ along $f|_{V,U}$ such that
$$X|_{V,U}(f|_U) = X(f)|_V, \quad f \in C^\infty(M) \ .$$
This field will be called the *restriction of $X$ to $V$ and $U$*.

(3) Let $\{U_i\}_{i \in I}$ and $\{V_i\}_{i \in I}$ be open coverings of $M$ and $N$, respectively, such that
$$f(V_i) \subseteq U_i \quad \forall i \in I \ .$$
For each $i \in I$, let $X_i$ be a vector field along the restriction $f|_{V_i, U_i}$. If the vector fields $X_i$ agree on the intersections, *i.e.*,
$$X_i|_{V_i \cap V_j, U_i \cap U_j} = X_j|_{V_i \cap V_j, U_i \cap U_j} \quad \forall i, j \in I \ ,$$
then there is a unique vector field $X$ along $f$ such that
$$X|_{V_i, U_i} = X_i \quad \forall i \in I \ .$$

(4) In local coordinates, $X$ may be described by

$$X(a)(y_1,\ldots,y_n) = \sum_{i=1}^{m} \alpha_i(y_1,\ldots,y_n) \frac{\partial a}{\partial x_i}(f(y_1,\ldots,y_n)).$$

(5) Let $Y$ be a vector field on $N$, $Z$ a vector field on $M$, and $U \subseteq M$, $V \subseteq N$ open submanifolds such that $f(V) \subseteq U$. Then

- if $Y$ projects through $f$ onto $X$, then $Y|_V$ projects onto $X|_{V,U}$ through $f|_{V,U}$;
- if $X$ projects through $f$ into $Z$, then $X|_{V,U}$ projects into $Z|_U$ through $f|_{V,U}$;
- if $Y$ projects through $f$ into $Z$, then $Y|_V$ projects into $Z|_U$ through $f|_{V,U}$.

## 0.4.16 Local Vector Fields

Working on a manifold $M$, one sometimes encounters vector fields that are defined only locally, *i.e.*, on an open $U \subseteq M$. Note that if $X$ is a vector field along the embedding $i : U \hookrightarrow M$, its restriction $Y = X|_{U,U}$ is a vector field on $U$. By the definition of restrictions, $Y$ is the unique vector field on $U$ projecting onto $X$. This gives a one-to-one correspondence between vector fields along $i$ and vector fields on $U$.

**Definition.** A *local vector field on $M$* is a vector field along the embedding $U \hookrightarrow M$ of an open submanifold, and it is sometimes identified with the corresponding vector field on $U$.

## 0.4.17 Splitting of Vector Fields on a Product Manifold

**Proposition.** Let $M$ and $N$ be smooth manifolds, one of them possibly with boundary, let $M \times N$ be their product and denote by $\pi_M : M \times N \to M$ and $\pi_N : M \times N \to N$ the projection maps. For every pair $(X_M, X_N)$ of vector fields respectively along $\pi_M$ and along $\pi_N$, there exists exactly one vector field $X$ on $M \times N$ that projects onto $X_M$ through $\pi_M$ and onto $X_N$ through $\pi_N$.

**Proof.** The result is true locally because of the coordinate description given in n. 0.4.15, (4). The global case follows with the help of a gluing procedure. □

**Corollary.** For every pair $(Y_M, Y_N)$ of vector fields, respectively on $M$ and on $N$, there exists exactly one vector field on $M \times N$ that is compatible with $Y_M$ with respect to $\pi_M$ and with $Y_N$ with respect to $\pi_N$.

**Proof.** It suffices to set $X_M = \pi_M^* \circ Y_M$, $X_N = \pi_N^* \circ Y_N$. □

### 0.4.18

The modules $D(M)_{\pi_M}$ and $D(N)_{\pi_N}$ of vector fields along the projection maps will also be respectively denoted by

$$D(M)_N \quad \text{and} \quad D(N)_M .$$

**Proposition.** There exists a decomposition

$$D(M \times N) = D(M)_N \oplus D(N)_M$$

with universal epimorphisms

$$\pi_{D(M)_N} : D(M \times N) \to D(M)_N, \quad X \mapsto X \circ \pi_M^* ,$$

$$\pi_{D(N)_M} : D(M \times N) \to D(N)_M, \quad X \mapsto X \circ \pi_N^* .$$

**Proof.** It is basically a reformulation of Proposition 0.4.17. □

### 0.4.19

A vector field $X \in D(M \times N)$ lies on the image of the natural monomorphism

$$\iota_{D(M)_N} : D(M)_N \hookrightarrow D(M \times N) = D(M)_N \oplus D(N)_M$$

if and only if it belongs to the kernel of the natural epimorphism

$$\pi_{D(N)_M} : D(M \times N) \to D(N)_M, \quad X \longmapsto X \circ \pi_N^* ,$$

i.e.,

$$X \circ \pi_N^* = 0 .$$

Therefore, $X \in \text{Im}\left(\iota_{D(M)_N}\right)$ if and only if $X$ vanishes on the image of the homomorphism

$$\pi_N^* : C^\infty(N) \to C^\infty(M \times N)$$

that defines $C^\infty(M \times N)$ as a $C^\infty(N)$-algebra. Hence $X \in \text{Im}\left(\iota_{D(M)_N}\right)$ if and only if $X$ is a $C^\infty(N)$-derivation of $C^\infty(M \times N)$ into itself:

$$\text{Im}\left(\iota_{D(M)_N}\right) = D_{C^\infty(N)}\left(C^\infty(M \times N)\right) .$$

The module $\mathrm{D}_{\mathrm{C}^\infty(N)}(\mathrm{C}^\infty(M \times N))$ of all derivations of the $\mathrm{C}^\infty(N)$-algebra $\mathrm{C}^\infty(M \times N)$ will be also denoted by
$$\mathrm{D}_N(M \times N) \,.$$
Summing up, there are natural isomorphisms
$$\mathrm{D}(M)_N \xrightarrow{\sim} \mathrm{D}_N(M \times N) \quad \text{and} \quad \mathrm{D}(N)_M \xrightarrow{\sim} \mathrm{D}_M(M \times N)$$
and an internal decomposition
$$\mathrm{D}(M \times N) = \mathrm{D}_N(M \times N) \oplus \mathrm{D}_M(M \times N) \,.$$

**0.4.20**

The decompositions
$$\mathrm{D}(M \times N) = \mathrm{D}(M)_N \oplus \mathrm{D}(N)_M$$
and
$$\mathrm{D}(M \times N) = \mathrm{D}_N(M \times N) \oplus \mathrm{D}_M(M \times N)$$
precisely express the intuitive fact that every vector field on a product may be decomposed into a horizontal and a vertical component. Moreover, n. 0.4.19 says that there are two natural formalizations of the concept of a 'horizontal' (respectively, vertical) vector field. The former is: a vector field along the projection on the horizontal (resp., vertical) factor. The other is: a vector field projecting into the zero vector field of the vertical (resp., horizontal) factor.

**0.4.21**

In this book, *interval* means a connected subset of $\mathbb{R}$. Intervals may be naturally considered as smooth manifolds, possibly with boundary ([27]). If $M$ is a manifold, possibly with boundary, by a *(smooth) curve in* $M$ we mean a smooth map $\mathbb{I} \to M$, with $\mathbb{I}$ being a nonempty interval, not reduced to a singleton.

**Definition.** Let $\mathbb{I}$ be a nonempty interval, not reduced to a singleton. The vector field
$$\mathrm{C}^\infty(\mathbb{I}) \to \mathrm{C}^\infty(\mathbb{I}), \quad f \mapsto f'$$
where $f'$ indicates the derivative of $f$, will be called the *standard vector field on* $\mathbb{I}$.

---

[27] For open intervals, see n. 0.2.19; details about the other cases are left to the reader.

Of course, when speaking about the standard vector field on an interval $\mathbb{I}$, it will be always tacitly implied that $\mathbb{I}$ is nonempty and not reduced to a singleton. Since it is convenient to think of $\mathbb{R}$ as the timeline, the standard vector field on $\mathbb{I}$ will generally be denoted by $d/dt$.

### 0.4.22 *Trajectories*

Let $X$ be a vector field on a smooth manifold $M$, possibly with boundary.

**Definition.** A *(smooth) trajectory of* $X$ (also called a *(smooth) integral curve*) is a curve

$$\gamma : \mathbb{I} \to M \;,$$

such that $X$ is $\gamma$-compatible with the standard vector field $d/dt$ on $\mathbb{I}$.

Plainly, a trajectory $\gamma : \mathbb{I} \to M$ is said to be *maximal* if it cannot be prolonged, *i.e.*, there are no trajectories $\gamma' : \mathbb{I}' \to M$ such that

$$\mathbb{I}' \supsetneq \mathbb{I} \quad \text{and} \quad \gamma = \gamma'|_{\mathbb{I}} \;.$$

### 0.4.23

Let $f : N \to M$ be a smooth map and $X$ and $Y$ be vector fields respectively on $M$ and $N$, that are compatible with respect to $f$. From the definition of trajectories and the transitivity of compatibility condition it follows that

$$\gamma \text{ is a trajectory of } Y \Longrightarrow f \circ \gamma \text{ is a trajectory of } X \;.$$

### 0.4.24

The standard vector field $d/dt$ on $\mathbb{R}$ is clearly compatible with itself with respect to the translation map

$$\tau_s : \mathbb{R} \to \mathbb{R}, \quad t \mapsto t + s$$

by $s \in \mathbb{R}$. Hence, if $\gamma : \mathbb{I} \to M$ is a trajectory of $X \in D(M)$, and if, by abuse of notation, the restriction

$$\tau_s^{-1}(\mathbb{I}) \to \mathbb{I}$$

of $\tau_s$ is denoted again by $\tau_s$, then from n. 0.4.23 it immediately follows that $\gamma \circ \tau_s$ is a trajectory of $X$ as well. Since the inverse $\tau_{-s}$ of $\tau_s$ is again a translation, it easily follows that

$$\gamma \text{ is a trajectory} \iff \gamma \circ \tau_s \text{ is a trajectory} \;,$$

and

$$\gamma \text{ is maximal} \iff \gamma \circ \tau_s \text{ is maximal}.$$

## 0.5 Differential Forms

### 0.5.1 *Differential Forms with Values in a Module*

Let $k$ be a field, $A$ a commutative $k$-algebra, and $P$ an $A$-module.

**Definition.** A multilinear alternating function

$$\underbrace{\mathrm{D}(A) \times \cdots \times \mathrm{D}(A)}_{s \text{ factors}} \to P$$

of $A$-modules will be called a *differential s-form on $A$ with values in $P$*, or also, shortly, a *$P$-valued s-form*.

The $A$-module of all $P$-valued differential $s$-forms will be denoted by $\Lambda^s(P)$. The graded module with graded components $\Lambda^s(P)$ will be denoted by $\Lambda^\bullet(P)$.

A *(ordinary) differential form on $A$* will be a differential form on $A$ with values in $A$ itself. A differential form on a smooth manifold $M$ will be a differential form on the algebra $C^\infty(M)$. The $C^\infty(M)$–module $\Lambda^\bullet(C^\infty(M))$ of all differential forms on $M$ will be denoted simply by $\Lambda^\bullet(M)$ (and its graded components by $\Lambda^s(M)$).

### 0.5.2 *Cotangent Bundle*

Let $M$ be a manifold and $A = C^\infty(M)$. Arguing as in n. 0.4.9, with [Nestruev (2003), 11.6 (III)] in place of [Nestruev (2003), 11.6 (II)] and using [Nestruev (2003), 11.37, 11.39] instead of [Nestruev (2003), Exercise 9.40], one deduces that $\Lambda^1(M)$ is projective, finitely generated, and determines an equidimensional pseudobundle $\pi_{\Lambda^1(M)} : |\Lambda^1(M)| \to M$ which is, therefore, a vector bundle. In this book, the *cotangent bundle of $M$* will be, by definition, $\pi_{\Lambda^1(M)}$.

### 0.5.3

By the above construction, the fiber at $m \in M$ of the cotangent bundle of $M$ is naturally isomorphic to the cotangent space $T_m^\vee M \stackrel{\text{def}}{=} (T_m M)^\vee$, and the value at $m$ of a 1-form $\omega$, considered as a section of the tangent bundle, corresponds to the unique tangent covector $\omega_m : T_m M \to \mathbb{R}$ such that

$$\omega_m(X_m) = \omega(X)(m), \quad X \in \mathrm{D}(M) . \tag{0.2}$$

Moreover, $T_m^\vee M$ is an $\mathbb{R}$-vector space

$$\mathbb{R}_m \otimes_A \Lambda^1(M)$$

obtained from $\Lambda^1(M)$ by extension of scalars via $m : A \to \mathbb{R}$, with universal homomorphism $\Lambda^1(M) \to T_m^\vee M$ given by

$$\omega \mapsto \omega_m .$$

Finally, note that if it is assumed that

$$T_m M = \mathbb{R}_m \otimes_A \mathrm{D}(M)$$

(see n. 0.4.10) and

$$\mathbb{R}_m = \mathbb{R}_m \otimes_A A ,$$

then (0.2) leads to

$$\omega_m = \mathrm{id}_{\mathbb{R}_m} \otimes_A \omega ,$$

i.e., $\omega_m$ is the homomorphism obtained from $\omega$ by extension of scalars via $m : \mathrm{C}^\infty(M) \to \mathbb{R}$.

Thus, a 1-form $\omega$ is geometrically described by a family

$$\{\omega_m\}_{m \in M}$$

of tangent covectors.

**0.5.4**

For higher degrees, the geometric description is analogous. According to n. 0.1.5, (3), given $\omega \in \Lambda^s(M)$, for each point $m \in M$ there is a unique alternating $s$-linear form

$$\omega_m : T_m M \times \cdots \times T_m M \to \mathbb{R}$$

such that

$$\omega_m(X_{1m}, \ldots, X_{sm}) = \omega(X_1, \ldots, X_s)(m) \quad X_1, \ldots, X_s \in \mathrm{D}(M) . \tag{0.3}$$

Therefore, every $s$-form $\omega$ is geometrically described by a family

$$\{\omega_m\}_{m \in M}$$

of alternating $s$-linear forms on the tangent spaces.

### 0.5.5 The Ordinary Differential

Let $k$ be a field and $A$ be a commutative $k$-algebra. For each $a \in A$ define
$$\mathrm{d}\,a : \mathrm{D}(A) \to A, \quad X \mapsto X(a)$$
and
$$\mathrm{d} : A \to \Lambda^1(A) \quad a \mapsto \mathrm{d}\,a.$$

**Definition.** The function d will be called the *ordinary differential on $A$*.

If $A = \mathrm{C}^\infty(M)$, with $M$ a smooth manifold, then d will be also called the *ordinary differential on $M$*.

### 0.5.6 Universal Property of First Order Differential Forms

It can be immediately checked that the ordinary differential d on a $k$-algebra is a derivation of $A$ into the $A$-module $\Lambda^1(A)$. The following proposition asserts that the ordinary differential $\mathrm{d} : \mathrm{C}^\infty(M) \to \Lambda^1(M)$ on a manifold $M$, possibly with boundary, is a universal derivation into a geometric module.

**Proposition.** For every geometric $\mathrm{C}^\infty(M)$–module $P$ and for every derivation
$$X : \mathrm{C}^\infty(M) \to P$$
of $\mathrm{C}^\infty(M)$ into $P$, there exists exactly one $\mathrm{C}^\infty(M)$–module homomorphism
$$h_X : \Lambda^1(M) \to P$$
such that
$$X = h_X \circ \mathrm{d}\,.$$

**Proof.** See [Nestruev (2003), Theorem 11.43] (take into account the definitions of [Nestruev (2003), 11.42]). □

### 0.5.7 Action of Smooth Maps on 1-Forms

Let $f : N \to M$ be a smooth map and denote by
$$\mathrm{d}_M : \mathrm{C}^\infty(M) \to \Lambda^1(M) \quad \text{and} \quad \mathrm{d}_N : \mathrm{C}^\infty(N) \to \Lambda^1(N)$$
the ordinary differentials.

Since $\mathrm{d}_N$ is a derivation of $\mathrm{C}^\infty(N)$ into the $\mathrm{C}^\infty(N)$-module $\Lambda^1(N)$, the function $\mathrm{d}_N \circ f^*$ is a derivation of $\mathrm{C}^\infty(M)$ into $\Lambda^1(N)$, considered as a

$C^\infty(M)$–module (by restriction of scalars via $f^*$). According to n. 0.5.2, $\Lambda^1(N)$ is projective and finitely generated as a $C^\infty(N)$–module; hence it is geometric. According to n. 0.3.3, it is also geometric as a $C^\infty(M)$–module. Therefore, by Proposition 0.5.6, there is exactly one $C^\infty(M)$–module homomorphism

$$\Lambda^1(f^*) : \Lambda^1(M) \to \Lambda^1(N),$$

such that

$$\Lambda^1(f^*) \circ d_M = d_N \circ f^*.$$

### 0.5.8  Geometric Description of $\Lambda^1(f^*)$

A geometric description for $\Lambda^1(f^*)$, with $f : N \to M$ being a smooth map, is provided by statement (3) of the last Exercise in [Nestruev (2003), 11.45] (where $\Lambda^1(f^*)$ is denoted by $f^*$ for simplicity):

$$\left(\Lambda^1(f^*)(\omega)\right)_n = \omega_{f(n)} \circ d_n f, \qquad n \in N, \omega \in \Lambda^1(M). \tag{0.4}$$

(This fact also will come out later as a particular case of a much more general statement.)

### 0.5.9

**Proposition.** Let $f : N \to M$ be a smooth map and consider the homomorphism

$$\nu : D(M) \to D(M)_f, \quad X \mapsto f^* \circ X.$$

(In other words, the map $\nu$ associates to each vector field $X$ on $M$ the vector field along $f$ that projects into $X$.) Then $D(M)_f$ is a $C^\infty(N)$–module obtained from $D(M)$ by extension of scalars via $f^*$, with universal homomorphism $\nu$.

**Proof.** Proposition 0.5.6 and n. 0.3.3 imply that

$$i : \mathrm{Hom}_{C^\infty(M)}\left(\Lambda^1(M), C^\infty(N)\right) \xrightarrow{\sim} D(M)_f, \quad h \mapsto h \circ d,$$

with d being the ordinary differential on $M$, is a $C^\infty(N)$–module isomorphism.

According to n. 0.5.2, $\Lambda^1(M) = D(M)^\vee$ is projective and finitely generated. Therefore n. 0.1.5, (7) implies

$$\mathrm{Hom}_{C^\infty(M)}\left(\Lambda^1(M), C^\infty(N)\right) = \left(D(M)^\vee\right)^\vee \otimes C^\infty(N).$$

The result now easily follows from n. 0.1.5, (6). $\square$

As a consequence, if $\pi_T : TM \to M$ is the tangent bundle of $M$, then the pseudobundle on $N$ defined by the module $\mathrm{D}(M)_f$ is an induced bundle $f^*(\pi_T)$. Accordingly, a vector field along $f$ may be viewed as a section of the induced bundle $f^*(\pi_T)$.

**Example.** Let $E$ be a vector space considered as a smooth manifold according to n. 0.2.19, $f : N \to E$ a smooth map and recall that $TE$ is a standard trivial bundle with standard fiber $E$ (see n. 0.4.11). From the definition of a standard trivial bundle (n. 0.3.15) and n. 0.3.7 it follows that the induced from $TE$ by $f$ bundle is a standard trivial bundle with standard fiber $E$ as well. Hence a vector field along $f$, when regarded as a section of this bundle, gives rise to a smooth function $N \to E$, simply by composition with the trivializing morphism.

If $\alpha : \mathbb{I} \to E$ is a smooth curve and $\mathrm{d}/\mathrm{d}t$ is the standard vector field on $\mathbb{I}$, the function $\mathbb{I} \to E$ that corresponds to $(\mathrm{d}/\mathrm{d}t) \circ \alpha^*$ is the *derivative* of $\alpha$. It will be denoted by $\alpha'$. Note also that $\alpha'$ is characterized by

$$\alpha'^*(\varphi) = \frac{\mathrm{d}}{\mathrm{d}t}(\alpha^*(\varphi)), \quad \varphi \in E^\vee,$$

from which it can be easily recognized that when $E = \mathbb{R}^n$, $\alpha'$ is nothing but the usual component-wise derivative of $\alpha$.

### 0.5.10

Let $A$ be a $k$-algebra, $k$ being a field, and $P$ an $A$-module. Consider the bilinear function

$$t : \Lambda^s(A) \times P \to \Lambda^s(P)$$

defined by

$$t(\omega, p)(X_1, \ldots, X_s) = \omega(X_1, \ldots, X_s)p,$$

$$\omega \in \Lambda^s(A), \ p \in P, \ X_1, \ldots, X_s \in \mathrm{D}(A).$$

**Proposition.** If either $\mathrm{D}(A)$ or $P$ is projective and finitely generated, then $(\Lambda^s(P), t)$ is a tensor product:

$$\Lambda^s(P) = \Lambda^s(A) \otimes P.$$

*Proof.* By the universality of exterior powers, there are natural isomorphisms $\Lambda^s(P) \cong \mathrm{Hom}(\bigwedge^s \mathrm{D}(A), P)$ and $\Lambda^s(A) \cong (\bigwedge^s \mathrm{D}(A))^\vee$. Therefore the result easily comes from n. 0.1.5, (12) and (7). □

By n. 0.4.9, the above result applies to $A = C^\infty(M)$, with $M$ being a smooth manifold, possibly with boundary. Therefore, for every $C^\infty(M)$-module $P$,

$$\Lambda^s(P) = \Lambda^s(M) \otimes P \,.$$

## 0.5.11 Exterior Differential

Let $k$ be a field and $A$ a commutative $k$-algebra. Define a $k$-linear function

$$d_s : \Lambda^s(A) \to \Lambda^{s+1}(A)$$

by setting

$$d_s(\omega)(X_1, \ldots, X_{s+1}) = \sum_i (-1)^{i+1} X_i \left( \omega\left(X_1, \ldots, \widehat{X_i}, \ldots, X_{s+1}\right)\right)$$
$$+ \sum_{i<j} (-1)^{i+j} \omega\left([X_i, X_j], X_1, \ldots, \widehat{X_i}, \ldots, \widehat{X_j}, \ldots, X_{s+1}\right)$$
$$\omega \in \Lambda^s(M), X_1, \ldots X_{s+1} \in D(A) \,,$$

where $\widehat{t}$ denotes the omission of a term $t$.

**Definition.** The *exterior differential on $A$* in this book will be the first degree graded $k$-module endomorphism

$$d : \Lambda^\bullet(A) \to \Lambda^\bullet(A)$$

with degree $s$ component $d_s$ for each $s$ ([28]).

Note that the degree 0 component is the ordinary differential on $A$. Sometimes, by abuse of notation, each graded component will be denoted simply by d.

In the case when $A = C^\infty(M)$, with $M$ being a smooth manifold, the exterior differential on $A$ will be also called the *exterior differential on* M.

For a 'conceptual definition' of the exterior differential see [Vinogradov (2001), 1.1.6].

---

[28] For an arbitrary algebra $A$, the adjective 'exterior' could be a bit misleading. Indeed, generally $\Lambda^\bullet(A)$ is not an exterior algebra (but it will be such in situations of central interest for this book).

## 0.5.12 de Rham Complex

Let $k$ be a field and $A$ a commutative $k$-algebra.

**Proposition.** The exterior differential d on $A$ satisfies
$$\mathrm{d} \circ \mathrm{d} = 0 \,,$$
i.e.,
$$(\Lambda^\bullet(A), \mathrm{d})$$
is a complex.

**Proof.** A proof may be done by writing down the definition of $\mathrm{d}^2 = \mathrm{d} \circ \mathrm{d}$ on homogeneous forms, being careful in doing the simplifications. It results in a rather long (but somewhat interesting) calculation, which is left to the reader. □

## 0.5.13 Wedge Product

Let $A$ be a commutative $k$-algebra, $k$ being a field, $S_{r+s}$ the group of permutations of $\{1, \ldots, r+s\}$, with $r$, $s$ being nonnegative integers, and define the following subset
$$S_{r,s} \stackrel{\text{def}}{=} \{\sigma \in S_{r+s} : \sigma(1) < \cdots < \sigma(r) \text{ and } \sigma(r+1) < \cdots < \sigma(r+s)\} \,.$$
Then, the $A$-module $\Lambda^\bullet(A)$ turns into a graded $A$-algebra once equipped with the (distributive) operation $\wedge$ determined by
$$(\omega \wedge \varkappa)(X_1, \ldots, X_{r+s})$$
$$= \sum_{\sigma \in S_{r,s}} (-1)^{|\sigma|} \omega\left(X_{\sigma(1)}, \ldots, X_{\sigma(r)}\right) \varkappa\left(X_{\sigma(r+1)}, \ldots, X_{\sigma(r+s)}\right) \,,$$
$$\omega \in \Lambda^r(A), \varkappa \in \Lambda^s(A), X_1, \ldots, X_{r+s} \in \mathrm{D}(A) \,,$$
with $(-1)^{|\sigma|}$ being the parity of $\sigma$.

Moreover, consider the zeroth degree homomorphism of graded $A$-modules
$$\alpha : \bigwedge\nolimits^\bullet \Lambda^1(A) \to \Lambda^\bullet(A)$$
with graded components $\alpha_n$ determined by the condition:
$$\alpha_n (\omega_1 \wedge \cdots \wedge \omega_n)(X_1, \ldots, X_n) = \left|(\omega_i(X_j))_{i,j \in \{1, \ldots, n\}}\right| \,,$$
$$X_1, \ldots, X_n \in \mathrm{D}(A), \omega_1, \ldots, \omega_n \in \Lambda^1(A)$$

(here $\wedge$ denotes the exterior product in $\bigwedge^\bullet \Lambda^1(A)$). From the generalized Laplace expansion of determinants (see [Mac Lane and Birkhoff (1967), Chap. XVI, Sect. 7, Theorem 12 (p. 564)]) it follows that $\alpha$ is a graded algebra homomorphism too.

**Proposition.** If $D(A)$ is projective and finitely generated, then $\alpha$ is an isomorphism.

**Proof.** It follows from n. 0.1.5, (11) (details are left to the reader). $\square$

**Corollary.** If $M$ is a manifold, possibly with boundary, then $\Lambda^\bullet(M)$ is an exterior algebra of the $C^\infty(M)$–module $\Lambda^1(M)$. Moreover, for all $X_1, \ldots, X_n \in D(M)$ and $\omega_1, \ldots, \omega_n \in \Lambda^1(M)$,

$$(\omega_1 \wedge \cdots \wedge \omega_n)(X_1, \ldots, X_n) = \left| (\omega_i(X_j))_{i,j \in \{1,\ldots,n\}} \right|. \quad (0.5)$$

**Proof.** It immediately follows from the proposition, because $D(M)$ is projective and finitely generated by n. 0.4.9. $\square$

The operation $\wedge$ will be called *wedge product in* $\Lambda^\bullet(A)$. In conformity with the proposition, if $D(A)$ is projective and finitely generated then $\wedge$ may also be called *exterior product in* $\Lambda^\bullet(A)$. In particular, an exterior product is defined on $\Lambda^\bullet(M)$, with $M$ being a manifold.

Note that, since the characteristic of the algebra $C^\infty(M)$ is 0, the formula

$$(\omega \wedge \varkappa)(X_1, \ldots, X_{r+s})$$
$$= \sum_{\sigma \in S_{r,s}} (-1)^{|\sigma|} \omega\left(X_{\sigma(1)}, \ldots, X_{\sigma(r)}\right) \varkappa\left(X_{\sigma(r+1)}, \ldots, X_{\sigma(r+s)}\right)$$

may be also written

$$(\omega \wedge \varkappa)(X_1, \ldots, X_{r+s})$$
$$= \frac{1}{r!s!} \sum_{\sigma \in S_{r+s}} (-1)^{|\sigma|} \omega\left(X_{\sigma(1)}, \ldots, X_{\sigma(r)}\right) \varkappa\left(X_{\sigma(r+1)}, \ldots, X_{\sigma(r+s)}\right).$$

The present definition of the wedge product agrees with the one that may be generally found in the literature (see, *e.g.*, [Berger and Gostiaux (1988), 0.1.4]). In some texts (*e.g.*, [Singer and Thorpe (1976), Sect. 5.2 (p. 120)]) a different definition is assumed by replacing $r!s!$ by $(r+s)!$. The resulting algebra structure on $\Lambda^\bullet(M)$ is different but, anyway, isomorphic.

### 0.5.14 Leibnitz Rule for the Exterior Differential

**Proposition.** Let $d : \Lambda^\bullet(A) \to \Lambda^\bullet(A)$ be the exterior differential on a commutative $k$-algebra $A$, $k$ being a field. For all $\omega_r \in \Lambda^r(A)$ and $\varkappa \in \Lambda^\bullet(A)$,
$$d(\omega_r \wedge \varkappa) = (d\omega_r) \wedge \varkappa + (-1)^r \omega_r \wedge d\varkappa.$$

**Proof.** Basically, a proof may be done by an easy (but cumbersome) calculation, based on some preliminaries about permutations. The details are left to the reader. □

In other words, the exterior differential is a graded derivation of $\Lambda^\bullet(A)$.

### 0.5.15

The universal property of the ordinary differential $d : C^\infty(M) \to \Lambda^1(M)$ easily implies that the $C^\infty(M)$-module $\Lambda^1(M)$ is generated by the image of d, i.e., by elements of the form $da$ with $a \in C^\infty(M)$ ([29]).

It follows, more generally, that for all $s \in \mathbb{N}_0$, the $C^\infty(M)$-module $\Lambda^s(M)$ is generated by elements of the form
$$da_1 \wedge \cdots \wedge da_s,$$
with $a_1, \cdots, a_s \in C^\infty(M)$. In other words, every $s$-form $\omega_s$ may be written as a sum
$$\omega_s = \sum_i a_i \, da_{i1} \wedge \cdots \wedge da_{is},$$
with the $a$'s in $C^\infty(M)$. Propositions 0.5.12 and 0.5.14 imply
$$d\omega_s = \sum_i da_i \wedge da_{i1} \wedge \cdots \wedge da_{is}.$$

Let $U$ be open in $\mathbb{R}^n$. From the local description of smooth vector fields (see n. 0.4.1) it easily follows that
$$da = \frac{\partial a}{\partial x_1} dx_1 + \cdots + \frac{\partial a}{\partial x_n} dx_n, \quad a \in C^\infty(U). \tag{0.6}$$
Therefore $dx_1, \ldots, dx_n$ generate $\Lambda^1(U)$ as a $C^\infty(U)$-module and $\Lambda^\bullet(U)$ as a $C^\infty(U)$-algebra. This gives the usual description in local coordinates of an $s$-form $\omega_s$ on a manifold $M$:
$$\sum_{i_1 < \cdots < i_s} a_{i_1 \ldots i_s} dx_{i_1} \wedge \cdots \wedge dx_{i_s}.$$

---

[29] Actually, in the present setting, the logical dependence between these facts is inverted, because the proof of [Nestruev (2003), Theorem 11.43] uses [Nestruev (2003), Corollary 11.49].

Accordingly, the local description of the exterior differential $d\omega_s$ is obtained from a rearrangement of

$$\sum_{i_1<\cdots<i_s} d\, a_{i_1\ldots i_s} \wedge d\, x_{i_1} \wedge \cdots \wedge d\, x_{i_s}$$

after substitution of the various $d\, a_{i_1\ldots i_s}$ with the corresponding expressions given by (0.6).

### 0.5.16 Action of Smooth Maps on de Rham Complexes

Let $f : N \to M$ be a smooth map between manifolds, possibly with boundary.

By n. 0.5.7, there exists exactly one $C^\infty(M)$–module homomorphism

$$\Lambda^1(f^*) : \Lambda^1(M) \to \Lambda^1(N)$$

such that

$$\Lambda^1(f^*) \circ d_M = d_N \circ f^*, \qquad (0.7)$$

with $d_M$, $d_N$ being the (zeroth components of) exterior differentials. Since $\Lambda^\bullet(M)$ is an exterior algebra of $\Lambda^1(M)$ by Corollary 0.5.13, $\Lambda^1(f^*)$ induces a $C^\infty(M)$–algebra homomorphism of $\Lambda^\bullet(M)$ into the exterior algebra of the $C^\infty(M)$–module obtained from $\Lambda^1(N)$ by restriction of scalars. Upon composing with the natural homomorphism of n. 0.1.5, (4) one gets a $C^\infty(M)$–algebra homomorphism

$$\Lambda^\bullet(f^*) : \Lambda^\bullet(M) \to \Lambda^\bullet(N),$$

which is the unique one with first degree component $\Lambda^1(f^*)$.

Using the (global) description of differential forms recalled in n. 0.5.15, one also easily deduces that

$$\Lambda^\bullet(f^*) \circ d_M = d_N \circ \Lambda^\bullet(f^*).$$

In conclusion, $\Lambda^\bullet(f^*) : \Lambda^\bullet(M) \to \Lambda^\bullet(N)$ is both a cochain homomorphism and a graded homomorphism of $C^\infty(M)$–algebras, when $\Lambda^\bullet(N)$ is considered so by restriction of scalars via $f^*$. Moreover, $\Lambda^\bullet(f^*)$ is characterized by these conditions. It will be called the (de Rham) *cochain homomorphism induced by* $f$.

### 0.5.17 Geometric Description of $\Lambda^\bullet(f^*)$

Let $f : N \to M$ be a smooth map. A geometric pointwise description of $\Lambda^\bullet(f^*)$ is given by

$$(\Lambda^\bullet(f^*)(\omega))_n(\xi_1, \ldots, \xi_s) = \omega_{f(n)}(\mathrm{d}_n f(\xi_1), \ldots, \mathrm{d}_n f(\xi_s)),$$

$$n \in N, \ \omega \in \Lambda^s(M), \ \xi_1, \ldots, \xi_s \in T_n N \ .$$

Indeed, note that it suffices to check it in the case when

$$\omega = \omega_1 \wedge \cdots \wedge \omega_s$$

with $\omega_i \in \Lambda^1(M)$ for all $i$, and that by n. 0.4.10 vector fields $Y_1, \ldots, Y_s$ on $N$ may be chosen such that

$$Y_{1n} = \xi_1, \ \ldots, \ Y_{sn} = \xi_s$$

and vector fields $X_1, \ldots, X_s$ on $M$ such that

$$X_{1 f(n)} = \mathrm{d}_n f(\xi_1), \ \ldots, \ X_{s f(n)} = \mathrm{d}_n f(\xi_s) \ .$$

With these assumptions, the required formula comes from an easy calculation based on (0.3), p. 53, (0.4), p. 55 and (0.5), p. 59.

### 0.5.18 Insertion Operator

Let $k$ be a field, $A$ a commutative $k$-algebra, $X \in \mathrm{D}(A)$ and $P$ an $A$-module. If $\omega \in \Lambda^s(A)$, with $s \geq 1$, then

$$\omega_X(X_1, \ldots, X_{s-1}) \stackrel{\mathrm{def}}{=} \omega(X, X_1, \ldots, X_{s-1}), \quad X_1, \ldots, X_{s-1} \in \mathrm{D}(A) \ ,$$

defines a form $\omega_X \in \Lambda^{s-1}(P)$. If $\omega \in \Lambda^0(P) \cong P$, set $\omega_X = 0$ by definition.

**Definition.** The $(-1)$-th degree graded $A$-module endomorphism

$$\Lambda^\bullet(P) \to \Lambda^\bullet(P)$$

with degree $s$ components

$$\Lambda^s(P) \to \Lambda^{s-1}(P), \quad \omega \mapsto \omega_X$$

is called the *insertion operator of $X$ into* $\Lambda^\bullet(P)$.

In the case when $P = A$, the insertion operator is usually denoted by $\mathrm{i}_X$; for an arbitrary $P$ the notation $\bar{\mathrm{i}}_{X,P}$, or often simply $\bar{\mathrm{i}}_X$, will be used. The form $\bar{\mathrm{i}}_X(\omega) = \omega_X$ is sometimes also denoted by

$$X \lrcorner \omega \ .$$

If $M$ is a smooth manifold, possibly with boundary, and $X$ a vector field on $M$, the *insertion operator of $X$* is the insertion operator of $X$ into $\Lambda^\bullet(M)$ (*i.e.*, it is assumed that $P = A = \mathrm{C}^\infty(M)$).

### 0.5.19 Leibnitz Rule for Insertion Operators

The insertion operator of $X \in D(A)$ into $\Lambda^\bullet(A)$ satisfies the following Leibnitz Rule.

**Proposition.** For all $\omega_r \in \Lambda^r(A)$ and $\varkappa \in \Lambda^\bullet(A)$,

$$i_X(\omega_r \wedge \varkappa) = i_X(\omega_r) \wedge \varkappa + (-1)^r \omega_r \wedge i_X(\varkappa) \ .$$

*Proof.* The situation is similar to (but fortunately easier than) that of Proposition 0.5.14. A proof may consist of a calculation together with some manipulation about permutations. □

In other words, $i_X$ is a graded derivation.

### 0.5.20

Let $M$ be a manifold, $X$ a vector field on $M$, $d : C^\infty(M) \to \Lambda^1(M)$ the ordinary differential on $M$, and $i_X : \Lambda^1(M) \to C^\infty(M)$ the first degree component of the insertion operator of $X$. Then for all $a \in C^\infty(M)$, by definition of d and $i_X$,

$$i_X(d\,a) = (d\,a)(X) = X(a) \ .$$

Therefore

$$i_X \circ d = X \ .$$

By the universal property of d (Proposition 0.5.6), the above equality characterizes $i_X$ among $C^\infty(M)$-module homomorphisms $\Lambda^1(M) \to C^\infty(M)$.

### 0.5.21

Let $f : N \to M$ be a smooth map, $X$ a vector field on $M$, $Y$ a vector field on $N$, and let

$$i_X : \Lambda^1(M) \to C^\infty(M), \ i_Y : \Lambda^1(N) \to C^\infty(N)$$

be the first degree components of the insertion operators of $X$ and $Y$, respectively.

**Exercise.** Show that $X$ and $Y$ are compatible with respect to $f$ if and only if the diagram

$$\begin{array}{ccc} \Lambda^1(M) & \xrightarrow{\Lambda^1 f^*} & \Lambda^1(N) \\ {\scriptstyle i_X}\downarrow & & \downarrow{\scriptstyle i_Y} \\ C^\infty(M) & \xrightarrow{f^*} & C^\infty(N) \end{array}$$

is commutative.

**Hint.** Compose with the ordinary differential on $M$ and take into account n. 0.5.20.

**Proposition.** The vector fields $X$ and $Y$ are $f$-compatible if and only if the diagram

$$\begin{array}{ccc} \Lambda^\bullet(M) & \xrightarrow{\Lambda^\bullet f^*} & \Lambda^\bullet(N) \\ {\scriptstyle i_X}\downarrow & & \downarrow{\scriptstyle i_Y} \\ \Lambda^\bullet(M) & \xrightarrow{\Lambda^\bullet f^*} & \Lambda^\bullet(N) \end{array}$$

is commutative.

*Proof.* It reduces to extend the 'only if' assertion of the preceding Exercise to every degree. By means of the description of $\omega \in \Lambda^s(M)$ given in n. 0.5.15, it suffices to make straightforward use of Proposition 0.5.19. □

Now let $X_1, \ldots, X_n \in D(M)$, $Y_1, \ldots, Y_n \in D(N)$ and $\omega \in \Lambda^n(M)$.

**Corollary.** If $X_i$ and $Y_i$ are $f$-compatible for all $i \in \{1, \ldots, n\}$, then
$$(\Lambda^\bullet f^*)(\omega)(Y_1, \ldots, Y_n) = f^*(\omega(X_1, \ldots, X_n)) .$$

*Proof.* It suffices to apply $n$ times the Proposition. □

### 0.5.22 Differential Forms Along Maps

Let $f : N \to M$ be a smooth map and consider the $C^\infty(M)$-module structure on $C^\infty(N)$ given by $f^*$.

**Definition.** A *(differential) form on $M$ along $f$* is a differential form on $C^\infty(M)$ with values in the $C^\infty(M)$-module $C^\infty(N)$.

The graded $C^\infty(N)$-module of all differential forms on $M$ along $f$ will be denoted by $\Lambda^\bullet(M)_f$, and its degree $s$ component by $\Lambda^s(M)_f$.

**0.5.23**

Ordinary differential forms on a manifold may be understood as differential forms along the identity. Note also that if $\omega$ is a form along $f : N \to M$ and $g : V \to N$ is a smooth map, then the composition

$$g^* \circ \omega$$

is a form along $f \circ g$.

**0.5.24**

**Proposition.** The module $\Lambda^s(M)_f$ of differential $s$-forms along a smooth map $f : N \to M$ is a module obtained from $\Lambda^s(M)$ by extension of scalars via $f^*$, with universal homomorphism

$$\nu_s : \Lambda^s(M) \to \Lambda^s(M)_f, \quad \omega \mapsto f^* \circ \omega .$$

*Proof.* Immediate from Proposition 0.5.10. □

As a consequence, $\Lambda^\bullet(M)_f$ is a graded $C^\infty(N)$–algebra obtained from $\Lambda^\bullet(M)$ by extension of scalars via $f^*$, with universal homomorphism $\nu \stackrel{\text{def}}{=} \bigoplus_{s \in \mathbb{N}_0} \nu_s$.

**0.5.25**

**Example.** Let $f : N \to M$ be a smooth map and $\pi_{T^\vee} : T^\vee M \to M$ be the cotangent bundle of $M$. By Proposition 0.5.24, the vector bundle on $N$, defined by the module $\Lambda^1(M)_f$ of all 1-forms along $f$, is an induced bundle $f^*(\pi_{T^\vee})$. If $f$ is replaced by a vector bundle $\pi : E \to M$ (understood as an object of $\mathfrak{VB}_g$), a 1-form along $\pi$ corresponds to what in [Nestruev (2003), 11.26] is called a *horizontal 1-form* on $E$.

**0.5.26**

**Proposition.** Let $f : M \to N$ be a smooth map. Then $\Lambda^\bullet(M)_f$ is an exterior algebra of the $C^\infty(N)$–module $\Lambda^1(M)_f$.

*Proof.* By Corollary 0.5.13, $\Lambda^\bullet(M)$ is an exterior algebra of $\Lambda^1(M)$ and, by n. 0.5.24, $\Lambda^\bullet(M)_f$ is an algebra obtained from $\Lambda^\bullet(M)$ by extension of scalars via $f^*$. Hence, it suffices to invoke n. 0.1.5, (5). □

## 0.5.27 Wedge Product of Differential Forms Along Maps

According to Proposition 0.5.26, $\Lambda^\bullet(M)_f$ carries an exterior product $\wedge$, that will be also called *wedge product*.

The explicit expression of the wedge product in $\Lambda^\bullet(M)_f$ is similar to the expression of the wedge product in $\Lambda^\bullet(M)$ (see n. 0.5.13). To see this, let $\omega \in \Lambda^r(M)_f, \varkappa \in \Lambda^s(M)_f$ and denote by

$$\nu : \Lambda^\bullet(M) \to \Lambda^\bullet(M)_f$$

the universal homomorphism into the scalar extension. Since $\Lambda^\bullet(M)_f$ is generated by the image of $\nu$, it may be assumed that

$$\omega = \sum_{i=1}^t a_i \nu(\omega_i), \quad \varkappa = \sum_{j=1}^u b_j \nu(\varkappa_j) .$$

According to Proposition 0.5.24, $\nu(\omega_i) = f^* \circ \omega_i$, $\nu(\varkappa_j) = f^* \circ \varkappa_j$. Therefore, from the expression of the wedge products $\omega_i \wedge \varkappa_j$ in $\Lambda^\bullet(M)$, by a straightforward calculation it is easily deduced that

$$(\omega \wedge \varkappa)(X_1, \ldots, X_{r+s})$$
$$= \sum_{\sigma \in S_{r,s}} (-1)^{|\sigma|} \omega\left(X_{\sigma(1)}, \ldots, X_{\sigma(r)}\right) \varkappa\left(X_{\sigma(r+1)}, \ldots, X_{\sigma(r+s)}\right) ,$$

where $S_{r,s}$ is, as usual, the subset of the group $S_{r+s}$ of permutations of $\{1, \ldots, r+s\}$ given by

$$S_{r,s} = \{\sigma \in S_{r+s} : \sigma(1) < \cdots < \sigma(r) \text{ and } \sigma(r+1) < \cdots < \sigma(r+s)\} .$$

## 0.5.28 Insertion Along Maps

Since differential forms along $f : N \to M$ are, in particular, forms with values in a module, according to Definition 0.5.18 for each vector field $X$ on $M$, an insertion operator

$$\bar{i}_X : \Lambda^\bullet(M)_f \to \Lambda^\bullet(M)_f ,$$

which is clearly $C^\infty(N)$–linear, is defined. It is easy to define, more generally, insertion operators of vector fields *along f*. Indeed, according to Proposition 0.5.9, the function

$$\nu : D(M) \to D(M)_f, \quad X \to f^* \circ X$$

is the universal homomorphism into the scalar extension $D(M)_f$ of $D(M)$. Therefore, according to n. 0.1.5, (3), a differential $s$-form $\omega$ along $f$ naturally corresponds to the $s$-linear alternating form

$$\omega_f : D(M)_f \times \cdots \times D(M)_f \to C^\infty(N)$$

characterized by
$$\omega_f(f^* \circ X_1, \ldots, f^* \circ X_s) = \omega(X_1, \ldots, X_s), \quad X_1, \ldots, X_s \in D(M).$$
It is sometimes convenient to identify $\omega$ with $\omega_f$, so that it will make sense
$$\omega(Y_1, \ldots, Y_s) \in C^\infty(N), \quad \omega \in \Lambda^s(M)_f, Y_1, \ldots, Y_s \in D(M)_f.$$
Now, let $Y$ be a vector field along $f$. For each $\omega \in \Lambda^s(M)_f$, if $s \geq 1$, set
$$\omega_Y(X_1, \ldots, X_{s-1}) \stackrel{\text{def}}{=} \omega_f(Y, f^* \circ X_1, \ldots, f^* \circ X_{s-1}),$$
$$X_1, \ldots, X_{s-1} \in D(M);$$
if $s = 0$, set $\omega_Y = 0$.

**Definition.** The $(-1)$-th degree graded $C^\infty(N)$–module homomorphism
$$i_Y : \Lambda^\bullet(M)_f \to \Lambda^\bullet(M)_f$$
with degree $s$ components
$$\Lambda^s(M)_f \to \Lambda^{s-1}(M)_f, \quad \omega \mapsto \omega_Y$$
will be called the *insertion operator of $Y$ into* $\Lambda^\bullet(M)_f$.

Sometimes, the form $\omega_Y = i_Y(\omega)$ will be also denoted by
$$Y \lrcorner \omega.$$

**0.5.29**

If $N = M$ and $f = \text{id}_M$, then $i_Y$ coincides with the ordinary insertion operator of the ordinary vector field $Y$ into ordinary differential forms on $M$.

**0.5.30** *Leibnitz Rule for Insertions Along Maps*

**Proposition.** Let $Y$ be a vector field along a smooth map $f : N \to M$. For all $\omega_r \in \Lambda^r(M)_f$ and $\varkappa \in \Lambda^\bullet(M)_f$,
$$i_Y(\omega_r \wedge \varkappa) = i_Y(\omega_r) \wedge \varkappa + (-1)^r \omega_r \wedge i_Y(\varkappa).$$

*Proof.* The description of the wedge product in $\Lambda^\bullet(M)_f$ given in n. 0.5.27 still holds if the ordinary vector fields $X_1, \ldots, X_{r+s}$ are replaced by vector fields along $f$ (and forms along $f$ are understood as forms on $D(M)_f$ according to n. 0.5.28). To see this, it suffices to decompose the arguments as linear combinations of ordinary vector fields with coefficients in $C^\infty(N)$. Using this description, the proof becomes formally identical to that of Proposition 0.5.19. □

## 0.5.31

Let $f : N \to M$ be a smooth map and recall that $D(M)_f$ and $\Lambda^s(M)_f$ are, respectively, obtained by extension of scalars via $f^*$ with universal homomorphisms

$$D(M) \to D(M)_f, \quad X \mapsto X \circ f^*$$

and

$$\Lambda^s(M) \to \Lambda^s(M)_f, \quad \omega \mapsto \omega \circ f^*.$$

By the definition of insertion operators along $f$, for all $X \in D(M)$ and $\omega \in \Lambda^s(M)$,

$$f^* \circ (X \lrcorner \omega) = (f^* \circ X) \lrcorner (f^* \circ \omega).$$

This shows that the bilinear map

$$D(M)_f \times \Lambda^s(M)_f \to \Lambda^{s-1}(M)_f, \quad (Y, \varkappa) \mapsto Y \lrcorner \varkappa$$

coincides with the bilinear map obtained from

$$D(M) \times \Lambda^s(M) \to \Lambda^{s-1}(M), \quad (X, \varkappa) \mapsto X \lrcorner \omega$$

by extension of scalars via $f^*$ (see n. 0.1.5, (2)).

Now let $g : V \to N$ be a smooth map. If $Y \in D(M)_f$ and $\varkappa \in \Lambda^s(M)_f$ then

$$g^* \circ Y \in D(M)_{f \circ g}, \quad g^* \circ \varkappa \in \Lambda^s(M)_{f \circ g}$$

(see nn. 0.4.13 and 0.5.23). Moreover, the homomorphisms

$$D(M)_f \to D_{f \circ g}(M), \quad Y \mapsto g^* \circ Y$$

and

$$\Lambda^s(M)_f \to \Lambda^s(M)_{f \circ g}, \quad \varkappa \mapsto g^* \circ \varkappa$$

both satisfy the universal property of scalar extension via $g^*$, because this is true for the analogous homomorphisms from $D(M)$ and $\Lambda^s(M)$. As a consequence, the bilinear map

$$D(M)_{f \circ g} \times \Lambda^s(M)_{f \circ g} \to \Lambda^{s-1}(M)_{f \circ g}, \quad (Z, \rho) \mapsto Z \lrcorner \rho$$

coincides with the bilinear map obtained from

$$D(M)_f \times \Lambda^s(M)_f \to \Lambda^{s-1}(M)_f, \quad (Y, \varkappa) \mapsto Y \lrcorner \varkappa$$

by extension of scalars via $g^*$. In other words, the following formula holds:

$$(g^* \circ Y) \lrcorner (g^* \circ \varkappa) = g^* \circ (Y \lrcorner \varkappa), \quad Y \in D(M)_f, \varkappa \in \Lambda^s(M)_f. \quad (0.8)$$

### 0.5.32

**Proposition.** Let $f : N \to M$ be a smooth map. For all $\omega \in \Lambda^1(M)$ and $Z \in D(N)$,

$$\Lambda^1(f^*)(\omega)(Z) = (Z \circ f^*) \lrcorner (f^* \circ \omega).$$

**Proof.** According to n. 0.5.31, for all $a \in C^\infty(M)$ and $X \in D(M)$

$$(f^* \circ X) \lrcorner (f^* \circ da) = f^* \circ (X \lrcorner da) = f^* \circ X(a).$$

Notice that here $X(a)$ is interpreted as a 0-form, so that $f^* \circ X(a)$ is nothing but $f^*(X(a))$. Therefore,

$$(f^* \circ X) \lrcorner (f^* \circ da) = (f^* \circ X)(a).$$

In other words, the $C^\infty(N)$-module homomorphism

$$D(M)_f \to C^\infty(N), \quad Y \mapsto Y \lrcorner (f^* \circ da)$$

coincides with

$$D(M)_f \to C^\infty(N), \quad Y \mapsto Y(a)$$

when $Y$ is in the image of the universal homomorphism $D(M) \to D(M)_f$, $X \mapsto f^* \circ X$. Therefore, they coincide for all $Y$:

$$Y \lrcorner (f^* \circ da) = Y(a), \quad Y \in D(M)_f, a \in C^\infty(M). \tag{0.9}$$

Now consider the $C^\infty(M)$-module homomorphism

$$\varphi : \Lambda^1(M) \to \Lambda^1(N)$$

that, with $\omega \in \Lambda^1(M)$, associates the 1-form on $N$ given by

$$Z \mapsto (Z \circ f^*) \lrcorner (f^* \circ \omega).$$

For all $a \in C^\infty(M)$ and $Z \in D(N)$,

$$\varphi(da)(Z) = (Z \circ f^*) \lrcorner (f^* \circ da) \stackrel{(0.9)}{=} Z(f^*(a)) = (d_N f^*(a))(Z),$$

with $d_N$ being the ordinary differential on $N$. This shows that $\varphi$ fulfills the condition

$$\varphi \circ d = d_N \circ f^*$$

that characterizes $\Lambda^1(f^*)$. □

Note that the above proposition, together with (0.8) and n. 0.4.10, implies the geometric description mentioned in n. 0.5.8 (compose the formula in the statement with $n : C^\infty(N) \to \mathbb{R}$).

Recall that $\Lambda^1(M)_f$ is obtained from $\Lambda^1(M)$ by extension of scalars via $f^*$ (see Proposition 0.5.24). Hence, to the $C^\infty(M)$–module homomorphism $\Lambda^1 f^*$ naturally corresponds a $C^\infty(N)$–module homomorphism

$$(\Lambda^1 f^*)_f : \Lambda^1(M)_f \to \Lambda^1(N) \ .$$

**Corollary.** For all $\varkappa \in \Lambda^1(M)_f$ and $Z \in D(N)$

$$(\Lambda^1 f^*)_f(\varkappa)(Z) = (Z \circ f^*) \lrcorner \varkappa \ .$$

**Proof.** By the Proposition, the composition of the universal homomorphism

$$\Lambda^1(M) \to \Lambda^1(M)_f, \quad \omega \mapsto f^* \circ \omega$$

with the $C^\infty(N)$–module homomorphism given by

$$Z \mapsto (Z \circ f^*) \lrcorner \varkappa$$

is precisely $\Lambda^1 f^*$. □

Consider now the $C^\infty(N)$–module homomorphism

$$(\Lambda^\bullet f^*)_f : \Lambda^\bullet(M)_f \to \Lambda^\bullet(N)$$

corresponding to $\Lambda^\bullet f^*$.

**Exercise.** Show that

$$Z \lrcorner (\Lambda^\bullet f^*)_f(\varkappa) = (\Lambda^\bullet f^*)_f\big((Z \circ f^*) \lrcorner \varkappa\big), \quad \varkappa \in \Lambda^\bullet(M)_f, Z \in D(N) \ .$$

**Hint.** Argue by induction using the Leibniz rule for insertions (see n. 0.5.30).

### 0.5.33 Local Differential Forms

Let $U$ be an open submanifold of a manifold $M$ and $\varkappa$ be an $s$-form along the embedding $i : U \hookrightarrow M$. Repeated applications of Exercise 0.5.32 lead to

$$(\Lambda^\bullet i^*)_i(\varkappa)\,(X_1|_U, \ldots, X_s|_U) = \varkappa(X_1, \ldots, X_s), \quad X_1, \ldots, X_s \in D(M) \ .$$

Moreover, according to n. 0.4.16, there exists a natural identification $D(M)_i = D(U)$, which shows that $D(U)$ is generated as a $C^\infty(U)$–module by the restrictions of vector fields on $M$ (by Proposition 0.5.9). This implies that $\omega = (\Lambda^\bullet i^*)_i(\varkappa)$ is the unique $s$-form on $U$ such that

$$\omega\left(X_1|_U, \ldots, X_s|_U\right) = \varkappa\left(X_1, \ldots, X_s\right), \quad X_1, \ldots, X_s \in \mathrm{D}(M).$$

Conversely, given an $s$-form $\omega$ on $U$, the above equality immediately defines an $s$-form $\varkappa$ along $i$.

In conclusion, there exists a natural one-to-one correspondence between $s$-forms along $i$ and $s$-forms on $U$. This identification is compatible with wedge products, as it easily follows form n. 0.5.27.

**Definition.** A *local (differential) form* on $M$ is a form along the embedding $U \hookrightarrow M$ of an open submanifold, and it is sometimes identified with the corresponding $s$-form on $U$.

### 0.5.34 Splitting of Differential Forms on a Product Manifold

Let $M$ and $N$ be smooth manifolds, at least one of them without boundary, $P$ a $C^\infty(M \times N)$–module, and

$$\pi_M : M \times N \to M, \quad \pi_N : M \times N \to N$$

the projection maps.

**Proposition.** If $P_M$ and $P_N$ respectively denote the modules obtained from $P$ by restriction of scalars via $\pi_M^*$ and $\pi_N^*$, then

$$\Lambda^1(P) = \Lambda^1(P_M) \oplus \Lambda^1(P_N).$$

*Proof.* From Propositions 0.4.18 it follows that

$$\mathrm{Hom}_{C^\infty(M\times N)}\left(\mathrm{D}(M\times N), P\right)$$
$$= \mathrm{Hom}_{C^\infty(M\times N)}\left(\mathrm{D}(M)_N, P\right) \oplus \mathrm{Hom}_{C^\infty(M\times N)}\left(\mathrm{D}(N)_M, P\right).$$

Therefore, it suffices to note that by Proposition 0.5.9 there are natural isomorphisms

$$\mathrm{Hom}_{C^\infty(M\times N)}(\mathrm{D}(M)_N, P) \cong \mathrm{Hom}_{C^\infty(M)}(\mathrm{D}(M), P_M)$$

and

$$\mathrm{Hom}_{C^\infty(M\times N)}(\mathrm{D}(N)_M, P) \cong \mathrm{Hom}_{C^\infty(N)}(\mathrm{D}(N), P_N). \qquad \square$$

## 0.5.35

When $P = C^\infty(M \times N)$, Proposition 0.5.34 leads to a decomposition of $\Lambda^1(M \times N)$. To write it in a more expressive form, let us introduce the following notation.

**Definition.** Let $\pi_M : M \times N \to M$ be the projection onto $M$ of a product $M \times N$ of smooth manifolds. The $C^\infty(M \times N)$–module $\Lambda^\bullet(M)_{\pi_M}$ of all differential forms on $M$ along $\pi_M$ will be also denoted by

$$\Lambda^\bullet(M)_N$$

and its degree $s$ component $\Lambda^s(M)_{\pi_M}$ by $\Lambda^s(M)_N$.

In this notation, Proposition 0.5.34 gives in particular

$$\Lambda^1(M \times N) = \Lambda^1(M)_N \oplus \Lambda^1(N)_M \ .$$

From the construction of the above decomposition and Corollary 0.5.32, it easily follows that the natural monomorphisms

$$\Lambda^1(M)_N \hookrightarrow \Lambda^1(M \times N) \quad \text{and} \quad \Lambda^1(N)_M \hookrightarrow \Lambda^1(M \times N)$$

into the direct sum are nothing but the homomorphisms that naturally correspond to $\Lambda^1(\pi_M^*)$ and $\Lambda^1(\pi_N^*)$, once $\Lambda^1(M)_N$ and $\Lambda^1(N)_M$ are considered as scalar extensions from $\Lambda^1(M)$ and $\Lambda^1(N)$, respectively.

## 0.5.36 *Splitting of Derivations*

**Proposition.** In the notation of n. 0.5.34, if $P$ is geometric then there is a decomposition

$$D(P) = D(P_M) \oplus D(P_N)$$

such that the natural epimorphisms onto the summands are respectively given by

$$X \mapsto X \circ \pi_M^* \quad \text{and} \quad X \mapsto X \circ \pi_N^* \ .$$

**Proof.** It easily follows from Proposition 0.5.6 and n. 0.5.35. □

## 0.5.37 *Splitting of Tangent Vectors*

When $P = \mathbb{R}$ with the $C^\infty(M \times N)$–module structure given by $(m,n) : C^\infty(M \times N) \to \mathbb{R}$, n. 0.3.3 and Proposition 0.5.36 leads to the decomposition

$$T_{(m,n)} M \times N = T_m M \oplus T_n N \ ,$$

with natural epimorphisms given by

$$\mathrm{d}_{(m,n)}\,\pi_M \quad \text{and} \quad \mathrm{d}_{(m,n)}\,\pi_N\,.$$

It is also easy to recognize that the natural monomorphisms of the summands are nothing but

$$\mathrm{d}_m\, i_n \quad \text{and} \quad \mathrm{d}_n\, j_m\,,$$

with $i_n : M \to M \times N$ and $j_m : N \to M \times N$ being the embeddings at $m$ and $n$, respectively (it suffices to check the four compositions with the natural epimorphisms, taking into account what are the compositions of $i_n$, $j_m$ with $\pi_M$ and $\pi_N$).

**Exercise.** Show that a morphism of vector bundles is regular if and only if it is regular as a morphism in $\mathfrak{VB}_g$ (according to n. 0.3.26).

**Hint.** Take into account the Inverse Function Theorem (see, *e.g.*, [Nestruev (2003), 6.21] or [Berger and Gostiaux (1988), 0.2.22]); *cf.* also [Nestruev (2003), 11.30].

## 0.6 Lie Derivative

In this section, the notion of *Lie derivative* of a differential form along a vector field $X$ is reviewed. Although it could be directly introduced by means of the Cartan formula (see n. 0.6.20), it is also worth it to recall the description based on the flow of $X$.

### 0.6.1 *Flow Generated by a Vector Field*

The existence and uniqueness theorem for ordinary (smooth) differential equations may be stated in the following form. For every smooth vector field $X$ on a manifold $M$, possibly with boundary, and for all $m \in M$, there exists a unique maximal trajectory $\gamma : \mathbb{I} \to M$ of $X$ such that $\gamma(0) = m$. The smooth dependence of trajectories on the initial data is encoded by the *flow of X*. More precisely, if $M$ is without boundary, the flow of $X$ is a smooth map

$$\Phi : \mathbb{I}_M \to M\,,$$

where $\mathbb{I}_M$ is an open submanifold of $M \times \mathbb{R}$, such that for all $m \in M$, the map

$$\gamma_m : \mathbb{I}_m \to M, \quad \gamma_m(t) = \Phi(m, t)$$

(where $\mathbb{I}_m = \{t \in \mathbb{R} | (m,t) \in \mathbb{I}_M\}$) results in the unique maximal trajectory of $X$ such that $\gamma_m(0) = m$. For the proof of existence and uniqueness ([30]) of the flow, the reader is referred to [Berger and Gostiaux (1988), 3.5.7] (it is also convenient to have a look on [Arnold (2006), Chap. 4, 31.8, Solutions 1, 2 (p. 277)]).

### 0.6.2  Flows on a Manifold with Boundary

When $M$ has a nonempty boundary, some pathologies may arise. Of course, a map $\Phi : \mathbb{I}_M \to M$ still exists such that maximal trajectories are given by $\gamma_m(t) = \Phi(m,t)$, but the domain $\mathbb{I}_M$ may be not a submanifold of $M \times \mathbb{R}$. However, when $\mathbb{I}_M$ is identified with $\left| C^\infty(M \times \mathbb{R}) \right|_{\mathbb{I}_M} |$ (through the homeomorphism $\mu$ examined in n. 0.2.20), $\Phi$ is smooth in the sense that it is the dual map of an algebra homomorphism. Equivalently, this means that $a \circ \Phi \in C^\infty(M \times \mathbb{R})|_{\mathbb{I}_M}$ for all $a \in C^\infty(M)$.

### 0.6.3  Relative Intervals

In some sense, the flow is a 'universal trajectory': a single maximal trajectory is to a single point $m$ as the flow is to the whole of $M$. Accordingly, it may be said that the subset $\mathbb{I}_M$ of $M \times \mathbb{R}$ is to $M$ as an interval is to a single point. By this reason, it is expressive to call $\mathbb{I}_M$ a *relative interval* over $M$, though this term is commonly used with a different meaning in Physics. A precise definition is stated below: it will be limited to the case when $\mathbb{I}_M$ is a manifold.

Let $M$ be a smooth manifold and, for all $m \in M$, consider the embedding at $m$ into the product $M \times \mathbb{R}$:

$$\mathbb{R} \to M \times \mathbb{R}, \quad t \mapsto (m,t).$$

**Definition.** A *relative interval over* $M$ will be, in this book, a submanifold $\mathbb{I}_M$ of $M \times \mathbb{R}$ such that for all $m \in M$ the inverse image through the embedding at $m$ is an interval.

A relative interval will be said to be *open* if it is such as a submanifold of $M \times \mathbb{R}$. For each $m \in M$, restricting the embedding at $m$ one gets a

---

[30] Since, in our setting, a product is defined up to natural diffeomorphisms, the uniqueness holds only in view of the natural identification of $M \times \mathbb{R}$ with the set-theoretic product.

smooth map
$$j_m : \mathbb{I}_m \to \mathbb{I}_M .$$
Similarly, for all $t \in \mathbb{R}$, consider the embedding at $t$ into the product $M \times \mathbb{R}$,
$$M \to M \times \mathbb{R}, \quad m \mapsto (m,t),$$
and let $M_t$ be the inverse image of $\mathbb{I}_M$ through this map. If $M_t$ is a submanifold of $M$ (it is certainly the case if $\mathbb{I}_M$ is open or, more generally, open in $M \times \mathbb{I}$, with $\mathbb{I}$ being an interval), then the restriction map
$$i_t : M_t \to \mathbb{I}_M$$
is defined. The maps $j_m$ and $i_t$ will be also called the *embeddings into the relative interval* $\mathbb{I}_M$ *at* $m$ and *at* $t$, respectively.

Finally, let $\partial/\partial t$ be the unique vector field on $M \times \mathbb{R}$ that is compatible with the zero vector field with respect to the projection map $\pi_M : M \times \mathbb{R} \to M$ and with the standard vector field $\mathrm{d}/\mathrm{d}\,t$ on $\mathbb{R}$ with respect to the projection map $\pi_\mathbb{R} : M \times \mathbb{R} \to \mathbb{R}$ (see Corollary 0.4.17). If $\mathbb{I}_M$ is open, then the restriction on $\mathbb{I}_M$ of $\partial/\partial t$ is defined. Note that if $\mathbb{I}_M = M \times \mathbb{I}$, with $\mathbb{I}$ being a nonempty interval not reduced to a singleton ([31]), the restriction of $\partial/\partial t$ may be defined as well ([32]). Thus, if $\mathbb{I}_M$ is open in $M \times \mathbb{I}$, the restriction of $\partial/\partial t$ on $\mathbb{I}_M$ will be called the *standard vector field* on the relative interval $\mathbb{I}_M$ ([33]).

A relative interval gives rise to a family of intervals $\mathbb{I}_m$. In view of the identification of $M \times \mathbb{R}$ with the set-theoretic product, it will be assumed that a relative interval that determines a fixed family (if it exists) is unique.

Note that the standard vector field $\mathrm{d}/\mathrm{d}\,t$ on the interval $\mathbb{I}_m$ is $j_m$-compatible with the standard vector field on $\mathbb{I}_M$. Although this fact may be trivially proved with the help of local coordinates, it is worth mentioning the following algebraic trick, which will be useful in similar situations. Suppose first that $\mathbb{I}_m = \mathbb{I}$ and $\mathbb{I}_M = M \times \mathbb{I}$. From the definitions of $j_m$ and of the standard vector fields it immediately follows that
$$j_m^* \circ \frac{\partial}{\partial t} \circ \pi_M^* = 0, \quad j_m^* \circ \frac{\partial}{\partial t} \circ \pi_\mathbb{I}^* = \frac{\mathrm{d}}{\mathrm{d}\,t}$$

---

[31] Thus $\mathbb{I}$ may include an endpoint and, in this case, we assume that $M$ is without boundary.

[32] Here, the uniquely determined vector field that is compatible with $\partial/\partial t$ through the embedding is meant. Namely, it is the unique vector field on $M \times \mathbb{I}$ that is compatible with the zero vector field with respect to the projection $M \times \mathbb{I} \to M$ and with the standard vector field $\mathrm{d}/\mathrm{d}\,t$ on $\mathbb{I}$ with respect to the projection $M \times \mathbb{I} \to \mathbb{I}$.

[33] It would be easy to see that the standard vector field may be defined whenever $\dim(\mathbb{I}_M) = \dim(M \times \mathbb{R})$, but all relative intervals that will be needed in the sequel are, in fact, open in $M \times \mathbb{I}$.

and

$$\frac{\mathrm{d}}{\mathrm{d}t} \circ j_m^* \circ \pi_M^* = 0, \quad \frac{\mathrm{d}}{\mathrm{d}t} \circ j_m^* \circ \pi_\mathbb{I}^* = \frac{\mathrm{d}}{\mathrm{d}t}.$$

Therefore, n. 0.3.3 and Proposition 0.5.36 (with $P$ equal to the $C^\infty(M \times \mathbb{I})$-module $C^\infty(\mathbb{I}_m)$ given by $j_m$), imply that $j_m^* \circ \partial/\partial t = \mathrm{d}/\mathrm{d}t \circ j_m^*$, i.e., $\mathrm{d}/\mathrm{d}t$ and $\partial/\partial t$ are $j_m$-compatible. When $\mathbb{I}_m$ and $\mathbb{I}_M$ are open submanifolds, respectively of $\mathbb{I}$ and $M \times \mathbb{I}$, it suffices to invoke, in addition, n. 0.4.15, (5).

### 0.6.4 Smooth Dependence on the Initial Data

**Theorem.** Let $X$ be a smooth vector field on a manifold $M$ without boundary and $f : N \to M$ be a smooth map. Then there exists a smooth map

$$\Phi_f : \mathbb{I}_N \to M,$$

where $\mathbb{I}_N$ is an open relative interval over $N$, such that, for all $n \in N$, the curve

$$\gamma_n = \Phi_f \circ j_n$$

is the unique maximal trajectory of $X$ such that $\gamma_n(0) = f(n)$, with $j_n : \mathbb{I}_n \to \mathbb{I}_N$ being the embedding at $n \in N$.

**Proof.** If $N = M$ and $f$ is the identity map, the statement results in the ordinary existence theorem of the flow

$$\Phi : \mathbb{I}_M \to M.$$

To find $\Phi_f$ in the general case, it suffices to consider the smooth map

$$f \times \mathrm{id}_\mathbb{R} : N \times \mathbb{R} \to M \times \mathbb{R},$$

set

$$\mathbb{I}_N = (f \times \mathrm{id}_\mathbb{R})^{-1}(\mathbb{I}_M)$$

and take

$$\Phi_f \stackrel{\mathrm{def}}{=} \Phi \circ (f \times \mathrm{id}_\mathbb{R})|_{\mathbb{I}_N, \mathbb{I}_M}.$$

$\square$

When $M$ has a nonempty boundary, the above result may be appropriately rephrased in view of n. 0.6.2.

### 0.6.5

In the notation of the preceding theorem, note that the standard vector fields $\partial/\partial t$ on $\mathbb{I}_N$ and $X$ are compatible with respect to $\Phi_f$. In particular, this is true when $\Phi_f$ coincides with the flow $\Phi$.

### 0.6.6  One-parameter Group Generated by a Vector Field

Let
$$\Phi : \mathbb{I}_M \to M$$
be the flow of a vector field $X$ on a manifold $M$ without boundary. For all $t \in \mathbb{R}$ let
$$i_t : M_t \to \mathbb{I}_M$$
be the embedding at $t$ into the relative interval $\mathbb{I}_M$ and set
$$\Phi_t = \Phi \circ i_t \,.$$
From n. 0.4.24, it easily follows that for all $s, t \in \mathbb{R}$ and $m \in M_s$, either
$$\Phi_t(\Phi_s(m)) = \Phi_{t+s}(m)$$
or both sides are undefined (that is, $\Phi_s(M) \not\subset M_t$ and $m \notin M_{t+s}$).

**Definition.** The family of smooth maps
$$\{\Phi_t\}_{t \in \mathbb{R}}$$
will be called the *one-parameter group generated by $X$*.

In the case when the domain $\mathbb{I}_M$ of the flow $\Phi$ is the whole of $M \times \mathbb{R}$, Proposition 0.6.6 actually asserts that
$$\{\Phi_t\}_{t \in \mathbb{R}} \,,$$
equipped with the operation given by map composition, is an abelian group; moreover, the $\Phi_t$'s are diffeomorphisms $M \to M$ (since they are invertible under composition). In the general case, the group operation coincides with map composition only in a local sense.

## 0.6.7

The standard vector field $\partial/\partial t$ and $X$ are compatible with respect to $\Phi$:

$$\frac{\partial}{\partial t} \circ \Phi^* = \Phi^* \circ X .$$

Upon composing with $i_{t_0}^*$, $t_0 \in \mathbb{R}$, and taking into account that, by definition, $\Phi \circ i_{t_0} = \Phi_{t_0}$, one deduces

$$i_{t_0}^* \circ \frac{\partial}{\partial t} \circ \Phi^* = \Phi_{t_0}^* \circ X .$$

In particular, since $\Phi_0 = \mathrm{id}_M$,

$$i_0^* \circ \frac{\partial}{\partial t} \circ \Phi^* = X .$$

These equalities are sometimes written

$$\left.\frac{\mathrm{d}}{\mathrm{d}t}\right|_{t=t_0} \Phi_t^* = \Phi_{t_0}^* \circ X$$

and

$$\left.\frac{\mathrm{d}}{\mathrm{d}t}\right|_{t=0} \Phi_t^* = X .$$

### 0.6.8 *Smooth Families of Smooth Maps*

Let $M$, $N$, and $T$ be smooth manifolds and $\{G_t\}_{t \in T}$ be a family of smooth maps $G_t : N \to M$, parameterized by the points of $T$. Suppose that at least one of $N$ and $T$ is without boundary ([34]).

**Definition.** The family $\{G_t\}_{t \in T}$ is said to be a *smooth family* if the map

$$G : N \times T \to M, \quad (n, t) \mapsto G_t(n)$$

is smooth.

It will be said that the smooth family $\{G_t\}_{t \in T}$ is *defined by* $G$, or that $G$ is the *corresponding map* of $\{G_t\}_{t \in T}$. It is convenient to explicitly remark that, for all $t \in T$,

$$G_t = G \circ i_t ,$$

with $i_t : N \to N \times T$ being the embedding at $t \in T$.

---

[34] The requirement on $N$ and $T$ is necessary to assure that $N \times T$ is a smooth manifold, possibly with boundary. This restriction could be avoided by using the dual space of the smooth envelope of $C^\infty(N) \otimes_\mathbb{R} C^\infty(T)$.

If $M = N$ and $G_t$ is a diffeomorphism for all $t \in T$, then $\{G_t\}_{t \in T}$ will be called a *smooth family of diffeomorphisms of $M$*.

More generally, let $U$ be an open submanifold of $N \times T$, let
$$H : U \to M$$
be a smooth map and, for all $t \in T$, set $N_t = i_t^{-1}(U)$ and
$$H_t : N_t \to M, \quad n \mapsto H(n, t).$$
Then $\{H_t\}_{t \in T}$ will be called a smooth family of *local smooth maps of $N$ into $M$*.

Finally, in the case when $T$ is a nonempty interval not reduced to a singleton and $U$ is a relative interval (over which the standard vector field $\partial/\partial t$ is defined), a smooth family of (possibly local) smooth maps will be also called a *(smooth) one-parameter family*.

### 0.6.9

Let $\{G_t\}_{t \in T}$ be a smooth family of smooth maps $N \to M$ defined by $G : N \times T \to M$ and set
$$\mathbb{G} \stackrel{\text{def}}{=} (G, \pi_T) : N \times T \to M \times T,$$
with $\pi_T : N \times T \to T$ being the projection map. Note that
$$\mathbb{G} \circ i_t = i_t \circ G_t.$$

If $\{H_t\}_{t \in T}$ is a smooth family of smooth maps $H_t : M \to V$ defined by $H : M \times T \to V$, then $\{H_t \circ G_t\}_{t \in T}$ is a smooth family because it corresponds to $H \circ \mathbb{G}$.

### 0.6.10 Smooth Families of Vector Fields

Let $M$ and $T$ be smooth manifolds, with at least one of them without boundary, $\pi : M \times T \to M$ the projection map onto $M$, and $X \in \mathrm{D}(M)_T$ a vector field along $\pi$. For all $t \in T$, let
$$i_t : M \to M \times T$$
be the embedding at $t$, and set
$$X_t = i_t^* \circ X.$$
By n. 0.4.13, $X_t$ is a vector field on $M$.

**Definition.** The family
$$\{X_t\}_{t \in T}$$
will be called a *smooth family of vector fields on $M$ defined by $X$*.

We shall also say that $X$ is the vector field along $\pi$ *corresponding to* $\{X_t\}_{t \in T}$.

More generally, let $U$ be an open submanifold of $M \times T$, $Y$ be a vector field along the restriction $U \to M$ of $\pi$, set $M_t = i_t^{-1}(U)$, and denote again by $i_t$ the restriction $M_t \to U$. Then the family
$$\{Y_t\}_{t \in T}$$
given by
$$Y_t = i_t^* \circ Y$$
will be called a *smooth family of local vector fields on $M$*, defined by $Y$ (*cf.* n. 0.4.16).

Finally, in the case when $T$ is a nonempty interval not reduced to a singleton and $U$ is a relative interval, a smooth family of (possibly local) vector field on $M$ will be sometimes called a *(smooth) time-dependent vector field*.

### 0.6.11 Time-dependent Vector Field Associated with a One-parameter Family of Diffeomorphisms

Let $\{G_t\}_{t \in \mathbb{I}}$ be a one-parameter family of diffeomorphisms defined by $G : M \times \mathbb{I} \to M$, with $\mathbb{I}$ being a nonempty interval, not reduced to a singleton, denote by $\pi_M : M \times \mathbb{I} \to M$ and $\pi_\mathbb{I} : M \times \mathbb{I} \to \mathbb{I}$ the projection maps and set $\mathbb{G} = (G, \pi_\mathbb{I}) : M \times \mathbb{I} \to M \times \mathbb{I}$. By definition,
$$\pi_M \circ \mathbb{G} = G, \quad \pi_\mathbb{I} \circ \mathbb{G} = \pi_\mathbb{I} .$$
Therefore, denoting the embedding at $t \in \mathbb{I}$ by $i_t : M \to M \times \mathbb{I}$ one gets
$$\pi_M \circ \mathbb{G} \circ i_t = G \circ i_t = G_t .$$
Since the maps $G_t$ are diffeomorphisms, $\mathbb{G}$ is a diffeomorphism ([35]). Therefore the operator
$$X \stackrel{\text{def}}{=} \left(\mathbb{G}^{-1}\right)^* \circ \frac{\partial}{\partial t} \circ G^*$$
is defined, with $\partial/\partial t$ being the standard vector field on $M \times \mathbb{I}$.

By n. 0.4.13, $X$ is a vector field along $G \circ \mathbb{G}^{-1}$. But, since $\pi_M \circ \mathbb{G} = G$,
$$G \circ \mathbb{G}^{-1} = \pi_M .$$
Hence, $X$ is a vector field along $\pi_M$.

---

[35] Argue as in Exercise 0.5.37.

**Definition.** The time-dependent vector field $\{X_t\}_{t\in \mathbb{I}}$ defined by $X = (\mathbb{G}^{-1})^* \circ \partial/\partial t \circ G^*$ is called the time-dependent vector field *associated with* $\{G_t\}_{t\in \mathbb{I}}$.

### 0.6.12

By definition,
$$\frac{\partial}{\partial t} \circ G^* = \mathbb{G}^* \circ X \ .$$

The above formula is sometimes written as
$$\frac{\mathrm{d}}{\mathrm{d}t} G_t^* = G_t^* \circ X_t \ .$$

More precisely, for all $t_0 \in \mathbb{I}$, let $i_{t_0} : M \to M \times \mathbb{I}$ be the embedding at $t_0$, and compose both sides of the above formula by $i_{t_0}^*$. Taking into account that
$$\mathbb{G} \circ i_{t_0} = i_{t_0} \circ G_{t_0} \ ,$$

one deduces

$$i_{t_0}^* \circ \frac{\partial}{\partial t} \circ G^* = i_{t_0}^* \circ \mathbb{G}^* \circ X = (\mathbb{G} \circ i_{t_0})^* \circ X = (i_{t_0} \circ G_{t_0})^* \circ X$$
$$= G_{t_0}^* \circ i_{t_0}^* \circ X = G_{t_0}^* \circ X_{t_0} \ .$$

This gives the fundamental formula
$$i_{t_0}^* \circ \frac{\partial}{\partial t} \circ G^* = G_{t_0}^* \circ X_{t_0} \ ,$$

which is also sometimes written
$$\left.\frac{\mathrm{d}}{\mathrm{d}t}\right|_{t=t_0} G_t^* = G_{t_0}^* \circ X_{t_0} \ .$$

### 0.6.13

Let $\{X_t\}_{t\in T}$ be a smooth family of vector fields corresponding to $X \in \mathrm{D}(M)_T$. Recall that there is a natural correspondence between vector fields along the projection $\pi_M : M \times T \to M$ and 'horizontal' vector fields on $M \times T$ (see n. 0.4.19). The vector field $\mathbb{X} \in \mathrm{D}_T(M \times N)$ corresponding to $X$ is characterized by

$$\mathbb{X} \circ \pi_T^* = 0, \quad \mathbb{X} \circ \pi_M^* = X \ ,$$

$\pi_T : M \times T \to T$ being the projection map onto $T$. If $i_t : M \to M \times T$ is the embedding at $t$, then
$$i_t^* \circ \mathbb{X} \circ \pi_T^* = 0, \quad i_t^* \circ \mathbb{X} \circ \pi_M^* = i_t^* \circ X = X_t .$$
But
$$X_t \circ i_t^* \circ \pi_T^* = 0, \quad X_t \circ i_t^* \circ \pi_M^* = X_t ,$$
because $\pi_T \circ i_t$ is a constant map and $\pi_M \circ i_t$ is the identity map. Therefore, from n. 0.3.3 and Proposition 0.5.36, it follows that $X_t$ and $\mathbb{X}$ are $i_t$–compatible for all $t \in T$. This gives an alternative description of the smooth family $\{X_t\}_{t \in T}$.

## 0.6.14

Suppose now that $\{X_t\}_{t \in \mathbb{I}}$ is the time-dependent vector field associated with a one-parameter family of diffeomorphisms $\{G_t\}_{t \in \mathbb{I}} : M \to M$. As before, set $\mathbb{G} = (G, \pi_{\mathbb{I}})$, with $G : M \times \mathbb{I} \to M$ being the map corresponding to the family and $\pi_{\mathbb{I}} : M \times \mathbb{I} \to \mathbb{I}$ the projection map.

**Exercise.** Show that if $\mathbb{X}$ is defined as in n. 0.6.13, then the standard vector field $\partial/\partial t$ on $M \times \mathbb{I}$ is $\mathbb{G}$-compatible with $(\partial/\partial t) + \mathbb{X}$.

**Hint.** Use Proposition 0.5.36 again.

## 0.6.15

Let $Y$ be a vector field on a manifold $M$ without boundary, $\Phi : \mathbb{I}_M \to M$ its flow, and $\{\Phi_t\}_{t \in \mathbb{R}}$ the generated one-parameter group.

Suppose first, for simplicity, that $\mathbb{I}_M = M \times \mathbb{R}$. It easily follows from the definitions that the time-dependent vector field $\{X_t\}_{t \in \mathbb{R}}$ associated with the one-parameter family $\{\Phi_t\}_{t \in \mathbb{R}}$ corresponds to $X = \pi_M^* \circ Y$, $\pi_M : M \times \mathbb{R} \to M$ being the projection map onto $M$. Therefore, if $i_t : M \to M \times \mathbb{R}$ is the embedding at $t$ then $X_t = i_t^* \circ \pi_M^* \circ Y = \mathrm{id}_M^* \circ Y = Y$ for all $t \in \mathbb{R}$. The vector field $\mathbb{X} \in \mathrm{D}_T(M \times N)$ that corresponds to $X$ coincides, in this case, with the unique vector field $\tilde{Y}$ that projects into $Y \in \mathrm{D}(M)$ and into $0 \in \mathrm{D}(\mathbb{R})$ (see Corollary 0.4.17).

Note that, according to Exercise 0.6.14, $\partial/\partial t$ is compatible with $(\partial/\partial t) + \tilde{Y}$ with respect to the map $\mathbb{G} = (\Phi, \pi_{\mathbb{R}})$. This holds also in the general case, i.e., when $\mathbb{I}_M$ is an open submanifold of $M \times \mathbb{R}$: the standard vector field $\partial/\partial t|_{\mathbb{I}_M}$ on $\mathbb{I}_M$ is compatible with $(\partial/\partial t) + \tilde{Y}$ with respect to $\mathbb{G} = (\Phi, \pi_{\mathbb{R}}) : \mathbb{I}_M \to M \times \mathbb{R}$.

### 0.6.16

The well-known intuitive description of the relationship between a vector field $X$ and the corresponding one-parameter group $\{\Phi_t\}_{t\in\mathbb{R}}$ is to say, in view of

$$\left.\frac{\mathrm{d}}{\mathrm{d}t}\right|_{t=0} \Phi_t^* = X$$

(see n. 0.6.7), that $X$ is an infinitesimal diffeomorphism that generates the group. In some sense, like a diffeomorphism $G : M \to M$ which acts on smooth functions via its algebraic counterpart $G^* : C^\infty(M) \to C^\infty(M)$, a vector field $X : C^\infty(M) \to C^\infty(M)$ may be understood as the algebraic counterpart of an infinitesimal diffeomorphism $M \to M$, which would be cumbersome to directly define in geometric terms.

Since a diffeomorphism $G : M \to M$ acts on differential forms via the corresponding de Rham complex homomorphism

$$\Lambda^\bullet G^* : \Lambda^\bullet(M) \to \Lambda^\bullet(M)$$

induced by $G$, it is natural to expect $X$ to give rise to a derivation

$$\Lambda^\bullet X : \Lambda^\bullet(M) \to \Lambda^\bullet(M) .$$

This derivation is called the *Lie derivative* with respect to $X$. The notation $L_X$ or $\mathcal{L}_X$ (instead of the more 'functorial' $\Lambda^\bullet X$) is generally used. A formalization of this concept is given in the following nn.

### 0.6.17 *Smooth Families of Differential Forms*

Let $M$ and $T$ be smooth manifolds, with at least one of them without boundary, and $\Omega \in \Lambda^s(M)_T$ an $s$-form on $M$ along the projection $\pi : M \times T \to M$. For all $t \in T$, set

$$\omega_t(X_1,\ldots,X_s) = i_t^*\left(\Omega(X_1,\ldots,X_s)\right), \quad X_1,\ldots,X_s \in \mathrm{D}(M) ,$$

that is, $\omega_t = i_t^* \circ \Omega$, with $i_t : M \to M \times T$ being the embedding at $t$ ([36]).

**Definition.** The family

$$\{\omega_t\}_{t\in T}$$

of $s$-forms on $M$ will be called a *smooth family defined by* $\Omega$. We shall also say that $\Omega$ is the form *corresponding to* $\{\omega_t\}_{t\in T}$.

---

[36] $\omega_t$ is an $s$-form on $M$, according to n. 0.5.23.

More generally, let $U$ be an open submanifold of $M \times T$, $K$ an $s$-form along the restriction $U \to M$ of $\pi$, set $M_t = i_t^{-1}(U)$, and denote again by $i_t$ the restriction $M_t \to U$. Then the family
$$\{\varkappa_t\}_{t \in T}$$
given by
$$\varkappa_t(X_1, \ldots, X_s) = i_t^*(K(X_1, \ldots, X_s)), \quad X_1, \ldots, X_s \in D(M),$$
will be said to be a *smooth family of local differential forms on $M$* (*cf.* n. 0.5.33), defined by $K$ (or, corresponding to $K$).

Finally, in the case when $T$ is a nonempty interval not reduced to a singleton and $U$ is a relative interval, then a family of (possibly local) differential forms is sometimes called a *(smooth) time-dependent $s$-form on $M$*.

**Exercise.** With $U$ and $i_t : M_t \to U$ being as above, let $\omega$ be an $s$-form on $U$. Show that
$$\{\Lambda^s \, i_t^*(\omega)\}_{t \in T}$$
is a smooth family (up to the natural identification introduced in n. 0.5.33).

**Hint.** Define an $s$-form $\Omega$ along $\pi|_U$ by
$$\Omega(X_1, \ldots, X_s) \stackrel{\text{def}}{=} \omega\left(\tilde{X}_1\Big|_U, \ldots, \tilde{X}_s\Big|_U\right), \quad X_1, \ldots, X_s \in D(M),$$
where $\tilde{X}$ denotes the unique vector field on $M \times T$ that projects onto $X \in D(M)$ through $\pi : M \times T \to M$ and onto $0$ through $\pi_T : M \times T \to T$. Then show that $\Lambda^s \, i_t^*(\omega)$ is identified with $i_t^* \circ \Omega$ according to n. 0.5.33.

## 0.6.18 Derivative of a Time-dependent Differential Form

Let $M$ be a smooth manifold, possibly with boundary, $\partial/\partial t$ the standard vector field on a relative interval $\mathbb{I}_M$ over $M$, and $\pi : \mathbb{I}_M \to M$ the projection map. Since
$$\frac{\partial}{\partial t} : C^\infty(\mathbb{I}_M) \to C^\infty(\mathbb{I}_M)$$
projects onto the zero vector field through $\pi$, it is a $C^\infty(M)$–module homomorphism when $C^\infty(\mathbb{I}_M)$ is considered as a $C^\infty(M)$–module via $\pi^*$. Therefore, for each differential $s$-forms $\Omega$ along $\pi$,
$$(X_1, \ldots, X_s) \mapsto \frac{\partial}{\partial t}(\Omega(X_1, \ldots, X_s))$$
gives again an $s$-form $\Omega'$ along $\pi$.

**Definition.** The time-dependent $s$-form $\{\omega'_t\}_{t\in \mathbb{I}}$ corresponding to $\Omega' = (\partial/\partial t)\circ\Omega$ will be called the *derivative* of the time dependent $s$-form $\{\omega_t\}_{t\in \mathbb{I}}$.

The local form $\omega'_{t_0}$, $t_0 \in \mathbb{R}$, will sometimes be denoted by

$$\left.\frac{d\,\omega_t}{d\,t}\right|_{t=t_0} \quad \text{or} \quad \left.\frac{d}{d\,t}\right|_{t=t_0} \omega_t .$$

It is interesting to describe the derivative of a time-dependent form obtained as in Exercise 0.6.17. To this end, it is useful to preliminarily point out the following Leibnitz rule.

**Exercise.** Let $\Omega \in \Lambda^r(M)_\pi$ and $K \in \Lambda^s(M)_\pi$ be differential forms along $\pi : \mathbb{I}_M \to M$, and

$$\{\omega_t\}_{t\in\mathbb{I}}, \quad \{\varkappa_t\}_{t\in\mathbb{I}},$$

the corresponding time-dependent local $s$-forms, with respective derivatives

$$\{\omega'_t\}_{t\in\mathbb{I}}, \quad \{\varkappa'_t\}_{t\in\mathbb{I}}.$$

Show that

$$\{\omega_t \wedge \varkappa_t\}_{t\in\mathbb{I}}$$

is smooth and that its derivative is

$$\{\omega'_t \wedge \varkappa_t + \omega_t \wedge \varkappa'_t\}_{t\in\mathbb{I}}.$$

**Hint.** Take into account n. 0.5.27.

**Proposition.** Let $\mathbb{I}_M$ be as above and $i_t : M_t \to \mathbb{I}_M$ as in n. 0.6.17 (with $U = \mathbb{I}_M$ open in $M \times \mathbb{I}$). If $\omega$ is an $r$-form on $\mathbb{I}_M$, then the derivative of the family

$$\{\Lambda^r\, i_t^*(\omega)\}_{t\in\mathbb{I}}$$

is

$$\{\Lambda^r\, i_t^*(\omega')\}_{t\in\mathbb{I}},$$

with

$$\omega' \stackrel{\text{def}}{=} \frac{\partial}{\partial t} \lrcorner\, d\omega + d\left(\frac{\partial}{\partial t} \lrcorner\, \omega\right).$$

**Proof.** The assertion is trivial for $r = 0$. Indeed, in this case $\omega$ is a smooth function on $\mathbb{I}_M$ and it can also be regarded as a 0-form along the projection $\pi : \mathbb{I}_M \to M$, that corresponds to the time-dependent form

$$\{i_t^*(\omega)\}_{t\in\mathbb{I}} = \{\Lambda^0 \, i_t^*(\omega)\}_{t\in\mathbb{I}} \,.$$

Then the derivative corresponds simply to $\partial \omega / \partial t$, and the required equality directly follows from the definitions of the ordinary differential and insertion operators (note that $\partial/\partial t \lrcorner \, \mathrm{d}\omega = \mathrm{d}\omega(\partial/\partial t) = \partial\omega/\partial t$ and $\partial/\partial t \lrcorner \, \omega = 0$).

Now consider the case when $\omega = \mathrm{d} a$, $a \in C^\infty(\mathbb{I}_M)$. According to Exercise 0.6.17, the derivative of $\{\Lambda^1 \, i_t^*(\mathrm{d} f)\}_{t\in\mathbb{I}}$ corresponds to the form along $\pi$ given by

$$X \mapsto \frac{\partial}{\partial t}\left(\mathrm{d} a\left(\tilde{X}|_{\mathbb{I}_M}\right)\right) = \frac{\partial}{\partial t}\left(\tilde{X}|_{\mathbb{I}_M}(a)\right),$$

where $\tilde{X} \in D(M \times \mathbb{I})$ projects into $X \in D(M)$ and $0 \in D(\mathbb{I})$. On the other hand,

$$\frac{\partial}{\partial t} \lrcorner \, \mathrm{d}(\mathrm{d} a) + \mathrm{d}\left(\frac{\partial}{\partial t} \lrcorner \, \mathrm{d} a\right) = \mathrm{d}\left(\frac{\partial a}{\partial t}\right)$$

corresponds to the form along $\pi$ given by

$$X \mapsto \mathrm{d}\left(\frac{\partial a}{\partial t}\right)\left(\tilde{X}|_{\mathbb{I}_M}\right) = \tilde{X}|_{\mathbb{I}_M}\left(\frac{\partial a}{\partial t}\right).$$

Hence, the required equality holds in this case too, because $\partial/\partial t$ clearly commutes with $\tilde{X}|_{\mathbb{I}_M}$ (e.g., by Corollary 0.4.17 and n. 0.4.4).

In the general case recall that, according to n. 0.5.15, $\omega$ may be written as a sum

$$\omega = \sum_i a_i \, \mathrm{d} a_{i1} \wedge \cdots \wedge \mathrm{d} a_{ir}\,.$$

Then the result follows from the Leibnitz rule stated in the preceding exercise and the fact that the operation

$$\omega \mapsto \omega' = \frac{\partial}{\partial t} \lrcorner \, \mathrm{d}\omega + \mathrm{d}\left(\frac{\partial}{\partial t} \lrcorner \, \omega\right)$$

also satisfies the Leibnitz rule (it is easily deduced by a straightforward calculation from Propositions 0.5.14 and 0.5.19). $\square$

### 0.6.19 Lie Derivative of Differential Forms

Let $X$ be a vector field and $\omega$ an $s$-form on a manifold $M$ without boundary. If $\{\Phi_t\}_{t\in\mathbb{R}}$ is the generated one-parameter group, then
$$\{\Lambda^s\,\Phi_t^*(\omega)\}_{t\in\mathbb{R}}$$
is a smooth time-dependent (local) form: it suffices to apply Exercise 0.6.17 to the form $\Lambda^s\,\Phi^*(\omega)$, where $\Phi:\mathbb{I}_M\to M$ is the flow of $X$. Moreover, since $\Phi_0$ is the identity map of $M$, the $s$-form
$$\left.\frac{\mathrm{d}}{\mathrm{d}t}\right|_{t=0}\Lambda^s\,\Phi_t^*(\omega)$$
is global, *i.e.*, it is actually an $s$-form on $M$.

**Definition.** In the above notation, the $s$-form
$$\mathcal{L}_X(\omega) \stackrel{\mathrm{def}}{=} \left.\frac{\mathrm{d}}{\mathrm{d}t}\right|_{t=0}\Lambda^s\,\Phi_t^*(\omega)$$
is called the *Lie derivative of $\omega$ along $X$*.

The notation $\mathcal{L}_X$ will refer more generally to the homogeneous operator $\Lambda^\bullet(M)\to\Lambda^\bullet(M)$ with components given by
$$\omega\mapsto\mathcal{L}_X(\omega)\,.$$

Since the operator determined by
$$\omega\mapsto\left.\frac{\mathrm{d}}{\mathrm{d}t}\right|_{t=0}\Lambda^s\,\Phi_t^*(\omega)\,,$$
may also be naturally denoted by
$$\left.\frac{\mathrm{d}}{\mathrm{d}t}\right|_{t=0}\Lambda^\bullet\,\Phi_t^*\,,$$
the above definition is summarized by
$$\mathcal{L}_X = \left.\frac{\mathrm{d}}{\mathrm{d}t}\right|_{t=0}\Lambda^\bullet\,\Phi_t^*\,.$$

### 0.6.20 Cartan Formula

**Proposition.** If $X$ is a vector field on a manifold $M$ without boundary, then
$$\mathcal{L}_X = [\mathrm{i}_X, \mathrm{d}]^{(\mathrm{gr})}\,.$$

**Proof.** Let $\Phi : \mathbb{I}_M \to M$ and $\{\Phi_t\}_{t \in \mathbb{R}}$ be the flow and the one-parameter group of $X$, and denote, as usual, by $i_0 : M \to \mathbb{I}_M$ the embedding at $t = 0$ and by $\partial/\partial t$ the standard vector field on $\mathbb{I}_M$. Denoting the exterior differential on $\mathbb{I}_M$ again by d, one may write

$$\mathcal{L}_X(\omega) = \left.\frac{\mathrm{d}}{\mathrm{d}t}\right|_{t=0} \Lambda^s\, \Phi_t^*(\omega)$$

$$\overset{\text{Prop. 0.6.18}}{=} \Lambda^s\, i_0^* \left( \mathrm{i}_{\frac{\partial}{\partial t}} \left( \mathrm{d}\left(\Lambda^s\, \Phi^*(\omega)\right)\right) + \mathrm{d}\left(\mathrm{i}_{\frac{\partial}{\partial t}}\left(\Lambda^s\, \Phi^*(\omega)\right)\right)\right)$$

$$\overset{\text{Prop. 0.5.21, nn. 0.6.5, 0.5.16}}{=} \Lambda^s\, i_0^* \left( \mathrm{i}_{\frac{\partial}{\partial t}} \left(\Lambda^s\, \Phi^*(\mathrm{d}\omega)\right) + \mathrm{d}\left(\Lambda^s\, \Phi^*\left(\mathrm{i}_X(\omega)\right)\right)\right)$$

$$\overset{\text{Prop. 0.5.21, nn. 0.6.5, 0.5.16}}{=} \Lambda^s\, i_0^* \left( \Lambda^s\, \Phi^*\left(\mathrm{i}_X(\mathrm{d}\omega)\right) + \Lambda^s\, \Phi^*\left(\mathrm{d}\left(\mathrm{i}_X(\omega)\right)\right)\right)$$

$$= \Lambda^s\, \Phi_0^*\left(\mathrm{i}_X(\mathrm{d}\omega) + \mathrm{d}\left(\mathrm{i}_X(\omega)\right)\right) = \mathrm{i}_X(\mathrm{d}\omega) + \mathrm{d}\left(\mathrm{i}_X(\omega)\right) = [\mathrm{i}_X, \mathrm{d}]^{(\mathrm{gr})}(\omega)$$

as required. □

The above result, called the *Cartan formula*, allows one to extend the notion of a Lie derivative to differential forms on arbitrary commutative algebras. In particular, a Lie derivative is also defined on manifolds with boundary. On the other hand, on a manifold $M$, the definition by means of the associated one-parameter group is not limited to the case of differential forms. For instance, the reader may try to define the Lie derivative of tensors of type $(p,q)$ (*i.e.*, sections of the bundle $\left| \mathrm{D}(M)^{\otimes p} \otimes \Lambda^1(M)^{\otimes q} \right|$).

### 0.6.21  Leibnitz Rule for Lie Derivatives

From Exercise 0.6.18, a Leibnitz rule immediately follows:

$$\mathcal{L}_X(\omega \wedge \varkappa) = \mathcal{L}_X(\omega) \wedge \varkappa + \omega \wedge \mathcal{L}_X(\varkappa), \quad \omega, \varkappa \in \Lambda^\bullet(M).$$

Note that it may be also deduced from the Cartan formula by means of a direct calculation based on Propositions 0.5.14 and 0.5.19 (*cf.* the end of the proof of Proposition 0.6.18). With this second argument, the Leibnitz rule is proved for manifolds with boundary too (and, more generally, for arbitrary commutative algebras).

### 0.6.22

**Exercise.** Prove that if $X \in \mathrm{D}(M)$ and $\omega \in \Lambda^s(M)$, then

(1)
$$\mathcal{L}_X(\omega)(X_1,\ldots,X_s) = X\big(\omega(X_1,\ldots,X_s)\big)$$
$$- \sum_{i=1}^{s} \omega\left(X_1,\ldots,[X,X_i],\ldots,X_s\right), \qquad X_1,\ldots,X_s \in \mathrm{D}(M)\,;$$

(2) $[\mathcal{L}_X, i_Y] = i_{[X,Y]}$, $Y \in \mathrm{D}(M)$.

**Hint.** Use the Cartan formula and the definition of the exterior differential to prove (1). Deduce (2) from (1).

Note that the above formulas hold for manifolds with boundary too (and, more generally, for arbitrary commutative algebras).

## Chapter 1

# Basic Differential Calculus on Fat Manifolds

Below fat manifolds are defined simply as vector bundles and at the first glance it might seem to be just an unprovoked change of terminology. Of course, it would not have much sense if so. What we really pursue with this renaming is to consider vector bundles as objects of *another* category by creating this way a new *conceptual environment* around them. By this reason a true understanding of what really are fat manifolds the reader will get after having read this chapter. Nevertheless, a few words explaining the idea in general outline would be healthy.

It is rather natural to treat a vector bundle as a manifold of *fat points*, namely, its fibers. This terminology stresses that we are dealing with some specific points that have a non trivial *internal structure*. If this structure is *classical*, *i.e.*, single points of a fiber can be observed and, therefore, distinguished one from another with classical means, then they form, according to J. Nestruev (see [Nestruev (2003), 10.1]), a smooth manifold. This way one gets nothing new with respect to the standard (classical) approach according to which fiber bundles are considered as smooth maps connecting smooth manifolds which are spectra of algebras of classical observables. From this point of view a fiber is simply an *association* of points put together on the basis of an *external* prescription. On the contrary, the impossibility to distinguish between points of a fiber with classical means makes their unity physically natural and *intrinsic*. So, in such a situation one must develop some *non-classical* means to deal with these fat points. It is natural to look for such means in the *fat* calculus, *i.e.*, a fat analogue of the standard calculus on smooth manifolds. First steps in constructing this analogue are done in this chapter. In doing that the adjective 'fat' plays a guiding role.

The situation when we are forced to renounce the classical observability of single points belonging to a fiber is what happens exactly in the theory of gauge fields and the related mathematics must respect this fact. We do not touch the physical aspect here but it was the main motivation in developing the *fat calculus*. In this book we present a naive and geometrically demonstrative approach to the subject which may be considered as an introduction to a more satisfactory one based on the theory of dioles (see [Vinogradov and Vinogradov (in preparation)]) and of iterated forms (see [Vinogradov and Vitagliano (2006)]).

## 1.1 Basic Definitions

From now on we fix a real vector space $F$ of a finite nonzero dimension.

### 1.1.1 Fat Manifolds

**Definition.** A *fat manifold* is a vector bundle over a manifold $M$, possibly with boundary, with general fiber $F$. Its fiber over a point $m \in M$ is called the *fat point over* $m$ and denoted by $\overline{m}$. In its turn $m$ is called the *base of* $\overline{m}$. The total space of a fat manifold with base $M$ will be usually indicated by $\overline{M}$. By *type* of a fat manifold we shall simply mean the dimension of the general fiber $F$.

Similarly to as it is common in the theory of fiber bundles we indicate only the total space $\overline{M}$ when referring to a fat manifold $\overline{M} \to M$.

According to Definition 0.3.4 the defining module of the vector bundle $\overline{M} \to M$ is denoted by $\Gamma\left(\overline{M}\right)$. We shall interpret its elements as sections of this bundle (*cf.* n. 0.3.2). If $s \in \Gamma\left(\overline{M}\right)$, then $s(m) \in \overline{m}$ stands for the value of $s$ at $m$.

### 1.1.2

Let $P$ be the module of smooth sections (*i.e.*, the defining module) of a fat manifold $\overline{M}$. According to Definition 0.3.4, $P$ is projective and finitely generated. Therefore the dual $P^\vee$ is projective and finitely generated as well. Moreover, $P^\vee$ determines an equidimensional pseudobundle because of n. 0.1.5, (8). By these reasons, there is a fat manifold with base $M$ and module of smooth sections $P^\vee$. We shall denote it by $\overline{M}^\vee$.

In a similar way, functors on modules often act on fat manifolds as well, and we shall denote this action by a superscript. For instance, $\overline{M}^{\text{End}}$ and

$\overline{M}^{S^r}$ are fat manifolds of different types defined by the modules $\mathrm{End}(P)$ and $\mathrm{S}^r(P)$, respectively.

### 1.1.3

Now we have to understand what would be morphisms of fat manifolds, or, in other words, fat analogues of smooth maps. It is clear that these should be morphisms of vector bundles, *i.e.*, smooth maps that send linearly fibers to fibers. These fiber-to-fiber maps are interpreted naturally as morphisms of 'internal structures' of the corresponding fat points. So, one can figure various reasonable options for such morphisms in conformity with the meaning given to words 'internal structure'. Accordingly, there are various natural possibilities to organize fat manifolds in a category. In this book we choose a simplest one that would correspond to the theory of one elementary particle. In such a context fiber-to-fiber maps must be isomorphisms in order to ensure the integrity of the 'intrinsic structure'. In the theory of gauge fields the 'intrinsic structure' of an elementary particle is fixed by means of its symmetry group, which, due to elementary character of the particle, must be an irreducible group of linear transformations of the fiber $F$. Our choice in this book corresponds to the group $\mathrm{GL}(F) \cong \mathrm{GL}(n, \mathbb{R})$ but can be easily generalized to other *structure groups* that emerge in situations when fat manifolds are supplied with additional geometric structures (see Sect. 1.7).

### 1.1.4 Fat Maps

Recall that a morphism of vector bundles is called *regular* if the associated fiber-to-fiber maps are isomorphisms (see Exercise 0.5.37). The above considerations lead us to the following definition.

**Definition.** A *(smooth) fat map* between fat manifolds is a regular morphism of vector bundles. If the base of a fat smooth map is denoted by $f$, then the fat map itself will be usually denoted by $\overline{f}$. A *fat diffeomorphism* is an invertible fat map (*i.e.*, the inverse is also a fat map).

This definition means that we are working in the category of fat manifolds (vector bundles) with the common standard fat point (fiber). Since in these notes we mainly consider the most general structure group $\mathrm{GL}(n, \mathbb{R})$, the isomorphism class of the common standard fiber is uniquely characterized by the dimension of the standard fiber (fat point), *i.e.*, by the type.

It is worth stressing that the concept of the type of a fat manifold becomes more delicate when the 'internal structure' of fat points in consideration is more rich. For instance, it could be an equivalence class of bilinear forms, or linear operators, *etc.*

Since fat maps are regular morphisms, a fat map is a fat diffeomorphism if and only if its base is a diffeomorphism.

### 1.1.5

Recall that a homomorphism of modules

$$\overline{f}^* : \Gamma\left(\overline{M}\right) \to \Gamma\left(\overline{N}\right)$$

is associated with a regular morphism

$$\overline{f} : \overline{N} \to \overline{M}$$

of vector bundles (see Definition 0.3.13). If the fiber is nonzero (as for fat manifolds), the correspondence

$$\overline{f} \leftrightarrow \overline{f}^*$$

is bijective.

This module homomorphism is the fat analogue of the homomorphism of smooth function algebras associated with a smooth map of manifolds.

### 1.1.6

By nn. 0.3.6 and 0.3.12 we have that $\Gamma\left(\overline{N}\right)$ is a $C^\infty(N)$–module obtained from $\Gamma\left(\overline{M}\right)$ by extension of scalars via $f^* : C^\infty(M) \to C^\infty(N)$. Taking into account our conventions about tensor products and scalar extensions (see n. 0.1.1) we have the following important fact.

Given a fat smooth map

$$\overline{f} : \overline{N} \to \overline{M},$$

we can assume that

$$\Gamma\left(\overline{N}\right) = C^\infty(N) \otimes_{C^\infty(M)} \Gamma\left(\overline{M}\right)$$

and that the associated homomorphism

$$\overline{f}^* : \Gamma\left(\overline{M}\right) \to C^\infty(N) \otimes_{C^\infty(M)} \Gamma\left(\overline{M}\right)$$

is the universal homomorphism

$$s \mapsto 1 \otimes s.$$

### 1.1.7 Fat Identity Maps

**Definition.** Let $\overline{M}$ and $\overline{M}'$ be fat manifold over the same base $M$. A fat map $\overline{M} \to \overline{M}'$ with base the identity map of $M$ will be called a *fat identity map*.

A fat identity map $\overline{M} \to \overline{M}$ is *not* necessarily the identity map of $\overline{M}$ in the category of fat manifolds. The name is due to the physical interpretation: note that when $\overline{M} = \overline{M}'$ a fat identity map is nothing else but a *gauge transformation*.

### 1.1.8 Fat Submanifolds

Let $\overline{M}$ be a fat manifold, $N$ a submanifold of $M$ and $i : N \hookrightarrow M$ the embedding map. Then $i$ induces from $\overline{M}$ a bundle $\overline{N}$ together with a fat map $\overline{i} : \overline{N} \hookrightarrow \overline{M}$ (see Definition 0.3.6).

**Definition.** The induced manifold $\overline{N}$ is a *fat submanifold of $\overline{M}$ over $N$*. The induced map $\overline{i}$ is the *fat embedding of $\overline{N}$*.

If $N$ is an open submanifold, then $\overline{N}$ is called an *open fat submanifold*. If $N$ is a closed submanifold, then $\overline{N}$ is called a *closed fat submanifold*.

For an arbitrary fat submanifold $\overline{N}$ we shall identify the fat point of $\overline{N}$ over a point $n \in N$ with the fat point of $\overline{M}$ over $i(n)$ via the isomorphism $\overline{i}_n$ (see n. 0.3.8). In particular, two fat submanifolds over the same submanifold are identified. For a given $s \in \Gamma\left(\overline{M}\right)$ the section $\overline{i}^*(s)$ is called the *restriction of $s$ to $\overline{N}$*. Sometimes we shall use $s|_{N,\overline{N}}$, or, simply, $s|_N$ for $\overline{i}^*(s)$.

### 1.1.9 Fat Curves

In order to develop a fat analogue of dynamics we must, first of all, know what are *fat trajectories* and, more generally, what *fat curves*. To do that it is sufficient to replace all 'normal' ingredients by the corresponding fat ones in the definition of a 'normal' curve. Since a (smooth) curve on $M$ is a (smooth) map $\gamma : \mathbb{I} \to M$ the result is the following.

**Definition.** A *fat interval* $\overline{\mathbb{I}}$ is a standard trivial bundle over an interval $\mathbb{I} \subseteq \mathbb{R}$ with the standard fiber $F$ (see Definition 0.3.15). A *fat curve* on a

fat manifold $\overline{M}$ is a fat map
$$\overline{\gamma} : \overline{\mathbb{I}} \to \overline{M}.$$
Sometimes $\overline{\gamma}$ will be called a *lift of $\gamma$*.

### 1.1.10

Any single fiber of a vector bundle is isomorphic to the standard one, but such an isomorphism is not canonical. In this connection it is worth to emphasize that fat points of a fat interval are identified canonically with the standard fiber $F$ (*cf.* Definition 0.3.16). Hence, a fat curve
$$\overline{\gamma} : \overline{\mathbb{I}} \to \overline{M}$$
allows one to identify fat points of $\overline{M}$ along $\gamma$ with the standard fiber $F$. Note that if $\gamma$ is not injective, then some fat points may admit several different identifications.

### 1.1.11  Fat Restriction

Let
$$\overline{\gamma} : \overline{\mathbb{I}} \to \overline{M}$$
be a fat curve and consider a fat interval $\overline{\mathbb{J}}$ such that $\mathbb{J} \subseteq \mathbb{I}$. If $i : \mathbb{J} \hookrightarrow \mathbb{I}$ is the inclusion map, a fat map
$$\overline{i} : \overline{\mathbb{J}} \hookrightarrow \overline{\mathbb{I}}$$
gives an inclusion of fat intervals, provided that it is *uniform* (see Definition 0.3.17).

**Definition.** A regular uniform morphism $\overline{i}$ over the inclusion map $i : \mathbb{J} \hookrightarrow \mathbb{I}$ will be called a *(fat) inclusion of $\overline{\mathbb{J}}$ into $\overline{\mathbb{I}}$*. The composition
$$\overline{\gamma} \circ \overline{i} : \overline{\mathbb{J}} \to \overline{M}$$
will be called a *(fat) restriction of $\overline{\gamma}$ to $\overline{\mathbb{J}}$*.

### 1.1.12  Fat Translation

Let $\overline{\mathbb{I}}$ be a fat interval. Denote by
$$\tau_t : \mathbb{R} \to \mathbb{R}, \qquad \lambda \mapsto \lambda + t$$
the translation by $t \in \mathbb{R}$ and put $\mathbb{J} = \tau_t^{-1}(\mathbb{I})$. Abusing the notation indicate the restriction $\mathbb{J} \to \mathbb{I}$ of $\tau_t$ again by $\tau_t$.

**Definition.** A regular uniform morphism of fat intervals

$$\overline{\tau_t} : \overline{\mathbb{J}} \to \overline{\mathbb{I}}$$

over the translation $\tau_t$ is called a *fat translation by t*.

### 1.1.13 The Total Space as a Manifold

An important feature of our definition of a vector bundle is that it does not require any smooth manifold structure on the total space. This is motivated by some 'observability' principles (see [Nestruev (2003), 10.1]). However, to make use of the smooth manifold structure on the total bundle is sometimes convenient, either by technical reasons or in order to compare our approach with the traditional one. For instance, it occurs in the proof of the existence theorem of *fat trajectories* and *flows* of fat fields.

The algebra $C^\infty(\overline{M})$ of smooth functions on the total space $\overline{M}$ considered as a smooth manifold (see n. 0.3.23) contains a copy of the dual module $P^\vee$ of $P = \Gamma(\overline{M})$. More precisely, the elements of $P^\vee$ may be interpreted as 'linear along fibers' functions:

$$\alpha(p(m)) = \alpha(p)(m), \quad \alpha \in P^\vee, p \in P, m \in M,$$

where on the left side $\alpha$ is considered as a function on $\overline{M}$ meanwhile on the right side it acts as a linear form on $P$ (see n. 0.3.24). The elements of $A = C^\infty(M)$ are identified with 'constant along fibers' functions:

$$f(p(m)) = f(m), \quad f \in A, p \in P, m \in M.$$

This way the $A$-module multiplication in $P^\vee$ is identified with the multiplication of linear along fibers functions by 'constant along fibers' ones.

### 1.1.14

**Exercise.** Let $\overline{M}$ be a fat manifold, $A = C^\infty(M)$, $P = \Gamma(\overline{M})$ and $f \in C^\infty(\overline{M})$. Show that

$$f \cdot P^\vee \subseteq P^\vee \iff f \in A.$$

**Hint.** Only the direct implication deserves a proof. To this end show, first, that on a real vector space the multiplication by a smooth function sends a given nonzero linear form to a linear form if and only if the smooth function is a constant. Then, note that for each fiber of $\overline{M} \to M$ there exists an $\alpha \in P^\vee$ that is different from zero on this fiber.

## 1.1.15

The notation $\overline{f}^*$, $\overline{f} : \overline{N} \to \overline{M}$ being a fat map, is ambiguous. In fact, it could indicate either the module homomorphism $\Gamma\left(\overline{M}\right) \to \Gamma\left(\overline{N}\right)$ or the algebra homomorphism $C^\infty\left(\overline{M}\right) \to C^\infty\left(\overline{N}\right)$ both associated to $\overline{f}$. As a rule in this book the first meaning will be assumed. In few occasions when the second interpretation is needed it will be explicitly mentioned.

## 1.1.16

In the sequel we shall need a characterization of smooth sections of a fat open submanifold that is similar to the well-known localization procedure in algebraic geometry. This is the subject of the following exercise.

**Exercise.** Let $\overline{N}$ be an open fat submanifold of a fat manifold $\overline{M}$ and $s : N \to \overline{N}$ a (not necessarily smooth) section of $\overline{N}$. Show that $s \in \Gamma\left(\overline{N}\right)$ (that is, $s$ is smooth) if and only if for all $n \in N$ there exists an open neighborhood $U_n \subseteq N$ and a smooth section $s_n \in \Gamma\left(\overline{M}\right)$ such that

$$s|_{U_n} = s_n|_{U_n}$$

(see n. 1.1.8).

**Hint.** The isomorphism $\Gamma\left(\overline{N}\right) = C^\infty(N) \otimes \Gamma\left(\overline{M}\right)$ reduces the direct implication to localization of functions.

## 1.2 The Lie Algebra of Der-operators

Now we are passing to *fat vector fields* and related matters.

### 1.2.1

Let $P, Q$ be modules over a commutative (unitary) $k$-algebra $A$, $k$ being a field. Consider the commutator map

$$\mathrm{Diff}_1(P, Q) \times A \to \mathrm{Diff}_0(P, Q) ,$$

defined by

$$(\square, a) \mapsto [\square, a] = \square(a \cdot) - a\square(\cdot)$$

(see n. 0.1.2). This map is linear over $A$ with respect to the first argument and a derivation with respect to the second. Hence it gives rise to an

$A$-module homomorphism
$$\Delta : \text{Diff}_1(P, Q) \to \text{D}(\text{Diff}_0(P, Q))$$
given by
$$\square \longmapsto (a \mapsto [\square, a]) \,.$$
Of course, $\text{Ker}\,\Delta = \text{Diff}_0(P, Q)$. Thus $\Delta$ is factorized to a monomorphism
$$\text{Smbl}_1(P, Q) \hookrightarrow \text{D}(\text{Diff}_0(P, Q)) \,, \tag{1.1}$$
where
$$\text{Smbl}_1(P, Q) \overset{\text{def}}{=} \frac{\text{Diff}_1(P, Q)}{\text{Diff}_0(P, Q)}$$
is the module of *first order symbols* (*cf.* [Nestruev (2003), 9.69]).

### 1.2.2

**Definition.** Let $P$, $Q$ be modules over a commutative $k$-algebra $A$, $k$ being a field. The *symbol* of an operator $\square \in \text{Diff}_1(P, Q)$ is the coset of $\square$ in
$$\frac{\text{Diff}_1(P, Q)}{\text{Diff}_0(P, Q)} \,,$$
denoted by $[\square]$.

The symbol $[\square]$ will be identified with the associated derivation
$$A \to \text{Diff}_0(P, Q) = \text{Hom}(P, Q)$$
defined earlier in (1.1).

In the case when $P = Q$ we have a derivation
$$[\square] \in \text{D}(\text{End}(P)) \,.$$
Consider also the map
$$\iota : A \to \text{End}(P), \qquad a \mapsto a\,\text{id}_P \,.$$
Obviously, $\iota$ is an $A$-module homomorphism whose image is the submodule of *scalar endomorphisms* of $P$. A natural homomorphism
$$\text{D}(\iota) : \text{D}(A) \to \text{D}(\text{End}(P)) \tag{1.2}$$
is defined by
$$X \mapsto \iota \circ X \,.$$

### 1.2.3

**Definition.** Let $k$ be a field, $A$ a commutative $k$-algebra and $P$ an $A$-module. A *der-operator in $P$* is a linear differential operator

$$\square : P \to P$$

of order $\leq 1$ such that its symbol lies in the image of homomorphism (1.2).

The above definition can be summarized by saying that a der-operator is a linear differential operator of order $\leq 1$ with scalar symbol.

The totality of all der-operators in $P$ has a natural $A$-module structure induced by the module structure in $\mathrm{Diff}_1(P,P)$, *i.e.*, that given by

$$(a\square)(p) = a\left(\square(p)\right), \quad a \in A, p \in P$$

(see n. 0.1.2). The $A$-module of all der-operators in $P$ is denoted by $\mathrm{Der}(P)$.

### 1.2.4 Leibnitz Rule for Der-operators

Consider an arbitrary operator $\square : P \to P$.

If $\square$ is a der-operator, then, by definition, there exists a derivation $X \in \mathrm{D}(A)$ such that

$$[\square](a) = X(a)\,\mathrm{id}_P, \quad a \in A\,.$$

Hence, for all $a \in A$ and $p \in P$,

$$X(a)p = X(a)\,\mathrm{id}_P(p) = [\square](a)(p) = [\square, a](p) = \square(ap) - a\square(p)\,, \quad (1.3)$$

that gives the following Leibnitz-like rule:

$$\square(ap) = X(a)p + a\square(p)\,. \tag{1.4}$$

Conversely, suppose that $\square$ is $k$-linear and satisfies (1.4) for some $X \in \mathrm{D}(A)$. Then the sequence (1.3) can be 'rearranged' in the following way:

$$[\square](a)(p) = [\square, a](p) = \square(ap) - a\square(p) = X(a)p = X(a)\,\mathrm{id}_P(p)\,.$$

This shows that $\square$ is a differential operator of order $\leq 1$, because $[\square, a]$ is an endomorphism of $P$ for all $a \in A$, and that the symbol $[\square]$ belongs to the image of (1.2), because the endomorphism $[\square, a] = [\square](a)$ is scalar. Hence $\square$ is a der-operator.

We conclude that an operator $\square : P \to P$ is a der-operator if and only if it is $k$-linear and satisfies (1.4).

### 1.2.5

**Definition.** A der-operator $\square$ in $P$ is said to be *over* a derivation $X \in D(A)$ if the image of $X$ in $D(\text{End}(P))$ according to (1.2) is $[\square]$, *i.e.*, Leibnitz rule (1.4) holds.

If $P$ is faithful, *i.e.*, there is no $f \neq 0$ such that $fp = 0$ for all $p \in P$, then every der-operator is over a unique derivation. Indeed, in this case the map

$$\iota : A \to \text{End}(P), \qquad a \mapsto a \, \text{id}_P$$

is injective and so is the map

$$D(A) \to D(\text{End}(P)).$$

A der-operator in $P$ over the zero derivation of $A$ is nothing but an endomorphism of $P$. Consequently, the difference of two der-operators over the same derivation is an endomorphism of $P$.

### 1.2.6

**Example.** Obviously, $\text{Der}(A) = \text{Diff}_1(A, A)$ and, in particular, $D(A) \subseteq \text{Der}(A)$.

### 1.2.7

**Example.** Let $E$ be a $k$-vector space of a finite dimension $m > 0$ and $X \in D(A)$. Consider the free module $P = A \otimes_k E$, which in a geometric situation is interpreted naturally as the module of smooth sections of a standard trivial vector bundle, and the operator

$$D_X = X \otimes_k \text{id}_E : P \to P.$$

Choose a basis $(e_1, \ldots, e_m)$ in $E$. Then for all $a \in A$ and $p = \sum a_i \otimes e_i \in P$ we have

$$D_X(ap) = D_X \left( \sum aa_i \otimes e_i \right) = \sum X(aa_i) \otimes e_i$$
$$= \sum (X(a) a_i + a X(a_i)) \otimes e_i = \left( X(a) \sum a_i \otimes e_i \right) + \left( a \sum X(a_i) \otimes e_i \right)$$
$$= X(a) p + a D_X(p).$$

Hence $D_X$ is a der-operator in $P$.

In coordinates $D_X$ is described as follows. Consider the isomorphism $P \cong A^m$ defined by

$$p \leftrightarrow (p_1, \ldots, p_m) \quad \Longleftrightarrow \quad p = \sum_{i=1}^{m} p_i \mathbf{e}_i$$

with $\mathbf{e}_i = 1 \otimes e_i$ and $p_i \in A$. Then up to this isomorphism

$$D_X (p_1, \ldots, p_m) = (X(p_1), \ldots, X(p_m)) \ .$$

**1.2.8**

Consider an arbitrary der-operator $\square$ in $P = A \otimes_k E$ over $X$. Since $\square$ and $D_X$ are over the same derivation $X$ their difference is an endomorphism $\Phi$ of $P$ (see n. 1.2.5) and, so,

$$\square = D_X + \Phi \ .$$

Conversely, every operator of this form is a der-operator.

**1.2.9**

In coordinates a der-operator $\square$ is described as follows. If $p = \sum p_i \mathbf{e}_i$, then

$$\square(p) = \sum_k X(p_k) \mathbf{e}_k + \sum_j p_j \square(\mathbf{e}_j) \ .$$

By putting

$$\square(\mathbf{e}_j) = \sum_k (\Gamma_\square)_j^k \mathbf{e}_k \ ,$$

we obtain

$$\square(p) = \sum_k \left( X(p_k) + \sum_j (\Gamma_\square)_j^k p_j \right) \mathbf{e}_k \ .$$

Represent now coordinate vectors as columns and introduce the matrix $\Gamma_\square$ with entries $(\Gamma_\square)_j^k$ (where $k$ is the row index), which represents $\Phi$ in the basis $(\mathbf{e}_1, \ldots, \mathbf{e}_m)$. In this notation the last formula reads as

$$\square \begin{pmatrix} p_1 \\ \vdots \\ p_m \end{pmatrix} = \underline{X} \begin{pmatrix} p_1 \\ \vdots \\ p_m \end{pmatrix} + \Gamma_\square \begin{pmatrix} p_1 \\ \vdots \\ p_m \end{pmatrix} ,$$

where $\underline{X}$ is the operator matrix

$$\underline{X} = \begin{pmatrix} X & & \mathbb{O} \\ & \ddots & \\ \mathbb{O} & & X \end{pmatrix}.$$

Conversely, every operator $\square$ represented coordinate-wisely in the form

$$\square = \underline{X} + \Gamma_\square$$

is a der-operator over $X$.

### 1.2.10

**Proposition.** If $\square_1$ and $\square_2$ are der-operators in an $A$-module $P$ over the derivations $X_1$ and $X_2$, respectively, then the commutator $[\square_1, \square_2]$ is a der-operator over $[X_1, X_2]$.

**Proof.** For all $a \in A$ and $p \in P$ we have

$$[\square_1, \square_2](ap) = \square_1(\square_2(ap)) - \square_2(\square_1(ap))$$
$$= \square_1(X_2(a)p + a\square_2(p)) - \square_2(X_1(a)p + a\square_1(p))$$
$$= X_1(X_2(a))p + X_2(a)\square_1(p) + X_1(a)\square_2(p) + a\square_1(\square_2(p))$$
$$- X_2(X_1(a))p - X_1(a)\square_2(p) - X_2(a)\square_1(p) - a\square_2(\square_1(p))$$
$$= [X_1, X_2](a)p + a[\square_1, \square_2](p) .$$

Therefore, $[\square_1, \square_2]$ satisfies the Leibnitz rule over $[X_1, X_2]$ (see Definition 1.2.5). $\square$

**Corollary.** Der $(P)$ is a Lie subalgebra of $\mathrm{Diff}_1(P, P)$.

### 1.2.11 *Der-operators Along Homomorphisms*

Let $\varphi : A \to B$ be a commutative $k$-algebra homomorphism, $k$ being a field, and $\overline{\varphi} : P \to Q$ an $A$-module homomorphism of an $A$-module into a $B$-module. Recall that the symbol of a differential operator

$$\square : P \to Q$$

of order $\leq 1$, with $Q$ considered as an $A$-module via $\varphi$, is identified with a derivation

$$A \to \mathrm{Hom}_A(P, Q)$$

(see n. 1.2.1).

Consider also the map

$$\iota_{\overline{\varphi}} : B \to \operatorname{Hom}_A(P, Q), \qquad b \mapsto b\overline{\varphi}.$$

Recall that $\mathrm{D}(A)_\varphi$ denotes the module of derivations along $\varphi$. A natural homomorphism

$$\mathrm{D}(\iota_{\overline{\varphi}}) : \mathrm{D}(A)_\varphi \to \mathrm{D}(A)_{\iota_{\overline{\varphi}} \circ \varphi}$$

is defined by

$$X \mapsto \iota_{\overline{\varphi}} \circ X.$$

**Definition.** A linear differential operator

$$\square : P \to Q$$

of order $\leq 1$ is called a *der-operator along $\overline{\varphi}$ over $X \in \mathrm{D}(A)_\varphi$* if

$$[\square] = \mathrm{D}(\iota_{\overline{\varphi}})(X).$$

All der-operators along $\overline{\varphi}$ constitute a $B$-submodule of $\operatorname{Diff}_1(P, Q)$ (the $B$-module structure being induced by that of $Q$), which will be denoted by $\operatorname{Der}(P)_{\overline{\varphi}}$.

As for ordinary der-operators, it is easy to see that a function

$$\square : P \to Q$$

is a der-operator along $\overline{\varphi}$ over a derivation $X$ along $\varphi$ if and only if it is $k$-linear and fulfills the following Leibnitz-like rule:

$$\square(ap) = X(a)\overline{\varphi}(p) + \varphi(a)\square(p), \quad a \in A, p \in P.$$

## 1.2.12

Like for ordinary der-operators it is easily proved that if the $B$-submodule generated by $\operatorname{Im}\overline{\varphi}$ is faithful, then every der-operator along $\overline{\varphi}$ is over a unique derivation along $\varphi$.

## 1.3 Fat Vector Fields

In this section der-operators are interpreted geometrically as fat fields.

## 1.3.1

**Definition.** Let $\overline{M}$ be a fat manifold. A *(smooth) fat vector field* $\overline{X}$ on $\overline{M}$ (or, for short, a *fat field*) is a der-operator in $\Gamma\left(\overline{M}\right)$. If $\overline{X}$ is a der-operator over a vector field $X$ on $M$, then $X$ is called the *base* of $\overline{X}$.

The totality of all fat fields on $\overline{M}$ will be denoted by $\overline{\mathrm{D}}(\overline{M})$, i.e., $\overline{\mathrm{D}}(\overline{M}) = \mathrm{Der}\left(\Gamma\left(\overline{M}\right)\right)$. As we have already seen in the previous section $\overline{\mathrm{D}}(\overline{M})$ is both a $\mathrm{C}^\infty(M)$–module and a Lie algebra.

Since the general fiber of $\overline{M}$ is nonzero, the module $\Gamma\left(\overline{M}\right)$ is faithful. Therefore, by n. 1.2.5 the base $X$ of a fat field $\overline{X}$ is uniquely determined.

### 1.3.2 Fat Tangent Vectors

Let $\overline{M}$ be a fat manifold and $m \in M$. Set $A = \mathrm{C}^\infty(M)$, $P = \Gamma(M)$ and recall that

$$\overline{m} = \frac{P}{\mu_m P} = \mathbb{R}_m \otimes P,$$

where $\mu_m = \mathrm{Ker}\, m$ is the ideal of functions vanishing at $m$ and $\mathbb{R}_m$ is the $A$-algebra given by $m : A \to \mathbb{R}$ (see n. 0.3.1 and Definition 0.3.4).

**Definition.** A *fat tangent vector* of $\overline{M}$ at $m$ (or at $\overline{m}$) is a der-operator along the evaluation map

$$\overline{h}_m : P \to \overline{m}, \quad p \mapsto p(m) = 1 \otimes p = p + \mu_m P.$$

## 1.3.3

In view of n. 1.2.12 every fat tangent vector $\overline{\xi}$ is over a unique ordinary tangent vector $\xi$. According to n. 1.2.11, a fat tangent vector is characterized as an $\mathbb{R}$-linear operator satisfying the following version of the Leibnitz rule:

$$\overline{\xi}(fp) = \xi(f)\, p(m) + f(m)\, \overline{\xi}(p).$$

### 1.3.4 Fat Tangent Space

The totality $\mathrm{Der}(P)_{\overline{h}_m}$ of all fat tangent vectors at $\overline{m}$, which is an $\mathbb{R}$-subspace of $\mathrm{Diff}_1\left(P,\overline{m}\right)$, will be called the *fat tangent space at $\overline{m}$* and denoted by $\overline{T}_m\overline{M}$.

## 1.3.5

According to the classical terminology, a field on a manifold is a rule associating with each point of the manifold a geometric object of a certain type. In that sense a fat vector field, as it was defined earlier, is a field of fat tangent vectors. In other words, with a fat vector field $\overline{X}$ on a fat manifold $\overline{M}$ and a point $m \in M$ (or, equivalently, a fat point $\overline{m}$ of $\overline{M}$) a fat tangent vector at $m$, denoted by $\overline{X}_m$, is associated. Namely, $\overline{X}_m$ is defined as

$$\overline{X}_m = \overline{h}_m \circ \overline{X}.$$

Moreover, it is easy to see that the correspondence

$$m \longmapsto \overline{X}_m$$

determines $\overline{X}$ uniquely. (But obviously not every field of fat tangent vectors gives a fat vector field, which is a smooth one.)

### 1.3.6 Locality of Fat Tangent vectors and Fat Fields

Fat tangent vectors and fat fields are differential operators and as such are local operators (*cf.* [Nestruev (2003), 9.61]). This means the following.

Let $\overline{\xi}$ be a be a fat tangent vector to $\overline{M}$ at a fat point $\overline{m}$ and $\overline{X}$ a fat field on $\overline{M}$. Assume that $s, t \in \Gamma\left(\overline{M}\right)$ are such that

$$s|_U = t|_U$$

with $U$ being a neighborhood $U$ of $m$. Then

$$\overline{\xi}(s) = \overline{\xi}(t) \quad \text{and} \quad \overline{X}(s)|_U = \overline{X}(t)|_U.$$

The reader is suggested to prove this locality property.

### 1.3.7 Fat Differential of a Fat Map

A fat map connecting two fat manifolds induces linear maps connecting fat tangent spaces at the corresponding points. The following rigorous definition mimics that of the usual differential.

Let $\overline{f} : \overline{N} \to \overline{M}$ be a fat map, $P = \Gamma\left(\overline{M}\right)$, $Q = \Gamma\left(\overline{N}\right)$, $n \in N$ and $f(n) = m$. Consider the restriction $\overline{f}_n : Q_n \to P_m$ of $\overline{f}$ to fibers $\overline{n} = Q_n$ and $\overline{m} = P_m$ and associate with a fat tangent vector to $\overline{N}$ at $\overline{n}$, $\overline{\xi} : Q \to Q_n$, the (differential) operator

$$\overline{\mathrm{d}}_n \overline{f}\left(\overline{\xi}\right) : P \to P_m, \quad \overline{\mathrm{d}}_n \overline{f}\left(\overline{\xi}\right) = \overline{f}_n \circ \overline{\xi} \circ \overline{f}^*.$$

By definition of $\overline{f}^*$ (see n. 0.3.13)
$$\overline{f}^*(p)(n) = \overline{f}_n^{-1}(p(f(n))), \quad p \in P,$$
so that
$$\overline{f}_n\left(\overline{f}^*(p)(n)\right) = p(m), \quad p \in P. \tag{1.5}$$
Moreover,
$$\overline{f}^*(ap) = f^*(a)\overline{f}^*(p), \quad a \in C^\infty(M), p \in P, \tag{1.6}$$
because $\overline{f}^*$ is a $C^\infty(M)$-module homomorphism and the $C^\infty(M)$-module structure on $Q$ is defined by means of $f^*$.

The operator $\overline{d}_n\overline{f}(\overline{\xi})$ is $\mathbb{R}$-linear, since $\overline{f}_n$, $\overline{\xi}$ and $\overline{f}^*$ are $\mathbb{R}$-linear, and for all $a \in C^\infty(M)$ and $p \in P$ we have

$$\overline{d}_n\overline{f}(\overline{\xi})(ap) = \overline{f}_n\left(\overline{\xi}\left(\overline{f}^*(ap)\right)\right) \stackrel{(1.6)}{=} \overline{f}_n\left(\overline{\xi}\left(f^*(a)\overline{f}^*(p)\right)\right)$$
$$= \overline{f}_n\left(\xi(f^*(a))\overline{f}^*(p)(n) + f^*(a)(n)\overline{\xi}\left(\overline{f}^*(p)\right)\right)$$
$$= \xi(f^*(a))\overline{f}_n\left(\overline{f}^*(p)(n)\right) + f^*(a)(n)\overline{f}_n\left(\overline{\xi}\left(\overline{f}^*(p)\right)\right)$$
$$\stackrel{(1.5)}{=} (d_n f(\xi)(a))p(m) + a(m)\left(\overline{d}_n\overline{f}(\overline{\xi})(p)\right).$$

In view of n. 1.3.3, this implies that $\overline{d}_n\overline{f}(\overline{\xi})$ is a fat tangent vector to $\overline{M}$ at $\overline{m}$ over the tangent vector $d_n f(\xi)$ to $M$. Finally the map
$$\overline{d}_n\overline{f}: \overline{T}_nN \to \overline{T}_mM, \quad \overline{\xi} \mapsto \overline{d}_n\overline{f}(\overline{\xi})$$
is $\mathbb{R}$-linear because $\overline{f}_n$ is $\mathbb{R}$-linear.

The above said leads us to the following definition.

**Definition.** The map $\overline{d}_n\overline{f}$ will be called the *fat differential of $\overline{f}$ at $\overline{n}$*.

### 1.3.8 Compatibility of Fat Fields

Another notion that naturally comes from the analogy with the standard case is that of compatibility of fat fields with respect to a fat map.

**Definition.** Suppose $\overline{f}: \overline{N} \to \overline{M}$ is a fat map and $\overline{X}$, $\overline{Y}$ are fat fields on $\overline{M}$, $\overline{N}$, respectively. Then $\overline{X}$ and $\overline{Y}$ are said to be *compatible with respect to $\overline{f}$* (shortly, $\overline{f}$-*compatible*) if
$$\overline{Y} \circ \overline{f}^* = \overline{f}^* \circ \overline{X},$$

*i.e.*, if the diagram

$$\begin{array}{ccc} \Gamma(\overline{M}) & \xrightarrow{\overline{f}^*} & \Gamma(\overline{N}) \\ \overline{X} \downarrow & & \downarrow \overline{Y} \\ \Gamma(\overline{M}) & \xrightarrow{\overline{f}^*} & \Gamma(\overline{N}) \end{array}$$

is commutative.

### 1.3.9

**Exercise.** Let $\overline{f} : \overline{N} \to \overline{M}$ be a fat smooth map and $\overline{X}$, $\overline{Y}$ fat fields on $\overline{M}$ and $\overline{N}$, respectively. Show that if $\overline{X}$ and $\overline{Y}$ are $\overline{f}$-compatible, then $X$ and $Y$ are $f$-compatible.

### 1.3.10 Restricted Fat Fields

**Exercise.** Let $\overline{X}$ be a fat field on a fat manifold $\overline{M}$, $\overline{N}$ an open fat submanifold of $\overline{M}$ and $\overline{i} : \overline{N} \hookrightarrow \overline{M}$ the corresponding fat embedding. Show that there is a unique fat field $\overline{X}|_N$ on $\overline{N}$ compatible with $\overline{X}$ with respect to $\overline{i}$.

**Hint.** Take into account nn. 1.1.16 and 1.3.6 (*cf.* Proposition 0.4.5).

**Definition.** The fat field $\overline{X}|_N$ will be called the *restriction of* $\overline{X}$ *to* $\overline{N}$.

As it results from the definitions of restriction and compatibility, the restriction of a fat field $\overline{X}$ is characterized by

$$\overline{X}|_N (s|_N) = \overline{X}(s)|_N, \quad s \in \Gamma(\overline{M}).$$

### 1.3.11

The following result concerning fat fields is, in fact, valid for arbitrary differential operators on a manifold $M$ and is a direct consequence of the fact that differential operators are localizable.

**Proposition.** Let $\overline{M}$ be a fat manifold, $\{U_i\}_{i \in \mathcal{I}}$ an open covering of $M$ and $\overline{U_i}$ the open fat submanifold over $U_i$. Suppose that for each $i$ $\overline{X_i}$ is a fat field on $\overline{U_i}$ and that for all $i, j \in \mathcal{I}$ the restrictions of $\overline{X_i}$ and $\overline{X_j}$ to

$\overline{U_i} \cap \overline{U_j}$ coincide. Then there is a unique fat field $\overline{X}$ on $\overline{M}$ such that its restriction to $\overline{U_i}$ is $\overline{X_i}$ for all $i$.

**Proof.** Given a section $s \in \Gamma(\overline{M})$ we have to define the value $\overline{X}(s)$ of $\overline{X}$ on $s$. Put $s_i = s|_{U_i}$, $t_i = \overline{X_i}(s_i)$, $\overline{X_{ij}} = \overline{X_i}|_{U_i \cap U_j} (= \overline{X_j}|_{U_i \cap U_j})$ and $s_{ij} = s|_{U_i \cap U_j}$. In view of n. 1.3.10, we have

$$t_i|_{U_i \cap U_j} = \overline{X_i}(s_i)|_{U_i \cap U_j} = \overline{X_{ij}}(s_{ij}) = \overline{X_j}(s_j)|_{U_i \cap U_j} = t_j|_{U_i \cap U_j} \quad (1.7)$$

for all $i, j \in \mathcal{I}$. For each $m \in M$ choose $i_m$ such that $m \in U_{i_m}$ and define the section

$$t : M \to \overline{M}, \qquad m \mapsto t_{i_m}(m) .$$

It follows from (1.7) that the section $t$ does not depend on the choice of indexes $i_m$ and that $t$ is smooth (see also Exercise 1.1.16). Thus we put

$$\overline{X}(s) \stackrel{\text{def}}{=} t .$$

It is straightforwardly verified that $\overline{X}$ is a fat field whose restriction to $\overline{U_i}$ is $\overline{X_i}$.

The uniqueness is obvious: if the restriction of a fat field $\overline{Y}$ to $\overline{U_i}$ is $\overline{X_i}$ for all $i \in \mathcal{I}$, then the restriction of $\overline{Y}(s)$ on $\overline{U_i}$ is $\overline{X_i}(s_i) = t_i$, and hence $\overline{Y}(s) = \overline{X}(s)$. □

**Definition.** In the above situation, the fat field $\overline{X}$ is said to be obtained by *gluing together* $\{\overline{X_i}\}_{i \in \mathcal{I}}$.

### 1.3.12

**Exercise.** Let $\overline{m}$ be a fat point of a fat manifold $\overline{M}$. Show that

$$\overline{T}_m \overline{M} = \mathbb{R}_m \otimes \overline{\mathrm{D}}(\overline{M}) ,$$

with $\mathbb{R}_m$ being as in n. 1.3.2, and

$$\overline{X}_m = 1 \otimes \overline{X}, \qquad \overline{X} \in \overline{\mathrm{D}}(\overline{M}) ,$$

**Hint.** Construct an isomorphism

$$\overline{T}_m \overline{M} \cong \frac{\overline{\mathrm{D}}(\overline{M})}{\mu_m \overline{\mathrm{D}}(\overline{M})} ,$$

with $\mu_m = \operatorname{Ker} m$ being the ideal of functions vanishing at $m$, such that $\overline{X}_m$ corresponds to the coset of $\overline{X}$.

In other words, the above exercise claims that $\overline{T}_m \overline{M}$ is obtained from $\overline{\mathrm{D}}(\overline{M})$ by extension of scalars via $m : C^\infty(M) \to \mathbb{R}$ with universal homomorphism

$$\overline{X} \mapsto \overline{X}_m .$$

### 1.3.13 Fat Fields Along Fat Maps

**Definition.** Let $\overline{f}$ be a fat map. A *fat (vector) field along $\overline{f}$* is a der-operator along $\overline{f}^*$.

The $C^\infty(N)$–module of all fat fields along $\overline{f} : \overline{N} \to \overline{M}$ will be denoted by $\overline{D}(\overline{M})_{\overline{f}}$.

Note that a fat tangent vector at $\overline{m}$ is a fat field along the inclusion
$$\overline{m} \hookrightarrow \overline{M}$$
according to the canonical identification $\Gamma(\overline{m}) = \overline{m}$ (see n. 0.3.15).

If $\overline{X}$ is a fat field along $\overline{f} : \overline{N} \to \overline{M}$ and $\overline{g} : \overline{V} \to \overline{N}$ is a fat map, then $\overline{g}^* \circ \overline{X}$ is a fat field along $\overline{f} \circ \overline{g}$. For this reason $\overline{X}$ can be described as a field that associates with $n \in N$ a fat tangent vector $\overline{X}_n$ to $\overline{M}$ at $m = f(n)$: consider the case when $\overline{g}$ is equal to the composition
$$\overline{m} \xrightarrow{\overline{f}_n^{-1}} \overline{n} \hookrightarrow \overline{N} \,.$$
Note also that in this case $\overline{g}^* = \overline{f}_n \circ \overline{h}_n$, where $\overline{h}_n : \Gamma(\overline{N}) \to \overline{n}$ is the evaluation map. Therefore the explicit formula for $\overline{X}_n$ is
$$\overline{X}_n = \overline{f}_n \circ \overline{h}_n \circ \overline{X} \,.$$

## 1.4 Fat Fields and Vector Fields on the Total Space

In this section we illustrate a natural correspondence arising between fat vector fields on a fat manifold and some special vector fields on its total space manifold.

### 1.4.1 Dual Der-Operators

Here, as before, $k$ is a field, $A$ a commutative $k$-algebra and $P$ an $A$-module.

**Exercise.** Let $P^\vee$ be the dual of $P$, $\overline{X}$ a der-operator in $P$ over a derivation $X$ and $\alpha \in P^\vee$. Show that
$$X \circ \alpha - \alpha \circ \overline{X} \in P^\vee$$
and that the map
$$\overline{X}^\vee : P^\vee \to P^\vee, \qquad \alpha \mapsto X \circ \alpha - \alpha \circ \overline{X}$$
is a der-operator in $P^\vee$ over $X$.

**Definition.** The der-operator $\overline{X}^\vee$ on $P^\vee$ is called the *dual der-operator of $\overline{X}$*.

### 1.4.2 Dual fat manifolds

**Exercise.** Show that

$$\overline{X} \longleftrightarrow \overline{X}^{\vee}$$

gives a bijection between fat fields on $\overline{M}$ and fat fields on $\overline{M}^{\vee}$ (that is defined in n. 1.1.2).

**Hint.** By definition, $P$ is projective and finitely generated and, so, one may use the natural isomorphism $P \cong (P^{\vee})^{\vee}$ (see n. 0.1.5, (6)).

### 1.4.3

Let $\pi : \overline{M} \to M$ be the projection of $\overline{M}$ onto its base and $A = C^{\infty}(M)$. Recall that elements of $P^{\vee}$ are interpreted naturally as linear along fibers of $\pi$ functions on $\overline{M}$ (see n. 1.1.13).

**Exercise.** Assume $\mathbb{X}$ to be a vector field on $\overline{M}$ such that

$$\mathbb{X}(P^{\vee}) \subseteq P^{\vee}, \quad \mathbb{X}(A) \subseteq A.$$

Show that the restriction of $\mathbb{X}$ to $P^{\vee}$ is a der-operator on $P^{\vee}$ whose base is the restriction of $\mathbb{X}$ to $A$ (the restrictions are intended on both the domain and the codomain). Moreover, show that $\mathbb{X}$ and $X$ are compatible with respect to $\pi$.

### 1.4.4

**Exercise.** In the notation of the previous Exercise show that

$$\mathbb{X}(P^{\vee}) \subseteq P^{\vee} \Longrightarrow \mathbb{X}(A) \subseteq A.$$

**Hint.** Let $a \in A$, $\alpha \in P^{\vee}$. Note that

$$\mathbb{X}(a)\alpha = \mathbb{X}(a\alpha) - a\mathbb{X}(\alpha)$$

and use Exercise 1.1.14.

Now one sees that in Exercise 1.4.3 the condition

$$\mathbb{X}(A) \subseteq A$$

may be removed.

### 1.4.5 Vector Fields Corresponding to Fat Fields

**Exercise.** Let $\overline{\chi}$ be a der-operator in $P^\vee$. Show that there exists a unique vector field $\mathbb{X}$ on $\overline{M}$ such that $\mathbb{X}(P^\vee) \subseteq P^\vee$ and such that its restriction to $P^\vee$ (on both the domain and the codomain) is $\overline{\chi}$.

**Hint.** Suppose first that $M$ is an open subset of some $\mathbb{R}^n$ and that $P^\vee$ is a free module. Let $x_1, \ldots, x_n$ be coordinate functions on $M$, take a basis $\alpha_1, \ldots, \alpha_m$ of $P^\vee$ and consider these elements as coordinate functions on the total space manifold $\overline{M}$. Deduce a coordinate expression for the required vector field from the decomposition of $\overline{\chi}$ obtained according to n. 1.2.9. For the general case, note that every fat manifold is locally (on the base) of the above type, and use a gluing procedure.

**Definition.** Let $\overline{X}$ be a fat field on a fat manifold $\overline{M}$, $P = \Gamma(\overline{M})$ and $\mathbb{X}$ the unique vector field on the total space manifold $\overline{M}$ that restricts to $\overline{X}^\vee$ on $P^\vee$ (see the above Exercise). Then $\mathbb{X}$ will be called the vector field on the total space manifold *associated* with the fat field $\overline{X}$.

The preceding exercises establish a one-to-one correspondence

$$\overline{X} \longleftrightarrow \mathbb{X}$$

between fat fields and vector fields on the total space leaving invariant $P^\vee$, i.e., $\mathbb{X}(P^\vee) \subseteq P^\vee$. This correspondence is in a sense functorial. For instance, compatibility of fat fields along a fat smooth map $\overline{f}$ implies compatibility of the associated vector fields along the map $\overline{f}$ considered as a smooth map between the total spaces. We deduce now this fact a consequence of a series of exercises.

### 1.4.6

Let $\overline{f} : \overline{N} \to \overline{M}$ be a fat map and set $P = \Gamma(\overline{M})$, $Q = \Gamma(\overline{N})$. Note that there is a fat map

$$\overline{f}^\vee : \overline{N}^\vee \to \overline{M}^\vee$$

over $f$ determined by

$$\left(\overline{f}^\vee\right)^* (\alpha) \left(\overline{f}^* (p)\right) = f^* (\alpha (p)), \quad p \in P, \alpha \in P^\vee$$

(see n. 0.1.5, (8)).

**Exercise.** Consider $\overline{f}$ as a smooth map between the total space manifolds and let $\overline{f}^* : C^\infty(\overline{M}) \to C^\infty(\overline{N})$ be the associated homomorphism (so that here $\overline{f}^*$ is *not* the homomorphism associated to $\overline{f}$ as a fat map; cf. n. 1.1.15). Show that $\overline{f}^*(P^\vee) \subseteq Q^\vee$ and that the restriction $P^\vee \to Q^\vee$ of $\overline{f}^*$ coincides with $\left(\overline{f}^\vee\right)^*$.

### 1.4.7

**Exercise.** Let $\overline{f} : \overline{N} \to \overline{M}$ be a fat map and $\overline{X}, \overline{Y}$, fat fields respectively on $\overline{M}, \overline{N}$. Show that

$$\overline{X}, \overline{Y} \text{ are } \overline{f}\text{-compatible} \iff \overline{X}^\vee, \overline{Y}^\vee \text{ are } \overline{f}^\vee\text{-compatible}.$$

### 1.4.8

**Exercise.** Let $\overline{M}$ be a fat manifold, $P = \Gamma(\overline{M})$, $g$ a smooth map of a smooth manifold $V$ into the total space $\overline{M}$, and $Z$, $W$ vector fields along $g$. Show that if $Z$ and $W$ coincide on $P^\vee \subseteq C^\infty(\overline{M})$ and on $C^\infty(M) \subseteq C^\infty(\overline{M})$, then they are equal.

### 1.4.9

**Exercise.** Let $\overline{f} : \overline{N} \to \overline{M}$ be a fat map, $P = \Gamma(\overline{M})$, $Q = \Gamma(\overline{N})$, and $\mathbb{X}$, $\mathbb{Y}$ vector fields on the total space manifolds $\overline{M}, \overline{N}$, respectively, such that

$$\mathbb{X}(P^\vee) \subseteq P^\vee, \quad \mathbb{Y}(Q^\vee) \subseteq Q^\vee.$$

Show that the restrictions

$$\overline{\chi} : P^\vee \to P^\vee, \quad \overline{v} : Q^\vee \to Q^\vee$$

of $\mathbb{X}$ and $\mathbb{Y}$, respectively, are fat fields compatible with respect to the fat map $\overline{f}^\vee$ if and only if $\mathbb{X}, \mathbb{Y}$ are compatible vector fields with respect to the smooth map $\overline{f}$.

**Hint.** According to Exercises 1.4.3 and 1.4.4, the operators $\overline{\chi}, \overline{v}$ are fat fields. Then use Exercises 1.3.9, 1.4.6 and 1.4.8.

### 1.4.10

**Proposition.** Let $\overline{f} : \overline{N} \to \overline{M}$ be a fat map and $\overline{X}, \overline{Y}$ fat fields respectively on $\overline{M}, \overline{N}$. Then $\overline{X}$ and $\overline{Y}$ are compatible with respect to the fat map $\overline{f}$

if and only if the associated vector fields on the total space manifolds are compatible with respect to the smooth map $\overline{f}$.

***Proof.*** Immediately from Exercises 1.4.7 and 1.4.9. □

## 1.5 Induced Der-operators

In the preceding section we discovered that every der-operator induces a dual der-operator. Here we see that this construction generalizes easily to modules of homomorphisms. By a similar procedure, we shall also see how to extend der-operators to tensor products and introduce pull-backs of fat fields as a particular case.

### 1.5.1 *Extension of Der-operators to Homomorphisms*

**Proposition.** Let $k$ be a field, $A$ a commutative $k$-algebra and $\square_1$, $\square_2$ der-operators in $A$-modules $P$, $Q$, respectively, over the same derivation $X$ of $A$. The formula

$$\square(\varphi) \stackrel{\text{def}}{=} \square_2 \circ \varphi - \varphi \circ \square_1, \quad \varphi \in \text{Hom}(P, Q)$$

defines a der-operator $\square$ over $X$ on $\text{Hom}(P, Q)$.

***Proof.*** For all $\varphi \in \text{Hom}(P, Q)$, $p \in P$ and $f \in A$ we have

$$\square(\varphi)(fp) = \square_2(\varphi(fp)) - \varphi(\square_1(fp))$$
$$= \square_2(f\varphi(p)) - \varphi(X(f)p + f\square_1(p))$$
$$= X(f)\varphi(p) + f\square_2(\varphi(p)) - X(f)\varphi(p) - f\varphi(\square_1(p))$$
$$= f[\square_2(\varphi(p)) - \varphi(\square_1(p))] = f(\square(\varphi))(p) .$$

Moreover, since $\square_1$, $\square_2$ and $\varphi$ are additive, $\square(\varphi) = \square_2 \circ \varphi - \varphi \circ \square_1$ is additive as well. Therefore $\square(\varphi)$ is an $A$-module homomorphism.

The operator $\square$ is linear over $k$ because $\square_2$ is linear over $k$ and, for all $\varphi \in \text{Hom}(P, Q)$ and $f \in A$,

$$\square(f\varphi) = \square_2 \circ f\varphi - f\varphi \circ \square_1 = X(f)\varphi + f\square_2 \circ \varphi - f\varphi \circ \square_1$$
$$= X(f)\varphi + f\square(\varphi) .$$

So, the assertion follows from n. 1.2.4. □

**Definition.** The so defined der-operator $\square$ is called the der-operator *induced by* $\square_1$ *and* $\square_2$ *on* $\text{Hom}(P, Q)$ and it will be denoted by $\text{Hom}(\square_1, \square_2)$.

## 1.5.2 Extension of Der-operators to Tensor Products

**Proposition.** Let $k$ be a field, $A$ a commutative $k$-algebra and $\square_1$, $\square_2$ der-operators in $A$-modules $P$, $Q$, respectively, over the same derivation $X$ of $A$. Then there exists a der-operator $\square : P \otimes Q \to P \otimes Q$ over $X$ determined by

$$\square(p \otimes q) = \square_1(p) \otimes q + p \otimes \square_2(q), \quad p \in P, q \in Q.$$

**Proof.** The formula in the statement determines a linear over $k$ operator

$$\square : P \otimes Q \to P \otimes Q$$

according to n. 0.1.4. A simple calculation shows that $\square$ satisfies the 'der-Leibnitz rule' (1.4), p. 100, and hence that it is a der-operator over $X$. $\square$

**Definition.** The so defined der-operator $\square$ is called the der-operator *induced by* $\square_1$ *and* $\square_2$ *on* $P \otimes Q$.

A natural notation $\square_1 \otimes \square_2$ for $\square$ is a little confusing because it could be understood as a $k$-endomorphism of $P \otimes_k Q$, or even as an $A$-endomorphism of $P \otimes_A Q$ when $X = 0$. By this reason we shall use $\square_1 \boxtimes \square_2$ instead.

## 1.5.3 Induced Fat Fields

**Proposition.** Let $\overline{f} : \overline{N} \to \overline{M}$ be a fat map and $\overline{X}$ a fat field on $\overline{M}$. If $Y$ is a vector field on $N$ $f$-compatible with $X$, then there exists a unique fat field $\overline{Y}$ over $Y$ on $\overline{N}$ that is $\overline{f}$-compatible with $\overline{X}$.

**Proof.** Let $A = C^\infty(M)$, $B = C^\infty(N)$, $P = \Gamma(\overline{M})$, $Q = \Gamma(\overline{N})$. Since $\overline{f}$ is a fat map, $Q = B \otimes_A P$ is a $B$-module obtained from $P$ by extension of scalars via $f^*$ and $\overline{f}^*$ is the universal homomorphism

$$P \to B \otimes_A P, \quad p \mapsto 1 \otimes p$$

(see n. 1.1.6). It results from $f$-compatibility of $X$ and $Y$ that for all $a \in A$ and $b \in B$ we have

$$Y(a \cdot b) = Y(f^*(a)b) = Y(f^*(a))b + f^*(a)Y(b)$$
$$= f^*(X(a))b + a \cdot Y(b) = X(a) \cdot b + a \cdot Y(b),$$

where the 'dot' stands for the $A$-module multiplication in $B$. Therefore, by n. 1.2.4, $Y$ is a der-operator over $X$ in the $A$-module $B$. Consider the der-operator $\overline{Y} = Y \boxtimes \overline{X}$ over $X$ in the $A$-module $Q = B \otimes_A P$ as defined in n. 1.5.2. By definition

$$\overline{Y}(b \otimes p) = Y(b) \otimes p + b \otimes \overline{X}(p), \quad b \in B, p \in P.$$

Once again in view of n. 1.2.4, we observe that the operator $\overline{Y}$ is a der-operator over $Y$ if $Q$ is considered as a $B$-module. In fact, $\overline{Y}$ is $k$-linear and it is sufficient to check the Leibnitz rule on the elements of the form $q = b \otimes p$ for arbitrary $b \in B$ and $p \in P$ as follows: for all $b' \in B$,

$$\begin{aligned}\overline{Y}(b'q) &= \overline{Y}(b'b \otimes p) = Y(b'b) \otimes p + b'b \otimes \overline{X}(p) \\ &= Y(b')b \otimes p + b'Y(b) \otimes p + b'\left(b \otimes \overline{X}(p_i)\right) \\ &= Y(b')(b \otimes p) + b'\left(Y(b) \otimes p + b \otimes \overline{X}(p)\right) \\ &= Y(b')q + b'\overline{Y}(q).\end{aligned}$$

This shows that $\overline{Y}$ is a fat field on $\overline{N}$. Moreover, it is $\overline{f}$-compatible with $\overline{X}$. Indeed, if $p \in P$ then

$$\overline{Y}\left(\overline{f}^*(p)\right) = \overline{Y}(1 \otimes p) = Y(1) \otimes p + 1 \otimes \overline{X}(p) = 1 \otimes \overline{X}(p) = \overline{f}^*\left(\overline{X}(p)\right),$$

i.e., $\overline{Y} \circ \overline{f}^* = \overline{f}^* \circ \overline{X}$.

Finally, if $\overline{Y}'$ is a fat field over $Y$ on $\overline{N}$ that is $\overline{f}$-compatible with $\overline{X}$, then for all $q = b \otimes p$, $b \in B$ and $p \in P$,

$$\begin{aligned}\overline{Y}'(q) &= \overline{Y}'(b(1 \otimes p)) = \overline{Y}'\left(b\overline{f}^*(p)\right) = Y(b)\overline{f}^*(p) + b\overline{Y}'\left(\overline{f}^*(p)\right) \\ &= Y(b)\overline{f}^*(p) + b\overline{f}^*\left(\overline{X}(p)\right) = Y(b) \otimes p + b \otimes \overline{X}(p) = \overline{Y}(q).\end{aligned}$$

This shows that

$$\overline{Y}' = \overline{Y},$$

i.e., $\overline{Y}$ is unique. □

**Definition.** The fat field $\overline{Y}$ is called the *fat field over $Y$ induced from $\overline{X}$ by $\overline{f}$*.

It is worth stressing that the induced fat field is defined only in the situation when $Y$ is $f$-compatible with $X$, the base of $\overline{X}$.

### 1.5.4

In the case when $f^*$ is surjective the construction of induced fat field is particularly simple. In such a situation the algebra $B$ is identified canonically with $A/I$ where $I = \operatorname{Ker} f^*$, $Q = B \otimes_A P$ with $P/IP$ and the fat map $\overline{f}^*$ reads as follows:

$$\overline{f}^*(p) = [p] ,$$

where $[p]$ stands for the coset $p + IP$. This shows that $\overline{f}^*$ is surjective and every element of $Q$ is of the form

$$1 \otimes p = [p] .$$

Note that

$$\overline{Y}([p]) = \overline{Y}(1 \otimes p) = Y(1) \otimes p + 1 \otimes \overline{X}(p) = 1 \otimes \overline{X}(p) = \left[\overline{X}(p)\right] .$$

So, in the considered case the induced fat field $\overline{Y}$ acts according to the formula

$$\overline{Y}([p]) = \left[\overline{X}(p)\right] .$$

Such a situation occurs if $f$ is the embedding of a closed submanifold (or a closed embedding, cf. n. 0.2.30) and $Y$ is tangent to the submanifold. In this particular case the induced fat field is nothing else but the *restriction* of $\overline{X}$ to $\overline{N}$.

### 1.5.5

Let $\overline{X}$ be a fat field on a fat manifold $\overline{M}$ and $\overline{N}$ an open fat submanifold of $\overline{M}$. Proposition 1.5.3 leads to an alternative and immediate solution of Exercise 1.3.10 concerning the restriction $\overline{X}|_{\overline{N}}$. Indeed, let $Y = X|_N$ (see Definition 0.4.5). By Proposition 1.5.3 there is a unique fat field $\overline{Y}$ on $\overline{N}$ over $Y$, compatible with $\overline{X}$ with respect to the fat embedding of $\overline{N}$. By Exercise 1.3.9 and Proposition 0.4.5, every fat field compatible with $\overline{X}$ via the fat embedding of $\overline{N}$ must have the base $Y$. Therefore $\overline{Y}$ is the unique fat field on $\overline{N}$ compatible with $\overline{X}$ with respect to the fat embedding of $\overline{N}$ in $\overline{M}$.

## 1.6 Fat Trajectories

In this section fat trajectories of fat fields are defined.

### 1.6.1 Fat Standard Model

The first step is to fix *the standard model*.

**Definition.** Let $\bar{\mathbb{I}}$ be a standard trivial fat manifold over an interval with standard fiber $E \cong F$. If $d/dt$ is the standard vector field on $\mathbb{I}$ (supposed nonempty nor reduced to a singleton; see n. 0.4.21), the fat field

$$D_{\frac{d}{dt}} = \frac{d}{dt} \otimes_{\mathbb{R}} \mathrm{id}_E$$

on $\bar{\mathbb{I}}$ (*cf.* Example 1.2.7) is called *standard* and denoted by $\overline{d/dt}$.

**Proposition.** Let $\overline{\tau_t} : \bar{\mathbb{J}} \to \bar{\mathbb{I}}$ be a fat translation by $t$ and $\overline{d/dt}$, $\overline{d/ds}$ standard fat fields on the fat intervals $\bar{\mathbb{I}}$ and $\bar{\mathbb{J}}$, respectively. Then $\overline{d/dt}$ and $\overline{d/ds}$ are compatible with respect to $\overline{\tau_t}$.

**Proof.** Recall that, by definition, $\overline{\tau_t} : \bar{\mathbb{J}} \to \bar{\mathbb{I}}$ is an uniform morphism over $\tau_t$ corresponding to a vector space isomorphism $\varphi : F \xrightarrow{\sim} F$ ($F$ is the standard fiber). Moreover,

$$\overline{\tau_t}^* = \tau_t^* \otimes_{\mathbb{R}} \varphi^{-1}$$

(see n. 0.3.17).

Standard vector fields $d/dt$ and $d/ds$ are $\tau_t$-compatible (see n. 0.4.24). Therefore, we have

$$\overline{\frac{d}{dt}} \circ \overline{\tau_t}^* = \left(\frac{d}{dt} \otimes_{\mathbb{R}} \mathrm{id}_F\right) \circ \left(\tau_t^* \otimes_{\mathbb{R}} \varphi^{-1}\right) = \left(\frac{d}{dt} \circ \tau_t^*\right) \otimes_{\mathbb{R}} \varphi^{-1}$$

$$= \left(\tau_t^* \circ \frac{d}{ds}\right) \otimes_{\mathbb{R}} \varphi^{-1} = \left(\tau_t^* \otimes_{\mathbb{R}} \varphi^{-1}\right) \circ \left(\frac{d}{ds} \otimes_{\mathbb{R}} \mathrm{id}_F\right) = \overline{\tau_t}^* \circ \overline{\frac{d}{ds}}. \quad \square$$

This shows that fat standard models are preserved under fat translations in full analogy with the classical case.

### 1.6.2

Recall that, in geometric terms, a fat interval $\bar{\mathbb{I}}$ is the projection $F \times \mathbb{I} \to \mathbb{I}$. It is easy to see that the vector field on $F \times \mathbb{I}$ associated with the standard fat field $\overline{d/dt}$ is nothing but the standard vector field $\partial/\partial t$ on $F \times \mathbb{I}$ (see n. 0.6.3). Note that the maximals trajectories of $\partial/\partial t$ are constant sections of the bundle and, on the other hand, constant sections $p$ are characterized by the following property:

$$\overline{\frac{d}{dt}}(p) = 0.$$

This last fact is generalized as follows.

### 1.6.3

**Proposition.** Let $X = d/dt$ be the standard vector field on an interval $\mathbb{I}$, $\bar{\mathbb{I}}$ a fat manifold over $\mathbb{I}$, $\overline{X}$ a fat field on $\bar{\mathbb{I}}$ over $X$ and $\mathbb{X}$ the associated with $\overline{X}$ vector field on the total space $\bar{\mathbb{I}}$. Then a section $p \in P = \Gamma(\bar{\mathbb{I}})$ is a trajectory of $\mathbb{X}$ if and only if
$$\overline{X}(p) = 0.$$

**Proof.** By definition, the section $p$ is a trajectory of $\mathbb{X}$ if and only if
$$\frac{d}{dt} \circ p^* = p^* \circ \mathbb{X}.$$
Note that $d/dt \circ p^*$ and $p^* \circ \mathbb{X}$ are vector fields along $p$. Thus, in view of Exercise 1.4.8, it suffices to check the equality on $C^\infty(\mathbb{I})$ and on $P^\vee$.

Let $f \in C^\infty(\mathbb{I})$. According to Exercise 1.4.3, $\mathbb{X}$ and $d/dt$ are compatible with respect to the projection $\bar{\mathbb{I}} \to \mathbb{I}$. So,
$$\mathbb{X}(f) = \frac{d}{dt}(f),$$
by identifying $f$ and $\pi^*(f)$. Moreover, since $p$ is a section, we have
$$p^*(f) = f$$
by applying the same identification. Hence
$$p^*(\mathbb{X}(f)) = p^*(\frac{d}{dt}(f)) = \frac{d}{dt}(f) = \frac{d}{dt}(p^*(f)).$$
Therefore, for every section $p$, $d/dt \circ p^*$ and $p^* \circ \mathbb{X}$ agree on $C^\infty(\mathbb{I})$ and it remains to show that $d/dt \circ p^*$ and $p^* \circ \mathbb{X}$ agree on $P^\vee$ if and only if $\overline{X}(p) = 0$.

Let $\alpha \in P^\vee$. To avoid a confusion denote by $a$ the smooth function
$$\bar{\mathbb{I}} \to \mathbb{R}$$
identified with $\alpha$. According to n. 1.1.13
$$a(p(t)) = \alpha(p)(t), \quad t \in \mathbb{I},$$
and, therefore,
$$p^*(a) = \alpha(p).$$
But the function on $\bar{\mathbb{I}}$ identified with $\overline{X}^\vee(\alpha) \in P^\vee$ is $\mathbb{X}(a)$. Thus we see that
$$p^*(\mathbb{X}(a)) = \overline{X}^\vee(\alpha)(p) = \frac{d}{dt}(\alpha(p)) - \alpha(\overline{X}(p)) = \frac{d}{dt}(p^*(a)) - \alpha(\overline{X}(p)).$$
This implies that $d/dt \circ p^*$ and $p^* \circ \mathbb{X}$ agree on $P^\vee$ if and only if
$$\alpha(\overline{X}(p)) = 0, \quad \alpha \in P^\vee.$$
This condition is equivalent to
$$\overline{X}(p) = 0$$
because $P$ is projective (see n. 0.1.5, (6)). $\square$

### 1.6.4 Fiber-wise Linearity of Fat Fields

The condition
$$\mathbb{X}(P^\vee) \subseteq P^\vee$$
characterizing vector fields associated with fat fields expresses a *fiber-wise linear* nature. In more geometric terms this means that 'every scalar multiple of a trajectory of $\mathbb{X}$ is again a trajectory of $\mathbb{X}$' and 'the sum of two trajectories of $\mathbb{X}$ is again a trajectory of $\mathbb{X}$'. The precise meaning of that is explained below.

First, some notational conventions. Let $\pi : E_\pi \to M$ be a vector bundle, $v, w \in E_\pi$ and $\lambda, \mu \in \mathbb{R}$. If $\pi(v) = \pi(w)$, then the linear combination $\lambda v + \mu w$ is well-defined since $v$ and $w$ belong to the same fiber, which is a vector space. Farther, if $S$ is a set and $f, g : S \to E_\pi$ are some maps such that $\pi \circ f = \pi \circ g$, then the linear combination
$$\lambda f + \mu g : S \to E_\pi$$
is defined by $(\lambda f + \mu g)(s) = \lambda f(s) + \mu g(s)$, $s \in S$. Obviously,
$$\pi \circ (\lambda f + \mu g) = \pi \circ f = \pi \circ g.$$

**Proposition.** Let $\overline{X}$ be a fat field on a fat manifold $\overline{M}$, $\mathbb{X}$ the associated vector field on $\overline{M}$ and $\pi : \overline{M} \to M$ the projection. If $\lambda \in \mathbb{R}$ and $\gamma, \gamma'$ are trajectories of $\mathbb{X}$ such that
$$\pi \circ \gamma = \pi \circ \gamma',$$
then $\lambda \gamma$ and $\gamma + \gamma'$ are trajectories of $\mathbb{X}$.

**Proof.** Put
$$\beta = \pi \circ \gamma = \pi \circ \gamma' : \mathbb{I} \to M$$
and consider the induced by $\beta$ bundle $\overline{\mathbb{I}} \to \mathbb{I}$ together with the induced map $\overline{\beta} : \overline{\mathbb{I}} \to \overline{M}$. Exercise 1.4.3 tells us that vector fields $\mathbb{X}$ and $X$ are $\pi$-compatible. Since the standard vector field $d/dt$ and $\mathbb{X}$ are $\gamma$-compatible, then $d/dt$ and $X$ are $(\pi \circ \gamma)$-compatible. In other words, $\beta = \pi \circ \gamma$ is a trajectory of $X$. Hence the induced by $\overline{\beta}$ from $\overline{X}$ fat field $\overline{Y}$ over $d/dt$ on $\overline{\mathbb{I}}$ is well defined. Denote by $\mathbb{Y}$ the associated with $\overline{Y}$ vector field on $\overline{\mathbb{I}}$.

Observe now that there exist smooth sections
$$\Delta, \Delta' : \mathbb{I} \to \overline{\mathbb{I}}$$
such that
$$\overline{\beta} \circ \Delta = \gamma, \quad \overline{\beta} \circ \Delta' = \gamma'$$

Indeed, the restriction $\overline{\beta}_t : \overline{t} \to \overline{\beta(\overline{t})} = \overline{\beta(t)}$ of $\overline{\beta}$ to the fiber $\overline{t}$ is an isomorphism of vector spaces and $\Delta$ is defined by putting $\Delta(t) = \overline{\beta}_t^{-1}(\gamma(t))$. Similarly for $\Delta'$. Moreover, $\mathbb{Y}$ and $\mathbb{X}$ are compatible in view of Proposition 1.4.10. This implies that $\Delta$ and $\Delta'$ are trajectories of the vector field $\mathbb{Y}$ on $\overline{\mathbb{I}}$.

Now, $\overline{Y}$ is a linear operator on $\Gamma(\overline{\mathbb{I}})$, thus the maps $\lambda\Delta$ and $\Delta + \Delta'$ are trajectories of $\mathbb{Y}$ as it results from Proposition 1.6.3. So, $\overline{\beta} \circ \lambda\Delta$ and $\overline{\beta} \circ (\Delta + \Delta')$ are trajectories of $\mathbb{X}$. Finally, it remains to observe that $\overline{\beta}$ is fiberwise linear, so that

$$\overline{\beta} \circ \lambda\Delta = \lambda(\overline{\beta} \circ \Delta) = \lambda\gamma \quad \text{and} \quad \overline{\beta} \circ (\Delta + \Delta') = \overline{\beta} \circ \Delta + \overline{\beta} \circ \Delta' = \gamma + \gamma'.$$

This concludes the proof. □

### 1.6.5

Let $X = \mathrm{d}/\mathrm{d}\,t$ be the standard vector field on an interval $\mathbb{I}$, $\overline{\mathbb{I}}$ a fat manifold over $\mathbb{I}$, $\overline{X}$ a fat field on $\overline{\mathbb{I}}$ over $X$ and $\mathbb{X}$ the associated with $\overline{X}$ vector field on the total space $\overline{\mathbb{I}}$. Proposition 1.6.3 tells that the kernel $\mathcal{F}_{\overline{X}}$ of $\overline{X}$ is composed of trajectories of $\mathbb{X}$ understood as sections of $\overline{\mathbb{I}}$. Since $\overline{X}$ is $\mathbb{R}$-linear, $\mathcal{F}_{\overline{X}}$ is a vector subspace of $P = \Gamma(\overline{\mathbb{I}})$. As such it can be restricted to any fat point of $\overline{t} \in \overline{\mathbb{I}}$ by means of the canonical evaluation homomorphism

$$\overline{h}_t : P \to \overline{t} = P_t = \frac{P}{\mu_t P} = \mathbb{R}_t \otimes P,$$

with $\mu_t = \mathrm{Ker}\,t$.

Denote the restriction of $\overline{h}_t$ to $\mathcal{F}_{\overline{X}}$ by $\overline{h}_{\overline{X},t}$. In other words, the linear map

$$\overline{h}_{\overline{X},t} : \mathcal{F}_{\overline{X}} \to \overline{t}$$

associates with a section $p \in \mathcal{F}_{\overline{X}}$ its value $p(t)$ at $t$.

The fact that the trajectory passing trough a given point is unique implies injectivity of $\overline{h}_{\overline{X},t}$. On the contrary, the existence theorem by itself does not guarantee surjectivity of $\overline{h}_{\overline{X},t}$. Namely, the maximal trajectory through an assigned $\mathbf{e} \in \overline{t}$, existence of which it establishes, is not guaranteed to be a global section of $\overline{\mathbb{I}}$, just a local one. So, some additional arguments are necessary.

**Proposition.** Through a given point of $\overline{\mathbb{I}}$ passes a trajectory of $\mathbb{X}$ that is a section of the bundle $\overline{\mathbb{I}} \to \mathbb{I}$.

***Proof.*** Let $t_1, t_2 \in \mathbb{I}$ be arbitrary. It is sufficient to show that any maximal trajectory of $\mathbb{X}$ starting from a point of the fiber $\overline{t_1}$ reaches the fiber $\overline{t_2}$. Observe, first of all, that the zero section of $\overline{\mathbb{I}}$ is a trajectory of $\mathbb{X}$. In other words, the trajectory starting from the origin of $\overline{t_1}$ reaches $\overline{t_2}$. Using some elementary general topology, we have that trajectories starting from points of a sufficiently small neighborhood, say, $U$, of the origin of $\overline{t_1}$ reach $\overline{t_2}$ according to the theorem of smooth dependence of solutions from initial data (see n. 0.6.4). But a vector $\mathbf{e} \in \overline{t_1}$ is a multiple of a vector belonging to $U$. So, the same multiple of the corresponding trajectories (see n. 1.6.4), starting, obviously, from $\mathbf{e}$ reaches $\overline{t_2}$ as well. □

**Corollary.** Let $\overline{X}, \overline{X}'$ be fat fields on $\overline{\mathbb{I}}$ over the standard vector field $X = \mathrm{d}/\mathrm{d}t$. If $\overline{X}$ and $\overline{X}'$ have the same kernel, that is,

$$\mathcal{F}_{\overline{X}} = \mathcal{F}_{\overline{X}'},$$

then they coincide.

***Proof.*** Let $\mathbb{X}$ and $\mathbb{X}'$ be the vector fields on $\overline{\mathbb{I}}$ associated with $\overline{X}$ and $\overline{X}'$, respectively. Since

$$\mathcal{F}_{\overline{X}} = \mathcal{F}_{\overline{X}'},$$

$\mathbb{X}$ and $\mathbb{X}'$ have the same maximal trajectories and hence they coincide. □

### 1.6.6

The above results are summarized in the following proposition.

**Proposition.** Let $X = \mathrm{d}/\mathrm{d}t$ be the standard vector field on an interval $\mathbb{I}$, $\overline{\mathbb{I}}$ a fat manifold over $\mathbb{I}$ and $\overline{X}$ a fat field on $\overline{\mathbb{I}}$ over $X$. Then for every fat point $\overline{t}$ of $\overline{\mathbb{I}}$, the evaluation map

$$\overline{h}_{\overline{X},t} : \mathcal{F}_{\overline{X}} \to \overline{t}$$

is a vector space isomorphism.

***Proof.*** Immediately from n. 1.6.5. □

**Example.** Let $\overline{\mathrm{d}/\mathrm{d}t}$ be the standard fat field on a fat interval $\overline{\mathbb{I}}$. Then, canonically,

$$\Gamma(\overline{\mathbb{I}}) = C^\infty(\mathbb{I}) \otimes_{\mathbb{R}} F$$

and the natural homomorphism

$$F \to \Gamma(\overline{\mathbb{I}}), \quad e \mapsto 1 \otimes e, \quad e \in F,$$

induces an isomorphism $\nu$ between $F$ and $\mathcal{F}_{\overline{d/dt}}$. In these terms the canonical identification of $F$ and $\bar{t}$, $t \in \mathbb{I}$, is given by the map $\overline{h}_{\overline{d/dt},t} \circ \nu$.

**Corollary.** Any basis $\mathbf{e}_1, \ldots, \mathbf{e}_n$ of the $\mathbb{R}$-vector space $\mathcal{F}_{\overline{X}}$ is simultaneously a free basis of the $\mathrm{C}^\infty(\mathbb{I})$-module $\Gamma(\overline{\mathbb{I}})$. In particular, the homomorphism of $\mathrm{C}^\infty(\mathbb{I})$-modules
$$\iota_{\overline{X}} : \mathrm{C}^\infty(\mathbb{I}) \otimes_{\mathbb{R}} \mathcal{F}_{\overline{X}} \to \Gamma(\overline{\mathbb{I}}), \quad a \otimes v \mapsto av,$$
with $a \in \mathrm{C}^\infty(\mathbb{I})$, $v \in \mathcal{F}_{\overline{X}}$, is an isomorphism.

**Proof.** It is sufficient to observe that, according to the above Proposition, the values of sections $\mathbf{e}_1, \ldots, \mathbf{e}_n$ at a point $m \in \mathbb{I}$ form a basis of $\overline{m}$. □

## 1.6.7

The previous results have the following important consequence. If a der-operator $\overline{X}$ over the standard vector field is assigned on a fat manifold $\overline{\mathbb{I}}$ over an interval, then $\overline{\mathbb{I}}$ is identified with a fat interval. This is due to the canonical identification between fats points and a fixed vector space, namely, $\mathcal{F}_{\overline{X}}$. The following result precisely states this fact.

**Proposition.** Let $X = \mathrm{d}/\mathrm{d}t$ be the standard vector field on an interval $\mathbb{I}$. If $\overline{\mathbb{I}}$ is a fat manifold over $\mathbb{I}$ and $\overline{X}$ a fat field on $\overline{\mathbb{I}}$ over $X$, then there exists a fat identity map $\overline{f} : \overline{\mathbb{I}}' \to \overline{\mathbb{I}}$ (see Definition 1.1.7), where $\overline{\mathbb{I}}'$ is a fat interval over $\mathbb{I}$, such that $\overline{X}$ is $\overline{f}$-compatible with the standard fat field on $\overline{\mathbb{I}}'$.

Moreover, a fat diffeomorphism $\overline{\mathbb{I}}'' \to \overline{\mathbb{I}}$ satisfies the same condition if and only if it is of the form $\overline{f} \circ \overline{u}$ with $\overline{u} : \overline{\mathbb{I}}'' \to \overline{\mathbb{I}}'$ being a uniform fat identity map between fat intervals.

**Proof.** Let $\zeta : F \to \mathcal{F}_{\overline{X}}$ be an isomorphism of vector spaces and $(v_1, \ldots, v_n)$ a basis of $F$. Then $(\zeta(v_1), \ldots, \zeta(v_n))$ is a basis of $\mathcal{F}_{\overline{X}}$ and hence a basis of the free module $\Gamma(\overline{\mathbb{I}})$ (see Corollary 1.6.6). So, the map
$$\iota_\zeta : \mathrm{C}^\infty(\mathbb{I}) \otimes_{\mathbb{R}} F \to \Gamma(\overline{\mathbb{I}}), \quad a \otimes v \mapsto a\zeta(v),$$
with $a \in \mathrm{C}^\infty(\mathbb{I})$, $v \in F$, is an isomorphism of $\mathrm{C}^\infty(\mathbb{I})$-modules. Hence the corresponding to $\iota_\zeta$ morphism $\overline{g}$ of vector bundles is a fat identity map sending $\overline{\mathbb{I}}$ to the fat interval $\overline{\mathbb{I}}'$ associated with the $\mathrm{C}^\infty(\mathbb{I})$-module $\mathrm{C}^\infty(\mathbb{I}) \otimes_{\mathbb{R}} F$. Moreover, $\overline{X}(a\zeta(v)) = (\mathrm{d}a/\mathrm{d}t)\zeta(v)$ because, by definition, $\zeta(v) \in \mathrm{Ker}\,\overline{X}$. We have
$$(\overline{g}^* \circ \overline{\mathrm{d}/\mathrm{d}t})(a \otimes v) = \overline{g}^*(\mathrm{d}a/\mathrm{d}t \otimes v) = \iota_\zeta(\mathrm{d}a/\mathrm{d}t \otimes v) = (\mathrm{d}a/\mathrm{d}t)\zeta(v)$$
$$= \overline{X}(a\zeta(v)) = (\overline{X} \circ \iota_\zeta)(a \otimes v) = (\overline{X} \circ \overline{g}^*)(a \otimes v).$$

Therefore, $\overline{g}^* \circ \overline{d/dt} = \overline{X} \circ \overline{g}^*$, or, equivalently, $\overline{f}^* \circ \overline{d/dt} = \overline{f}^* \circ \overline{X}$ for $\overline{f} = \overline{g}^{-1}$. This proves the first assertion of the Proposition.

Let now $\overline{\mathbb{I}}''$ be a fat interval over $\mathbb{I}$ and $\overline{h} : \overline{\mathbb{I}}'' \to \overline{\mathbb{I}}$ another fat diffeomorphism satisfying the conditions imposed on the fat identity map $\overline{f}$. Then standard fat fields on $\overline{\mathbb{I}}'$ and $\overline{\mathbb{I}}''$, both denoted $\overline{d/dt}$, are $\overline{u}$-compatible for $\overline{u} = \overline{g} \circ \overline{h} : \overline{\mathbb{I}}'' \to \overline{\mathbb{I}}'$. By this reason $\overline{u}$ sends $\mathcal{F}_{\overline{d/dt}}$ into $\mathcal{F}_{\overline{d/dt}}$ and, due to canonical identification of $F$ and $\mathcal{F}_{\overline{d/dt}}$ (see Example 1.6.6), gives rise to an automorphism $\tau$ of $F$. It is now obvious that $\overline{u}$ is an uniform fat identity map associated with $\tau$.

Conversely, the standard fat fields $\overline{d/dt}$ are compatible with respect to whatever uniform fat identity map $\overline{u}$ of fat intervals. Therefore, fat fields $\overline{d/dt}$ and $\overline{X}$ are $(\overline{f} \circ \overline{u})$-compatible for all such $\overline{u}$. □

### 1.6.8 Definition of Fat Trajectories

**Definition.** A *fat trajectory* (also called an *integral fat curve*) of a fat field $\overline{X}$ on a fat manifold $\overline{M}$ is a fat curve

$$\overline{\gamma} : \overline{\mathbb{I}} \to \overline{M}$$

such that $\overline{X}$ is $\overline{\gamma}$-compatible with the standard fat field $\overline{d/dt}$.

If $m = \gamma(t)$ for some $t \in \mathbb{I}$, then we say that the fat trajectory $\overline{\gamma}$ *passes through* the fat point $\overline{m}$ at the 'time' $t$.

A fat trajectory $\overline{\gamma} : \overline{\mathbb{I}} \to \overline{M}$ is said to be *maximal* if it cannot be prolonged. This means that if $\overline{\gamma}$ is the fat restriction of a fat trajectory $\overline{\gamma}' : \overline{\mathbb{I}'} \to \overline{M}$, then $\overline{\gamma}'$ coincides with $\overline{\gamma}$ up to an uniform fat identity map of the domains (that is, $\mathbb{I} = \mathbb{I}'$).

**Exercise.** Let $\overline{\gamma} : \overline{\mathbb{I}} \to \overline{M}$ be a fat trajectory of a fat field $\overline{X}$ over $X$. Show that $\gamma : \mathbb{I} \to M$ is a trajectory of $X$.

### 1.6.9

**Exercise.** Let $\overline{\gamma} : \overline{\mathbb{I}} \to \overline{M}$ be a fat trajectory of a fat field $\overline{X}$ and $\overline{\tau_t} : \overline{\mathbb{J}} \to \overline{\mathbb{I}}$ a fat translation by $t \in \mathbb{R}$. Show that $\overline{\gamma} \circ \overline{\tau_t}$ is a fat trajectory and that it is maximal if and only if $\overline{\gamma}$ is maximal.

**Hint.** Mimic n. 0.4.24.

### 1.6.10  Lifted Fat Trajectories

Now we prove that if $\overline{X}$ is a fat field, every trajectory of the base vector field $X$ can be lifted to a fat trajectory of $\overline{X}$.

**Proposition.** Let $\overline{X}$ be a fat field on a fat manifold $\overline{M}$ and $\gamma : \mathbb{I} \to M$ a trajectory of the base vector field $X$. Then there exists a fat trajectory

$$\overline{\gamma} : \overline{\mathbb{I}} \to \overline{M}$$

of $\overline{X}$ whose base is $\gamma$. Such a fat trajectory is unique up to an uniform fat identity map $\overline{\mathbb{I}}' \to \overline{\mathbb{I}}$.

**Proof.** Let $\overline{\mathbb{I}}_{\overline{X}}$ be the bundle induced by $\gamma$ from $\overline{M}$. Consider the induced map

$$\overline{\gamma}_{\overline{X}} : \overline{\mathbb{I}}_{\overline{X}} \to \overline{M}.$$

Since $\gamma$ is a trajectory of $X$, the standard vector field $d/dt$ on the base interval $\mathbb{I}$ is $\gamma$-compatible with $X$. Therefore, the induced by $\overline{\gamma}_{\overline{X}}$ from $\overline{X}$ fat field $\overline{Y}$ over $d/dt$ is defined. Consider then a fat interval $\overline{\mathbb{I}}$ and a fat identity map

$$\overline{f} : \overline{\mathbb{I}} \to \overline{\mathbb{I}}_{\overline{X}}$$

such that the standard fat field $\overline{d/dt}$ on it is $\overline{f}$-compatible with $\overline{Y}$. Such a fat map exists according to Proposition 1.6.7. The base of the fat map

$$\overline{\gamma} \stackrel{\text{def}}{=} \overline{\gamma}_{\overline{X}} \circ \overline{f}$$

is, obviously, $\gamma$ and $\overline{d/dt}$ is $\overline{\gamma}$-compatible with $\overline{X}$, i.e., $\overline{\gamma}$ is a fat trajectory of $\overline{X}$.

Conversely, if

$$\overline{\gamma}' : \overline{\mathbb{I}}' \to \overline{M}$$

is an arbitrary fat trajectory of $\overline{X}$ over $\gamma$, then, due to universal property of induced bundles (Proposition 0.3.14), there exists a unique fat identity map

$$\overline{f}' : \overline{\mathbb{I}}' \to \overline{\mathbb{I}}_{\overline{X}}$$

such that

$$\overline{\gamma}' = \overline{\gamma}_{\overline{X}} \circ \overline{f}'.$$

Since the base of $\overline{f}'$ is a diffeomorphism, $\overline{f}'$ itself is a fat diffeomorphism. Moreover, the fat field induced by $\overline{f}'$ from $\overline{Y}$ is equal, by definition of $\overline{Y}$, to the fat field induced by $\overline{\gamma}'$ from $\overline{X}$. But $\overline{\gamma}'$ is a fat trajectory and, so, this fat field is nothing but the standard fat field on $\overline{\mathbb{I}}$. In other words, the standard fat field on $\overline{\mathbb{I}}'$ is $\overline{f}'$-compatible with $\overline{Y}$. It follows now from Proposition 1.6.7 that $\overline{f}' = \overline{f} \circ \overline{u}$, $\overline{u} : \overline{\mathbb{I}}' \to \overline{\mathbb{I}}$ being an uniform fat identity map, and hence $\overline{\gamma}' = \overline{\gamma} \circ \overline{u}$.

Conversely, it is obvious that the composition of $\overline{\gamma}$ with a uniform fat identity map is a fat trajectory. □

**Exercise.** Let $\overline{\gamma}$ and $\overline{\gamma}'$ be fat trajectories of a fat field. Show that if the base of $\overline{\gamma}$ is a restriction of the base of $\overline{\gamma}'$, then $\overline{\gamma}$ is a fat restriction of $\overline{\gamma}'$.

Note that these facts, together with Exercise 1.6.8, imply that a fat trajectory of a fat field $\overline{X}$ is maximal if and only if its base trajectory is maximal.

### 1.6.11 *Existence and Uniqueness of Fat Trajectories*

**Theorem.** Let $\overline{X}$ be a fat field on a fat manifold $\overline{M}$ and $\overline{m}$ a fat point of $\overline{M}$. Then there exists a maximal fat trajectory

$$\overline{\gamma} : \overline{\mathbb{I}} \to \overline{M}$$

of $\overline{X}$ passing through $\overline{m}$ at a given 'time' $t_0 \in \mathbb{R}$. Such a trajectory is unique up to a uniform fat identity map of the domain fat intervals. A maximal fat trajectory passing through a fixed fat point (not necessarily at the same time) is unique up to a fat translation.

*Proof.* Immediately from the uniqueness of 'normal' maximal trajectories and the results of n. 1.6.10. □

### 1.6.12 *Parallel Translation Along Fat Trajectories*

As it was observed in n. 1.1.10, fat points lying on an embedded fat curve

$$\overline{\gamma} : \overline{\mathbb{I}} \to \overline{M}$$

are canonically identified with the standard fiber $F$ of $\overline{\mathbb{I}}$. When, as in the case of fat trajectories, the curve is defined up to an uniform fat diffeomorphism this canonical identification goes lost. Nevertheless, in such a situation fat points can be canonically identified each other.

More precisely, consider fat points $\overline{t}$ and $\overline{t}'$ of $\overline{\mathbb{I}}$, and their images

$$\overline{m} = \overline{\gamma}\left(\overline{t}\right), \quad \overline{m'} = \overline{\gamma}\left(\overline{t}'\right).$$

Fat points $\overline{m}$ and $\overline{m'}$ are identified with $F$ by means of the isomorphisms

$$c_t \circ \overline{\gamma}_t^{-1} : \overline{m} \to F \quad \text{and} \quad c_{t'} \circ \overline{\gamma}_{t'}^{-1} : \overline{m'} \to F,$$

with

$$c_t : \overline{t} \to F \quad \text{and} \quad c_{t'} : \overline{t'} \to F$$

being the identification isomorphisms (see Definition 0.3.16). These fat points are, therefore, identified by means of

$$\overline{\gamma}_{t'} \circ c_{t'}^{-1} \circ c_t \circ \overline{\gamma}_t^{-1} : \overline{m} \xrightarrow{\sim} \overline{m'}.$$

This identification does not change when passing to the curve $\overline{\gamma'} = \overline{\gamma} \circ \overline{f}$, $\overline{f}$ being a uniform fat diffeomorphism, as it can be easily checked. In particular, fat points belonging to embedded fat trajectories of a fat field $\overline{X}$ are canonically identified and the identification isomorphisms depend only on the base trajectories. The same identifications arise by means of the isomorphisms $\overline{h}_{\overline{Y},m}$ (see n. 1.6.5), with $\overline{Y}$ induced by $\overline{X}$ on the induced bundle by the base trajectory.

Now, fix $\overline{m} = \overline{\gamma}\left(\overline{t_0}\right)$ and denote by $\mathbf{T}_{t_0, t_1}$ the identification (along $\overline{\gamma}$) isomorphism between $\overline{m}$ and $\overline{m'} = \overline{\gamma}\left(\overline{t_1}\right)$. The family of isomorphisms $\mathbf{T}_{t_0,t_1}$'s may be viewed naturally as the *parallel translation* of $\overline{m}$ along $\overline{\gamma}$. Of course, this procedure makes sense even for trajectories that are not embedded (cycles), but in this case a fat point on $\overline{\gamma}$ may admit several different identifications with $\overline{m}$.

### 1.6.13  Fat Cylinders

As in the ordinary case, in order to construct the *fat flow* generated by a fat field and then to prove in this context the theorem of smooth dependence from initial data, a smooth family of fat intervals is to be organized into a suitable fat manifold, which will be called a *relative fat interval* (*cf.* Definition 0.6.3). Relative fat intervals are open fat submanifolds of *fat cylinders*. These are defined as follows.

Let $\overline{M}$ be a fat manifold, $T$ a manifold, possibly with boundary, and $\pi : M \times T \to M$ the projection onto $M$ ([1]). Consider the fat manifold

---

[1] We assume here that at least one of $M$ and $T$ is without boundary, in order the product to be really a manifold (possibly with boundary). This restriction could be easily leaved out, by means of obvious considerations that are left to the interested readers (*cf.* the footnote in Sect. 0.6.8).

$\overline{T_M}$ induced by $\pi$ from $\overline{M}$ and the induced map $\overline{\pi}: \overline{T_M} \to \overline{M}$. Denote by $i_t : M \to M \times T$, $j_m : T \to M \times T$ the canonical embeddings at $t \in T$ and at $m \in M$, respectively. According to Proposition 0.3.14 there is a uniquely determined fat map $\overline{i_t} : \overline{M} \to \overline{T_M}$ over $i_t$ such that $\overline{\pi} \circ \overline{i_t} = \mathrm{id}_{\overline{M}}$. Moreover, if $\overline{T}_m$ is a standard trivial fat manifold with standard fiber $\overline{m}$, then there is a uniquely determined fat map $\overline{j_m} : \overline{T}_m \to \overline{T_M}$ such that $\overline{\pi} \circ \overline{j_m}$ factors as

$$\overline{T}_m \xrightarrow{\tau} \overline{m} \hookrightarrow \overline{M},$$

where $\tau$ is the corresponding trivializing fat map (see n. 0.3.15).

**Definition.** The fat manifold $\overline{T_M}$ is called *fat $T$-cylinder over* $\overline{M}$, the map $\overline{\pi}$ *fat projection of* $\overline{T_M}$ and $\overline{i_t}$, $\overline{j_m}$ *fat embeddings into* $\overline{T_M}$ *at $t$ and at $m$*, respectively.

### 1.6.14 Relative Fat Intervals

In the case when $T$ is an interval $\mathbb{I}$, the fat cylinder $\overline{\mathbb{I}_M}$ may be viewed as a family of standard trivial fat manifolds $\overline{\mathbb{I}_m}$. These are not properly fat intervals, since the standard fibers are the fat points $\overline{m}$. However, they can be considered as such ones up to a uniform fat identity map, if an isomorphism $\overline{m} \cong F$ is fixed.

Thus a fat cylinder $\overline{\mathbb{I}_M}$ formalizes the idea of a smooth family of fat intervals over $\mathbb{I}$, parametrized by points of $M$. Similarly, the notion of a *relative fat interval* formalizes the idea of family of fat intervals of 'different lengths'.

**Definition.** A *relative fat interval* over a fat manifold $\overline{M}$ is a fat submanifold $\overline{\mathbb{I}_M}$ of a fat $\mathbb{R}$-cylinder $\overline{\mathbb{R}_M}$ provided that the base manifold $\mathbb{I}_M$ is a relative interval (see Definition 0.6.3).

The *fat projection* $\overline{\mathbb{I}_M} \to \overline{M}$ is the restriction of the fat projection $\overline{\mathbb{R}_M} \to \overline{M}$. The *fat embeddings at $t \in \mathbb{R}$ and at $m \in M$*, will be the restrictions (on both the domain and the range) of the corresponding fat embeddings into $\overline{\mathbb{R}_M}$.

Given two relative fat intervals over the same relative interval, there is a unique fat diffeomorphism $\overline{f}$ between them that is compatible with the projections onto $\overline{M}$. These fat intervals are identified by means of $\overline{f}$ and we may assume that there is a unique relative fat interval over a given relative interval.

### 1.6.15 Smooth Dependence on the Initial Data

Below we keep the notation of two preceding subsections. Like in the ordinary case (see n. 0.6.4), for simplicity here we assume that $\overline{M}$ is without boundary, that is, its base (or, equivalently, its total space) is without boundary.

**Theorem.** Let $\overline{X}$ be a fat field on $\overline{M}$ and $\overline{f}: \overline{N} \to \overline{M}$ a fat map. Then there exists a (essentially) unique open relative fat interval over $\overline{N}$ and a unique fat map

$$\overline{\Phi_{\overline{f}}}: \overline{\mathbb{I}_N} \to \overline{M},$$

such that for all $n \in N$ the fat map

$$\overline{\gamma_n} = \overline{\Phi_{\overline{f}}} \circ \overline{j_n},$$

is a maximal fat trajectory of $\overline{X}$ and

$$\overline{\Phi_{\overline{f}}} \circ \overline{i_0} = \overline{f}.$$

**Proof.** According to the ordinary theorem on smooth dependence (Theorem 0.6.4), there exists a unique relative interval $\mathbb{I}_N$ over $N$ and a unique smooth map

$$\Phi_f : \mathbb{I}_N \to M,$$

such that for all $n \in N$ the curve

$$\gamma_n = \Phi_f \circ j_n$$

is the unique maximal trajectory of $X$ with $\gamma_n(0) = f(n)$. Let $\overline{\mathbb{I}_N}$ be the relative fat interval with base manifold $\mathbb{I}_N$.

For each $n$, consider a maximal fat trajectory $\overline{\gamma_n}' : \overline{\mathbb{I}_n}' \to \overline{M}$ of $\overline{X}$ over $\gamma_n$, and the uniform fat identity map $\overline{u_n}: \overline{\mathbb{I}_n}' \to \overline{\mathbb{I}_n}$ corresponding to the isomorphism $\overline{f_n}^{-1} \circ (\overline{\gamma_n}')_0 : F \to \overline{n}$. Also put

$$\overline{\gamma_n} \stackrel{\text{def}}{=} \overline{\gamma_n}' \circ \overline{u_n}^{-1}.$$

Then there exists a unique map

$$\overline{\Phi_{\overline{f}}} : \overline{\mathbb{I}_N} \to \overline{M}$$

such that

$$\overline{\Phi_{\overline{f}}} \circ \overline{j_n} = \overline{\gamma_n}$$

for all $n$. Indeed, note that each vector in the total space $\overline{\mathbb{I}_N}$ can be represented as $\overline{j_n}(\mathbf{v})$ for a unique $\mathbf{v}$ in the total space $\overline{\mathbb{I}_n}$, with uniquely determined $n \in N$. The relation

$$\overline{\Phi_{\overline{f}}} \circ \overline{i_0} = \overline{f}$$

follows easily from definitions of $\overline{i_0}$ and $\overline{j_n}$ as well that the fat maps $\overline{\gamma_n}$ and, therefore, $\overline{\Phi_{\overline{f}}}$ are uniquely determined by this condition. It remains to show that $\overline{\Phi_{\overline{f}}}$ is a fat map.

Clearly, the map $\overline{\Phi_{\overline{f}}}$ sends linearly fat points into fat points; hence, according to n. 0.3.25, it suffices to prove that it is smooth as a map between the corresponding total spaces. Consider the vector field $\mathbb{X}$ on the total space $\overline{M}$ associated with $\overline{X}$. Note that the total space $\overline{\mathbb{I}_N}$ is an open submanifold of the total space $\overline{\mathbb{R}_N}$, which is a product $\overline{N} \times \mathbb{R}$. The total space $\overline{\mathbb{I}_n}$ is identified with $\overline{n} \times \mathbb{I}_n \subseteq \overline{N} \times \mathbb{R}$. This shows that

$$\mathbb{I}_n \to \overline{M}, \quad t \mapsto \overline{\gamma_n}(\mathbf{v}, t), \quad \mathbf{v} \in \overline{n}$$

is a trajectory of $\mathbb{X}$ just because $\overline{\gamma_n}$ is a fat trajectory of $\overline{X}$. But

$$\overline{\gamma_n}(\mathbf{v}, t) = \overline{\Phi_{\overline{f}}}\left(\overline{j_n}(\mathbf{v}, t)\right)$$

and $\overline{j_n}(\mathbf{v}, t)$ is nothing but the point $(\mathbf{v}, t)$ in the product $\overline{N} \times \mathbb{R}$. Therefore, the curve

$$\mathbb{I}_n \to \overline{M}, \quad t \mapsto \overline{\Phi_{\overline{f}}}(\mathbf{v}, t)$$

is a trajectory of $\mathbb{X}$ passing through $\overline{f}(\mathbf{v})$ at $t = 0$. Thus $\overline{\Phi_{\overline{f}}}$ is smooth by the ordinary theorem on smooth dependence. $\square$

**Definition.** The map $\overline{\Phi_{\overline{f}}}$ is called the *fat flow of* $\overline{X}$ if $\overline{f} = \mathrm{id}_{\overline{M}}$.

## 1.7 Inner Structures

In Differential Geometry, Field Theory, *etc.*, vector bundles are, as a rule, supplied with additional structures. These may be interpreted as inner structures of fat manifolds and the general theory should be refined suitably in order to take into account these features. In this section these topics are illustrated by two guiding examples: inner complex structures and inner metrics.

### 1.7.1 Principal Bundles

Up to now we ignored an important feature of the theory of vector bundles, namely, the notion of *structure group*. Besides Differential Geometry and Topology this concept is considered, non infrequently, to be fundamental in the theory of gauge fields. We do not share this last opinion by giving more credit to *symmetries* (see below). Nevertheless in this subsection this notion is sketched to allow the reader to judge about.

In the traditional approach the structure group is introduced by means of *principal bundles*. Such a (smooth) fiber bundle is defined by means of a free right action $T : G \times N \to N$ of a Lie group $G$ on a manifold $N$ whose orbits form a manifold, say, $M$, i.e., $M = N/G$ ([2]). If, additionally, $\rho$ is a left (linear) representation of $G$ in a vector space $V$, then the action $T_\rho$ on $N \times V$

$$(T_\rho)(g) : (x, v) \mapsto \bigl(T(g,x), \rho(g^{-1})(v)\bigr), \quad x \in N, v \in V, g \in G,$$

is, obviously, free. The trivial bundle $p : N \times V \to N$ is equivariant (*cf.* [Nestruev (2003), (10.5), p. 154]) with respect to actions $T_\rho$ and $T$ of $G$ on $N \times V$ and $N$, respectively. So, the quotient vector bundle

$$E = (N \times V)/G \to M = N/G$$

is well defined and called *associated* with the considered principal bundle $N \to M = N/G$. The Lie group $G$ is called the *structure group* of it. If a vector bundle is equivalent to an associated with a principal $G$-bundle one, then $G$ is called its structure group, too. In frames of this approach various features of a vector bundle are considered to be representations of those of the corresponding principal bundle.

In reality, the structure group appears as the symmetry group of an additional geometric structure (see nn. 1.7.4 and 1.7.7) the considered vector bundle is supplied. The point of view adopted here allows a direct natural approach to this kind of situations that simplifies much the theory. These points are illustrated by the subsequent exposition.

### 1.7.2 Inner Complex Structures

The first example we intend to discuss concerns complex structures in (real) vector bundles. From intuitive geometrical point of view such a structure

---

[2]The action is free when $g \neq h \Rightarrow T(g,x) \neq T(h,x)$ for *all* $x \in N$ ($g, h \in G$). For fundamentals about Lie groups we refer to [Lee (2003)] instead of [Singer and Thorpe (1976)].

assigns a complex structure to each fiber of the considered real vector bundle in a smooth manner. To this end it suffices to introduce a multiplication by the imaginary unit operator on each fiber. Having in mind the relationship between vector bundles and modules of their smooth sections this idea is formalized as follows.

**Definition.** Let $A$ be a commutative $\mathbb{R}$-algebra and $P$ an $A$-module. A *complex structure in $P$* is an endomorphism

$$J : P \to P$$

such that

$$J^2 = -\operatorname{id}_P .$$

An *inner complex structure* on a fat manifold $\overline{M}$ is a complex structure in the $\mathrm{C}^\infty(M)$-module $\Gamma\left(\overline{M}\right)$.

Note that if $\hat{A}$ is the $\mathbb{C}$-algebra obtained from $A$ by extension of scalars, then setting

$$(x + \mathrm{i}\, y)p = xp + J(yp), \quad p \in P ,$$

we get an $\hat{A}$-module structure in $P$.

Conversely, if an $\hat{A}$-module $P$ is given, then the multiplication by the imaginary unit is a complex structure on $P$, considered as an $A$-module by restriction of scalars.

To avoid confusions, we explicitly mention that a complex structure on the tangent bundle of a smooth manifold $M$ is usually called an *almost-complex structure on $M$*, because it does not necessarily come from some complex manifold structure on $M$.

### 1.7.3 Inner (pseudo-)metrics

A symmetric bilinear form $b : P \times P \to A$ on an $A$-module $P$ is called *nondegenerate* if the map

$$p \mapsto \varphi_p \in P^\vee = \operatorname{Hom}_A(P, A), \quad \varphi_p(q) = b(p, q), \quad p, q \in P ,$$

is an isomorphism $P \cong P^\vee$ of $A$-modules. In the case when $A = \mathrm{C}^\infty(M)$, a symmetric nondegenerate bilinear form $b$ on $P$ is called *positive* if $b(p,p)(m) \geq 0$, for all $p \in P, m \in M$.

An *inner bilinear form* on a fat manifold $\overline{M}$ is a bilinear form on the $\mathrm{C}^\infty(M)$-module $\Gamma\left(\overline{M}\right)$. Geometrically it means that each fat point $\overline{m}$ is supplied with an $\mathbb{R}$-bilinear form which depends smoothly on $\overline{m}$, *i.e.*, on $m$ (*cf.* [Nestruev (2003), 11.24]).

Indeed, let $b$ be such a form and $m \in M$. Since

$$\overline{m} = \mathbb{R}_m \otimes P$$

(see n. 1.3.2), n. 0.1.5, (2) shows that there exists a unique bilinear form $b_m$ on $\overline{m}$ such that

$$b_m\left(p_1(m), p_2(m)\right) = b\left(p_1, p_2\right)(m), \quad p_1, p_2 \in \Gamma(\overline{M}).$$

By n. 0.1.5, (3), if $b$ is symmetric (respectively, alternating), then $b_m$ is symmetric (respectively, alternating). This way an $\mathbb{R}$-bilinear form is assigned to any fat point so that one gets the 'field' $\{b_m\}_{m \in M}$ of bilinear forms on $M$.

An *inner pseudo-metric $g$ on $\overline{M}$* is a symmetric nondegenerate bilinear form on $\Gamma(\overline{M})$. If $g$ is positive, then it is called simply *inner metric on $\overline{M}$*. It is easy to see that the 'field' $\{g_m\}_{m \in M}$ consists of nondegenerate (positive) forms if $g$ is nondegenerate (positive).

An inner (pseudo-)metric $g$ on the tangent bundle of a smooth manifold $M$ is called a *(pseudo-)Riemmannian metric on $M$*, and $M$ together with $g$ is called a *(pseudo-)Riemannian manifold*.

### 1.7.4 Inner Symmetries

A *symmetry* of a bilinear form $b$ on an $A$-module $P$ is an automorphism $\varphi$ of $P$ such that

$$b(\varphi(p_1), \varphi(p_2)) = b(p_1, p_2), \quad p_1, p_2 \in P.$$

For fat manifolds, a fat identity map $\overline{f}$ is a *symmetry* of an inner bilinear form $g$ on $\overline{M}$ if

$$g(\overline{f}^*(p_1), \overline{f}^*(p_2)) = g(p_1, p_2), \quad p_1, p_2 \in \Gamma(\overline{M}),$$

that is, if $\overline{f}^*$ is a symmetry of $g$. The notion of symmetry of an inner (pseudo-)metric is a particular case of this definition.

Obviously, the totality of all symmetries of a bilinear form $b$ on $P$ is a group with respect to the operation of composition. It will be denoted by $O(P, b)$ and called *b-orthogonal group*. The group of all fat identity maps that are symmetries of an inner bilinear form $g$ on a fat manifold $\overline{M}$ will be denoted by $O\left(\overline{M}, g\right)$ and also called *g-orthogonal group*.

In many situations it is more convenient to use the infinitesimal version of the above notion of symmetry. As usual, it is understood to be the velocity at $\overline{f}_0 = \mathrm{id}$ of a time-dependent family $\overline{f}_t$ of ordinary symmetries. By deriving formally with respect to $t$ the equality

$$g\left(\overline{f}_t^*(p_1), \overline{f}_t^*(p_2)\right) = g(p_1, p_2), \quad p_1, p_2 \in \Gamma\left(\overline{M}\right),$$

one finds that

$$g(\varphi(p_1), p_2) + g(p_1, \varphi(p_2)) = 0, \qquad (1.8)$$

$\varphi = \mathrm{d}/\mathrm{d}t|_{t=0}\overline{f}_t^*$. This motivates to say that an endomorphism of the $C^\infty(M)$–module $\Gamma(\overline{M})$ is an *infinitesimal symmetry of $g$*, if it satisfies (1.8). Alternatively, infinitesimal symmetries are also called *$g$–skew-adjoint*.

It is easily verified that infinitesimal symmetries form a submodule o$\left(\overline{M}, g\right)$ of $\mathrm{End}\left(\Gamma(\overline{M})\right)$. Moreover, the commutator of two infinitesimal symmetries is again an infinitesimal symmetry so that o$\left(\overline{M}, g\right)$ is a *Lie subalgebra* of the Lie algebra $\mathrm{End}\left(\Gamma(\overline{M})\right)$.

It is worth noticing that this definition is a particular case of the following general algebraic one. An endomorphism $\varphi$ of an $A$-module $P$ is called an *infinitesimal symmetry* of a bilinear form $b$ on $P$ if

$$b(\varphi(p_1), p_2) + b(p_1, \varphi(p_2)) = 0, \quad p_1, p_2 \in P.$$

Infinitesimal symmetries of $b$ form a Lie subalgebra o$(P, b)$ of $\mathrm{End}(P)$.

Similar definitions can be done for other types of *internal structures* in fat manifolds. For instance, for complex structures it looks as follows.

Let $P$ be an $A$-module and $\psi$ an endomorphisms of $P$. An automorphism $\Phi$ of $P$ is called *symmetry of $\psi$* if $\Phi \circ \psi = \psi \circ \Phi$, i.e.,

$$[\Phi, \psi] = 0.$$

Obviously, the totality of all symmetries of $\psi$, i.e., of all automorphisms commuting with $\psi$, is a subgroup $\mathrm{GL}(P, \psi)$ of $\mathrm{Aut}\, P$. When $P = \Gamma(\overline{M})$, a fat identity map $\overline{f} : \overline{M} \to \overline{M}$ is a *symmetry of $\psi$* if the associated automorphism $\overline{f}^*$ is a symmetry of $\psi$. The group of all such symmetries of $\psi$ will be denoted by $\mathrm{GL}\left(\overline{M}, \psi\right)$. In particular, if $J$ is an inner complex structure in $\overline{M}$, then $\mathrm{GL}\left(\overline{M}, J\right)$ is the *symmetry group of $J$*.

The same heuristic considerations as before lead to call an endomorphism of $\Gamma(\overline{M})$ commuting with $\psi$ to be an *infinitesimal symmetry* of $\psi$. All such endomorphisms constitute a submodule in $\mathrm{End}\left(\Gamma(\overline{M})\right)$, denoted by $\mathrm{gl}\left(\overline{M}, \psi\right)$. Obviously, it is also a Lie subalgebra of the Lie algebra $\mathrm{End}\left(\Gamma\left(\overline{M}\right)\right)$. In particular, $\mathrm{gl}\left(\overline{M}, J\right)$ is the Lie algebra of all infinitesimal symmetries of the inner complex structure $J$. In general, when $\psi$ is an endomorphism of whatever $A$-module $P$, all endomorphims of $P$ commuting with it constitute a Lie subalgebra $\mathrm{gl}(P, \psi)$ of $\mathrm{End}(P)$.

### 1.7.5

A natural and important question is: to what extent an inner structure in a fat manifold is determined by its symmetry group (resp., its infinitesimal symmetry algebra). For pseudo-metrics and complex structures the answer is given below as a consequence of the following results.

**Proposition.** Let $g$ be an inner pseudo-metric on a fat manifold $\overline{M}$ and $b$ an inner symmetric bilinear form on it. Then
$$\mathrm{O}\left(\overline{M}, b\right) \supseteq \mathrm{O}\left(\overline{M}, g\right) \quad \Longleftrightarrow \quad b = \alpha g, \alpha \in \mathrm{C}^\infty(M).$$

**Proof.** Let $V$ be a finite dimensional vector space over $\mathbb{R}$ and $\beta : V \times V \to \mathbb{R}$ a symmetric nondegenerate bilinear form on it. Consider the connected component $\mathrm{O}_0(V, \beta)$ of the $\beta$-orthogonal group $\mathrm{O}(V, \beta)$ that contains the identity homomorphism. It is not difficult to see that if $\mathrm{O}_0(V, \beta) \subseteq \mathrm{O}_0(V, \beta')$, $\beta'$ being a symmetric bilinear form on $V$, then either $\beta' = \lambda \beta$, $0 \neq \lambda \in \mathbb{R}$, or $\beta' = 0$ (look at the values of the corresponding quadratic forms on orbits).

The inclusion $\mathrm{O}\left(\overline{M}, b\right) \supseteq \mathrm{O}\left(\overline{M}, g\right)$ implies $\mathrm{O}_0\left(\overline{m}, b_m\right) \supseteq \mathrm{O}_0\left(\overline{m}, g_m\right)$, $\forall m \in M$, because $g$ is nondegenerate (details are left to the reader). Hence $b_m = \lambda_m g_m$. It remains to put $\alpha(m) = \lambda_m$. It is easy to see that the so-defined $\alpha$ is smooth. $\square$

**Exercise.** Let $J$ be a complex structure on a fat manifold $\overline{M}$ and $\varphi$ an endomorphism of $\Gamma(\overline{M})$. Then
$$\mathrm{GL}\left(\overline{M}, \varphi\right) \supseteq \mathrm{GL}\left(\overline{M}, J\right) \quad \Longleftrightarrow \quad \varphi = a\,\mathrm{id} + bJ, a, b \in \mathrm{C}^\infty(M).$$

**Hint.** Reduce the proof to single fibers as in the proof of the preceding proposition, and use the fact that an endomorphism of a (complex) vector space commuting with all automorphisms is a (complex) scalar endomorphism.

Let $g$ and $g'$ be inner pseudo-metrics on a fat manifold $\overline{M}$ and $J$ and $J'$ inner complex structure in it.

**Corollary.** We have
$$\mathrm{O}\left(\overline{M}, g\right) = \mathrm{O}\left(\overline{M}, g'\right) \quad \Leftrightarrow \quad g' = \alpha g,$$
with an invertible $\alpha \in \mathrm{C}^\infty(M)$ and
$$\mathrm{GL}\left(\overline{M}, J\right) = \mathrm{GL}\left(\overline{M}, J'\right) \quad \Leftrightarrow \quad J' = \sigma J$$
with $\sigma$ being a continuous function on $M$ with values in $\{1, -1\}$. In particular, $J' = \pm J$ if $M$ is connected.

## 1.7.6

The results of the preceding n. remain valid when the symmetry groups figuring in it are replaced by the corresponding symmetry Lie algebras, i.e., O $(\overline{M}, g)$ by o $(\overline{M}, g)$, etc. Proofs are easier and left as an exercise for the reader.

For more general Lie groups, it suffices to recall that exponentials of endomorphisms that belongs to the symmetry algebra generate the corresponding symmetry group.

### 1.7.7 Relation with Principal Bundles

Principal bundles (see n. 1.7.1) arise naturally in connection with inner structures. We shall illustrate this topics here just to draw a comparison with the standard approach.

Let $g$ be an inner pseudo-metric on $\overline{M}$ and $\{g_m\}_{m \in M}$ the corresponding family of pseudo-metrics on fat points $\overline{m}$'s. If all $g_m$'s are of the same signature, say $(k, l)$, (it is automatic if $M$ is connected) then all groups O $(\overline{m}, g_m)$ are isomorphic to the orthogonal group $O(k, l)$. Consider now all $g_m$-orthonormal frames whose totality denote by $R_m$. The group $O(k, l)$ acts naturally on the right on $R_m$. Namely, if $\mathbf{e} \in R_m$, $\mathbf{e} = (e_1, \ldots, e_n)$, $e_i \in \overline{m}$, and $\theta = (\theta_{ij}) \in O(k, l)$, then $\theta$ sends $\mathbf{e}$ to the frame $\mathbf{e}' = (e'_1, \ldots, e'_n)$ with $e'_i = \sum_j \theta_{ji} e_j$. The manifold of all frames

$$R(\overline{M}) = \bigcup_{m \in M} R_m$$

projects naturally on $M$:

$$R(\overline{M}) \supseteq R_m \ni \mathbf{e} \mapsto m \in M$$

and fibers of the so-obtained fiber bundle are $R_m$'s. As we have already seen the group $O(k, l)$ acts, obviously, freely, on fibers of this bundles and hence it acts freely on $R(\overline{M})$. Thus

$$R(\overline{M}) \to M$$

is a principal bundle with the structure group $O(k, l)$. It is not difficult to show that the bundle $\overline{M} \to M$ is associated with this principal bundle (see n. 1.7.1).

The same procedure can be applied to an inner complex structure. The only thing to be modified with respect to the previous case is to make use of $J_m$-complex frames instead of $g_m$-orthogonal ones. The structure group of the so-obtained principal bundle is $GL(d, \mathbb{C})$.

The general case can be treated along essentially the same lines.

### 1.7.8 Fat Lie Algebras and Principal Bundles

Lie algebras $o\left(\overline{M}, g\right)$ and $\mathrm{gl}\left(\overline{M}, J\right)$ are projective modules over $C^\infty(M)$ and the corresponding to them vector bundles have $o(k,l)$ and $\mathrm{gl}(d,\mathbb{C})$ as general fibers, respectively. These bundles may be thought as fat manifolds with inner Lie algebra structures. This point of view leads to a general idea of a *fat Lie algebra* as a fiber bundle of Lie algebras over a manifold which are indistinguishable one from another. This means that fat points composing a fat Lie algebra should be isomorphic one to another and hence isomorphic to a Lie algebra $\mathfrak{g}$.

A delicate point here is that, from one side, there are many isomorphisms of $\mathfrak{g}$ to the fiber $\mathfrak{g}_m$, $m \in M$, of the considered fat Lie algebra and, from the other side, these isomorphisms should depend smoothly on $m$. This requires a due formalization.

The first step toward this end is to fix a set $S_m$ of admissible isomorphisms from $\mathfrak{g}$ onto $\mathfrak{g}_m$. If $\varphi \in S_m$, then

$$S_m = \{\psi = \varphi \circ a : a \in G_{m,\varphi}\}$$

with

$$G_{m,\varphi} = \{a \in \mathrm{Aut}\,\mathfrak{g} : \varphi \circ a \in S_m\}\,.$$

By passing, if necessary, to the subgroup of $\mathrm{Aut}\,\mathfrak{g}$ generated by $G_{m,\varphi}$, we may assume from the very beginning that $G_{m,\varphi}$ is a group. Obviously, in this case $G_{m,\varphi}$ does not depend on $\varphi$ and hence one can simply write $G_m$ by omitting $\varphi$. Note that $G_m$ acts naturally from the right on $S_m$, *i.e.*, $\psi \mapsto \psi \circ a$, $a \in G_m$.

Next, the group $G_m$ should not depend on $m \in M$. Otherwise, the situation cannot be treated as a *fat* one. Denote this common for all points of $M$ subgroup of $\mathrm{Aut}\,\mathfrak{g}$ by $G$. Finally, $G$ must be supposed a submanifold in $\mathrm{Aut}\,\mathfrak{g}$ in order to give a sense to various analytical questions appearing in this context, say, smooth dependence on $m$, *etc*. This motivates the assumption that $G$ is a closed Lie subgroup of $\mathrm{Aut}\,\mathfrak{g}$. Moreover, by the same reasons, the totality $S = \bigcup_{m \in M} S_m$ of all admissible isomorphisms of $\mathfrak{g}$ onto fibers of the considered fat Lie algebra should be supplied with a structure of a smooth manifold such that its natural projection $S \to M$, $S_m \ni \varphi \mapsto m \in M$, is smooth. Since $G_m$ acts freely and from the right on $S_m$, and $G = G_m$ for all $m \in M$, $G$ acts freely and from the right on $S$ and orbits of this action are the $S_m$'s.

Now, by putting together all above considerations we see that $S \to M$ is a principal bundle with the structure group $G$ which is a Lie subgroup of

Aut $\mathfrak{g}$. Thus a rigorous formalization of the idea of a fat Lie algebra requires a principal bundle as a necessary ingredient. Such a bundle formalizes the concept of the variety of all isomorphisms connecting the general fiber Lie algebra with single fat point Lie algebras. The fat Lie algebra corresponding to a principal bundle is the associated bundle with the general fiber $\mathfrak{g}$ on which the structure group $G \subseteq \mathrm{Aut}\,\mathfrak{g}$ acts naturally.

In the traditional approach $\mathfrak{g}$ is as a rule assumed to be the Lie algebra of $G$ on which $G$ acts via the adjoint representation.

We have used the term 'fat Lie algebra' with the only purpose to show that the notion of a principal bundle appears almost automatically in the context of 'fat' philosophy, and only in this subsection. This term is synonymous to 'gauge Lie algebra' which is commonly used and it would be unreasonable to try changing it. It is worth also stressing that the above considerations are in full parallel with motivations for introducing gauge fields.

The above is a geometrically intuitive definition of a gauge Lie algebra which will be made precise in n. 3.6.1.

### 1.7.9

Lie algebras o $(\overline{M}, g)$ and gl $(\overline{M}, J)$, as well as any other gauge Lie algebra, are $\mathrm{C}^\infty(M)$–Lie algebras, i.e., the Lie multiplication in them is $\mathrm{C}^\infty(M)$–bilinear. In the general algebraic context the corresponding notion is as follows.

**Definition.** Given a commutative $k$-algebra $A$, $k$ being a field, an $A$-module $P$ supplied with an $A$-bilinear skew-symmetric map

$$P \times P \xrightarrow{\langle\cdot,\cdot\rangle} P$$

is called a *Lie algebra over* $A$ (alternatively an *A-Lie algebra*) if the *Jacobi identity*

$$\langle p_1, \langle p_2, p_3 \rangle \rangle + \langle p_3, \langle p_1, p_2 \rangle \rangle + \langle p_2, \langle p_3, p_1 \rangle \rangle = 0$$

holds.

If $\mathcal{I}$ is an ideal in $A$ then, obviously, $\langle \mathcal{I}P, P \rangle \subseteq \mathcal{I}P$ and the Lie multiplication $\langle \cdot, \cdot \rangle$ induces naturally a Lie multiplication $\langle \cdot, \cdot \rangle_\mathcal{I}$ in the quotient module $P/\mathcal{I}P$ over the quotient algebra $A/\mathcal{I}$. In particular, this way one obtains a family of $k$-Lie algebras

$$|A| \ni h \mapsto \mathfrak{g}_h \stackrel{\mathrm{def}}{=} (P_h, \langle \cdot, \cdot \rangle_{\mu_h})$$

with $P_h$ and $\mu_h$ as in n. 0.3.1.

If $P = \Gamma(\overline{M})$, then all vector spaces $P_m = \overline{m}$ are isomorphic one to another, but it is no longer so for Lie algebras $\mathfrak{g}_m$'s, i.e., these algebras corresponding to different points of $M$ are not generally isomorphic.

**Example.** Let $\overline{M}$ be a trivial fat manifold of type 2, and $\mathbf{e}_1$, $\mathbf{e}_2$ be a basis of $\Gamma(\overline{M})$. Fix a function $f \in C^\infty(M)$ and define a $C^\infty(M)$–Lie algebra structure on $\Gamma(\overline{M})$ by putting $\langle \mathbf{e}_1, \mathbf{e}_2 \rangle = f\mathbf{e}_2$. Then the Lie algebra structure $\langle \cdot, \cdot \rangle_m$ in $\overline{m}$ is defined by the relation $\langle \mathbf{e}_1(m), \mathbf{e}_2(m) \rangle = f\mathbf{e}_2(m)$. So, the bidimensional algebra $\mathfrak{g}_m$ is abelian if and only if $f(m) = 0$. Since all bidimensional nonabelian algebras are isomorphic to each other, we see that $\mathfrak{g}_m$ and $\mathfrak{g}_{m'}$ are isomorphic either if $f(m) = f(m') = 0$ or $f(m) \neq 0$, $f(m') \neq 0$, and nonisomorphic otherwise.

This shows that the class of $A$-Lie algebras is wider than the class of gauge Lie algebras.

**Exercise.** Show that the described above algebra is a gauge Lie algebra in the sense of n. 1.7.8 if and only if either $f = 0$, or $f$ is nowhere zero on $M$, by describing explicitly the corresponding principal bundles.

Chapter 2

# Linear Connections

In order to be in conformity with the previous chapter and the underlying philosophy, the theory of linear connections in vector bundles should be developed exclusively in terms of the corresponding modules of sections. It means that any use of total spaces of vector bundles must be excluded. In the standard approach a connection is a *horizontal* distribution (or, something equivalent) on the total space (see [Singer and Thorpe (1976), 7.1 (p. 178)]) of the bundle. According to the point of view developed in [Nestruev (2003)] this presupposes that the sections can be observed point by point with the classical observation mechanism. So, no more of 'horizontal distributions' in a situation when the classical observation mechanism is not applicable. This is our stand point in this chapter.

The subject of observability is, in fact, a rather delicate and important matter which merits a more detailed discussion. It, however, goes much beyond frames of this book and we will limit ourself with these short general remarks.

## 2.1 Basic Definitions and Examples

In this section linear connections in modules and, consequently, on fat manifolds are defined.

### 2.1.1 *Linear Connections*

In itself the following definition looks rather natural: it just transforms 'normal' vector fields into fat vector fields. It is certainly satisfactory for fat manifolds, whereas for general modules it mainly plays, in a sense, a 'descriptive role'.

**Definition.** Let $A$ be a commutative $k$-algebra, $k$ being a field, and $P$ an $A$-module. We define a *linear connection in* $P$ as an $A$-module homomorphism
$$\nabla : \mathrm{D}(A) \to \mathrm{Der}(P)$$
such that $\nabla(X)$ is a der-operator over $X$ for all $X \in \mathrm{D}(A)$.

The image of $X \in \mathrm{D}(A)$ is usually denoted by $\nabla_X$, instead of $\nabla(X)$, and it is called the *covariant derivative along* $X$.

A *linear connection* on a fat manifold will be a linear connection on the corresponding module of smooth sections.

Denote by $\mathrm{End}_k(P)$ the group of all $k$-linear operators acting on $P$. In view of n. 1.2.4, a linear connection on $P$ may be seen as a map $\nabla : \mathrm{D}(A) \to \mathrm{End}_k(P)$ such that

(1) $\quad \nabla_{fX+gY} = f\nabla_X + g\nabla_Y, \quad f,g \in A, \ X,Y \in \mathrm{D}(A);$
(2) $\quad \nabla_X(fp) = X(f)p + f\nabla_X(p), \quad f \in A, X \in \mathrm{D}(A), p \in P.$

Accordingly, a linear connection on a fat manifold, that is a homomorphism of $A$-modules
$$\mathrm{D}(M) \to \overline{\mathrm{D}}\left(\overline{M}\right),$$
may be also viewed as a map
$$\mathrm{D}(M) \to \mathrm{D}\left(\overline{M}\right)$$
that sends a vector field $X$ on $M$ to a vector field on the total space $\overline{M}$ projecting onto $X$: namely, $X \mapsto \mathbb{X}$ where $\mathbb{X}$ is the associated to $\nabla_X$ vector field on $\overline{M}$ (see n. 1.4.5).

An arbitrary linear map $\mathrm{D}(M) \to \mathrm{D}\left(\overline{M}\right)$ that sends $X \in \mathrm{D}(M)$ to a vector field on $\overline{M}$ projecting onto $X$, but not necessarily associated with a fat vector field, may be taken as an alternative definition of a (not necessarily linear) *connection*. This approach is applicable as well to general smooth bundles, but it becomes unjustifiably cumbersome when dealing with *linear* connections and, by the reasons at the beginning of this chapter, is not very compatible with our goals.

Vector fields $\mathbb{X}$ corresponding to der-operators $\nabla_X$ generate a $C^\infty\left(\overline{M}\right)$-submodule in $\mathrm{D}\left(\overline{M}\right)$, say, $\mathrm{D}_\nabla\left(\overline{M}\right)$. The fact that any field $\mathbb{X}$ projects onto $X$ shows easily that it is a projective $C^\infty\left(\overline{M}\right)$-module of rank $n = \dim M$. So, this module may be interpreted as the module of smooth sections of an $n$-dimensional subbundle of the tangent bundle to $\overline{M}$, i.e., as an $n$-dimensional distribution on $\overline{M}$. The fiber of this bundle at a point $\mathbf{e} \in \overline{M}$ is generated by vectors of the form $\mathbb{X}_\mathbf{e}$ where fields $\mathbb{X}$'s are as above. These fibers project isomorphically onto the corresponding tangent spaces of $M$. By this reason this distribution is called *horizontal*. It is easy to see that it determines completely the original (non-linear) connection and hence can be taken as an alternative definition of it.

### 2.1.2

**Example.** Let $A$ be a commutative $k$-algebra, $k$ being a field, and $P = A$. A *tautological* linear connection $D$ in $P$ is defined due to a natural inclusion of $\mathrm{D}(A)$ in $\mathrm{Der}(P)$ (*cf.* Example 1.2.6). In other words,

$$D_X = X, \quad X \in \mathrm{D}(A).$$

$D$ is called the *trivial connection* in $A$. A more general version of this construction is as follows.

### 2.1.3

**Example.** With $k$ and $A$ being as above, let $E$ be a $k$-vector space of a finite dimension $m > 0$. Consider $P = A \otimes_k E$ and define a linear connection $D$ in the $A$-module $P$ by putting

$$D_X = X \otimes_k \mathrm{id}_E, \quad X \in \mathrm{D}(A)$$

(*cf.* Example 1.2.7).

**Definition.** The above defined linear connection $D$ will be called *trivial connection in* $P = A \otimes_k E$. If $\overline{M}$ is a standard trivial bundle, in particular, $\overline{M}$ may be a fat interval, then the trivial connection in $\Gamma\left(\overline{M}\right)$ will be also called the *trivial connection on* $\overline{M}$.

### 2.1.4 Christoffel Symbols

Let $P = A \otimes_k E$ and $D$ be as above and $\nabla$ a linear connection in $P$. Since $D_X$ and $\nabla_X$ are der-operators over the same derivation $X \in D(A)$ the difference $\Phi_X = \nabla_X - D_X$, according to n. 1.2.8, is an endomorphism of $P$. Hence $\nabla$ may be presented in the form $D + \Phi$ with $\Phi$ being the $A$-homomorphism $\mathrm{D}(A) \to \mathrm{End}(P)$, $X \mapsto \Phi_X$. If $(e_1, \ldots, e_m)$ is a basis of $E$, then $(1 \otimes_k e_1, \ldots, 1 \otimes_k e_m)$ is a basis of the free $A$-module $P$. In corresponding coordinates (see n. 1.2.9) we have

$$\nabla_X \begin{pmatrix} p_1 \\ \vdots \\ p_m \end{pmatrix} = \underline{X} \begin{pmatrix} p_1 \\ \vdots \\ p_m \end{pmatrix} + \Gamma_X \begin{pmatrix} p_1 \\ \vdots \\ p_m \end{pmatrix}$$

with $\Gamma_X$ being the representing $\Phi_X$ matrix.

Assume that $D(A)$ is also free and $\{\partial_i\}_{i=1,\ldots,n}$ is a basis of it. If $X \in D(A)$ corresponds to $(X_1, \ldots, X_n) \in A^n$ in this basis, then

$$\Gamma_X = \sum_i X_i \Gamma_{\partial_i} \Rightarrow \nabla_X = \sum_i X_i (\underline{\partial_i} + \Gamma_{\partial_i}).$$

Hence the linear connection $\nabla$ is determined by the entries

$$\Gamma_{ij}^k = (\Gamma_{\partial_i})_j^k$$

of the matrices $\Gamma_{\partial_i}$ (with $k$ being the row index).

Conversely, for every choice of the elements $\Gamma_{ij}^k \in A$ the formula

$$\nabla_X = \sum_i X_i (\underline{\partial_i} + \Gamma_{\partial_i}), \quad X = \sum_i X_i \partial_i \in D(A)$$

defines a linear connection in $P$.

**Definition.** The elements $\Gamma_{ij}^k \in A$ are called the *Christoffel symbols of* $\nabla$ with respect to the bases $(e_1, \ldots, e_m)$ and $(\partial_1, \ldots, \partial_n)$.

### 2.1.5

Every linear connection on a fat manifold $\overline{M}$ may be described locally as above. Indeed, it will be proved in n. 2.4.10 that a linear connection can be localized. If $\{U, (x_1, \ldots, x_n)\}$ is a trivializing local chart for the considered vector bundle, then one may take for $e_i$'s the corresponding local basis of sections and $\partial/\partial x_i$ for $\partial_i$.

### 2.1.6

**Example.** Let $k$ and $A$ be as above. Consider the $A$-module Diff $A$ of all $k$-linear differential operators $\square : A \to A$, with the 'left' module multiplication:

$$(f\square)(g) = f(\square(g)), \quad f, g \in A.$$

Put for any $X \in D(A)$ and $\square \in \text{Diff } A$

$$\nabla_X(\square) = X \circ \square.$$

Obviously, $\nabla_X$ is a $k$-linear map of Diff $A$ into itself.
Moreover,

$$\nabla_{fX+gY}(\Box) = (fX+gY) \circ \Box = f(X \circ \Box) + g(Y \circ \Box)$$
$$= f\nabla_X(\Box) + g\nabla_Y(\Box)$$

and

$$\nabla_X(f\Box) = X \circ f\Box = X(f)\Box + f(X \circ \Box) = X(f)\Box + f\nabla_X(\Box) .$$

Hence the so-defined map $\nabla$ is a linear connection in Diff $A$ (see n. 2.1.1).

### 2.1.7

Assume now that $A$ is a subalgebra of a $k$-algebra $B$ such that every $k$-derivation $X$ of $A$ extends to a unique derivation $\nabla_X$ of $B$. If $X$ and $Y$ are derivations of $A$ and $f, g \in A$, then, obviously, $f\nabla_X + g\nabla_Y$ is a derivation in $B$ whose restriction to $A$ is $fX + gY$. Hence by the uniqueness assumption $f\nabla_X + g\nabla_Y = \nabla_{fX+gY}$.

Since $\nabla_X$ is a derivation of $B$ extending $X$, for all $f \in A$ ($\subseteq B$) and $p \in B$, we have

$$\nabla_X(fp) = \nabla_X(f)p + f\nabla_X(p) = X(f)p + f\nabla_X(p) .$$

This shows, in view of of n. 2.1.1, that the extension operation

$$\nabla : X \longmapsto \nabla_X$$

is a linear connection in the $A$-module $B$.

**Example.** If $A$ has no zero divisors, then any $k$-derivation of $A$ extends uniquely to a derivation of its quotient field $B$. Namely,

$$\nabla_X\left(\frac{a}{b}\right) = \frac{X(a)b + aX(b)}{b^2}, \quad a, b \in A, b \neq 0 .$$

Thus, a linear connection accompanies naturally the quotient procedure.

### 2.1.8

Another situation of this kind is also well-known in algebra.

**Example.** Let $K$ be a field and $k$ a subfield of it. Consider a finite separable algebraic extension $L$ of $K$. If $X$ is a $k$-derivation of $K$, then it extends canonically to a derivation $\nabla_X$ of $L$. In fact, by the primitive element theorem, $L$ can be obtained by adding to $K$ a root $\theta$ of an irreducible polynomial

$p(x) \in K[x]$. Obviously, the value $\nabla_X(\theta)$ determines $\nabla_X$ completely and it can be found from the linear equation (over $L$)

$$\nabla_X(p(\theta)) = \nabla_X(0) = 0 \Leftrightarrow p'(\theta)\nabla_X(\theta) + q(\theta) = 0$$

with a certain $q(x) \in K[x]$. Note that $p'(\theta) \neq 0$ due to separability of the extension $K \subseteq L$.

### 2.1.9 Affine Space of Linear Connections

Let $\nabla$ be a linear connection in an $A$-module $P$, $A$ being an algebra over a field $k$. If $a \in A$ and $X \in D(A)$, then $a\nabla_X$ is a der-operator over $aX$. So, the module homomorphism $a\nabla : D(A) \to \mathrm{Der}(P)$ is not a linear connection, unless $aX = X$ for all $X \in D(A)$ (when $P = \Gamma(\overline{M})$ this happens if and only if $a = 1$ or $\dim M = 0$). Similarly, the sum of two linear connections is not, generally, a linear connection.

However, if $\nabla_1, \ldots, \nabla_n$ are linear connections in $P$ and $a_1, \ldots, a_n \in A$ are such that $a_1 + \cdots + a_n = 1$, then $a_1\nabla_1 + \cdots + a_n\nabla_n$ is a linear connection on $P$. This induces to think that linear connections in $P$ constitute an affine space. Indeed, if $\nabla$ and $\Delta$ are linear connections, then $\nabla_X - \Delta_X \in \mathrm{End}\,P$ for all $X \in D(A)$, since $\nabla_X$ and $\Delta_X$ have the common symbol. This is equivalent to say that $\nabla - \Delta \in \Lambda^1(\mathrm{End}(P))$ (see Definition 0.5.1), up to the obvious identification of $\Lambda^1(\mathrm{Der}(P))$ with the corresponding submodule of $\Lambda^1(\mathrm{End}(P))$ (due to $\mathrm{Der}(P) \subseteq \mathrm{End}(P)$). Conversely, it is obvious that $\nabla + \rho$, $\rho \in \Lambda^1(\mathrm{End}(P))$ is a linear connection in $P$. Hence linear connections in $P$ constitute an affine space modeled over the module $\Lambda^1(\mathrm{End}(P))$.

### 2.1.10 Lift of Tangent Vectors

The following exercise shows that a linear connection on a fat manifold admits a 'pointwise description'.

**Exercise.** Let $\nabla$ be a linear connection on a fat manifold $\overline{M}$ and $m \in M$. Consider the evaluation at $m$ map

$$\overline{h}_m : \Gamma(\overline{M}) \twoheadrightarrow \overline{m} = \frac{\Gamma(\overline{M})}{(\mathrm{Ker}\,m)\,\Gamma(\overline{M})}\,.$$

For an $X \in D(M)$ consider the tangent vector $X_m = m \circ X$ at $m$ and the fat tangent vector $\overline{X}_m = \overline{h}_m \circ \nabla_X$ at $\overline{m}$. Show that if $X, Y \in D(M)$, then

$$X_m = Y_m \Longrightarrow \overline{X}_m = \overline{Y}_m\,.$$

**Hint.** Make use of the following fact:
$$X_m = Y_m \Leftrightarrow X - Y \in (\operatorname{Ker} m)\, \mathrm{D}(M)\,.$$

For a tangent vector $\xi$ at $m$ put
$$\overline{\xi} = \overline{h}_m \circ \nabla_Z$$
with a field $Z \in \mathrm{D}(M)$ such that $\xi = Z_m$. It follows from the above exercise that $\overline{\xi}$ does not depend on the choice of $Z$ and is determined completely by $\xi$ and $\nabla$.

**Definition.** The so-defined fat tangent vector $\overline{\xi}$ is called the $\nabla$-*lift of* $\xi$.

The lift of $\xi$ by $\nabla$ will be sometimes denoted by $\nabla_\xi$.

In view of the above result a linear connection may be interpreted as a procedure associating a fat tangent vector to each tangent vector. Informally speaking, a connection allows to identify infinitesimally near fibers.

## 2.2  Parallel Translation

Once the notion of internal structure is formalized as that of fat point, an evolution of the internal state is represented as a time dependent family of fat points, *i.e.*, as a fat curve. Evolution of a fat point as a single whole is a 'normal' curve in the base of the considered fat manifold. From this point of view the role of a connection is that it completes the *external* evolutions, *i.e.*, a 'normal' curve, to the *internal* one, *i.e.*, a covering fat curve. Intuitively, it is quite clear. Indeed, a connections allows, as we have noticed earlier, to identify infinitesimally near along a certain direction fat points. So, when walking along a curve in the base of a fat manifold consecutively passed fat points are identified each other. In this section a due formalization of the above said is given.

### 2.2.1  *Related Linear Connections*

**Definition.** Let $\overline{f} : \overline{N} \to \overline{M}$ be a fat map and $\square$ and $\nabla$ be linear connections on $\overline{N}$ and $\overline{M}$, respectively. $\nabla$ and $\square$ are called $\overline{f}$-*related* if for every pair of $f$-compatible vector fields $Y \in \mathrm{D}(N)$, $X \in \mathrm{D}(M)$ the fat vector fields $\square_Y$ and $\nabla_X$ are $\overline{f}$-compatible.

The notion of related connections is not basic. It is convenient by some technical reasons.

### 2.2.2

**Exercise.** Let $\overline{\mathbb{I}}'$ be a standard trivial fat manifold over an interval $\mathbb{I}$, with the standard fat field $\overline{d/dt}$, $\nabla$ a linear connection on a fat manifold $\overline{\mathbb{I}}$ over $\mathbb{I}$ and

$$\overline{f} : \overline{\mathbb{I}}' \to \overline{\mathbb{I}}$$

a fat identity map. Show that the trivial connection $D$ on $\overline{\mathbb{I}}'$ is $\overline{f}$-related with $\nabla$ if and only if $\overline{d/dt}$ is $\overline{f}$-compatible with $\nabla_{d/dt}$.

**Hint.** Recall that every vector field on $\mathbb{I}$ is a multiple of $d/dt$ by a smooth function.

**Proposition.** Let $\overline{\mathbb{I}}$ be a fat manifold over an interval $\mathbb{I}$ and $\nabla$ a linear connection on $\overline{\mathbb{I}}$. Then there exists a fat identity map $\overline{f} : \overline{\mathbb{I}}' \to \overline{\mathbb{I}}$, $\overline{\mathbb{I}}'$ being a fat interval over $\mathbb{I}$, such that $\nabla$ is $\overline{f}$-related with the trivial connection on $\overline{\mathbb{I}}'$.

Moreover, $\overline{f}$ is unique up to an uniform fat identity map of $\overline{\mathbb{I}}'' \to \overline{\mathbb{I}}'$.

*Proof.* A direct consequence of Proposition 1.6.7 and the preceding Exercise. $\square$

### 2.2.3  A Property of Closed Embeddings

**Lemma.** Let $\overline{\gamma} : \overline{N} \to \overline{M}$ be a fat map over a closed embedding $\gamma : N \to M$ and $\nabla$ a linear connection on $\overline{M}$. Furthermore, consider fields $X, Y \in D(M)$ and $\overline{Z} \in \overline{D}(\overline{N})$ such that $Y$ is $\gamma$-compatible with $Z$ and $\nabla_X$ is $\overline{\gamma}$-compatible with $\overline{Z}$. Then $\nabla_Y$ is $\overline{\gamma}$-compatible with $\overline{Z}$.

*Proof.* By Exercise 1.3.9 $X$ is $\gamma$-compatible with $Z$. Since $X$ and $Y$ are both $\gamma$-compatible with $Z$ and $\gamma$ is a closed embedding, one has

$$Y = X + \sum f_i W_i$$

for some functions $f_1, \ldots, f_s \in \operatorname{Ker} \gamma^*$ and vector fields $W_1, \ldots, W_s \in D(M)$. Then

$$\overline{\gamma}^* \circ \nabla_Y = \overline{\gamma}^* \circ \nabla_{(X + \sum f_i W_i)} = \overline{\gamma}^* \circ \left( \nabla_X + \sum f_i \nabla_{W_i} \right)$$
$$= \overline{\gamma}^* \circ \nabla_X + \sum \gamma^*(f_i) \left( \overline{\gamma}^* \circ \nabla_{W_i} \right) = \overline{\gamma}^* \circ \nabla_X = \overline{Z} \circ \overline{\gamma}^* .$$
$\square$

### 2.2.4 Connection Induced by a Closed Embedding

**Proposition.** Let $\nabla$ be a linear connection on a fat manifold $\overline{M}$ and $\bar{i}$ : $\overline{N} \to \overline{M}$ a fat map whose base $i : N \to M$ is a closed embedding. Then there exists a unique linear connection $\nabla^{\overline{N}}$ on $\overline{N}$ that is $\bar{i}$-related with $\nabla$.

**Proof.** For each vector field $Y$ on $N$ there exists a vector field $X$ on $M$, $i$-compatible with $Y$. This is equivalent to the possibility to extend a vector field on a closed submanifold to the whole manifold (see nn. 0.2.30 and 0.4.8).

Now, according to Proposition 1.5.3, there exists a unique fat vector field on $\overline{N}$ over $Y$, $\bar{i}$-compatible with $\nabla_X$. Denote it by $\nabla_Y^{\overline{N}}$. Lemma 2.2.3 shows that this vector field does not depend on the choice of $X$. Hence the map

$$\nabla^{\overline{N}} : D(N) \to \overline{D}(\overline{N}), \quad Y \mapsto \nabla_Y^{\overline{N}}$$

is well-defined.

The map $\nabla^{\overline{N}}$ is linear. Indeed, consider a pair of $i$-compatible fields $Y_j \in D(N)$ and $X_j \in D(M)$, $j = 1, 2$. The field $f_1 Y_1 + f_2 Y_2$, $f_1, f_2 \in C^\infty(N)$, is, obviously, $i$-compatible with $g_1 X_1 + g_2 X_2$ assuming that functions $g_j$'s are extensions of $f_j$'s to $M$, i.e., $f_j = i^*(g_j)$. Similarly, fat vector fields $f_1 \nabla_{Y_1}^{\overline{N}} + f_2 \nabla_{Y_2}^{\overline{N}}$ and $g_1 \nabla_{X_1} + g_2 \nabla_{X_2} = \nabla_{g_1 X_1 + g_2 X_2}$ are $\bar{i}$-compatible and, therefore,

$$\nabla_{f_1 Y_1 + f_2 Y_2}^{\overline{N}} = f_1 \nabla_{Y_1}^{\overline{N}} + f_2 \nabla_{Y_2}^{\overline{N}}.$$

This proves that $\nabla_Y^{\overline{N}}$ is a linear connection. By definition, $\nabla_Y^{\overline{N}}$ is the unique linear connection $\bar{i}$-related with $\nabla$. □

Note that the hypothesis of the above Proposition is satisfied for fat submanifolds $\bar{i} : \overline{N} \to \overline{M}$.

**Definition.** The linear connection $\nabla^{\overline{N}}$ defined in the above proposition is called *induced by $\bar{i}$ from* $\nabla$. In the case when $\bar{i}$ is the embedding of a closed fat submanifold we also refer to $\nabla^{\overline{N}}$ as the *restriction of $\nabla$ to $\overline{N}$*.

The restriction will be denoted by $\nabla|_{\overline{N}}$.

### 2.2.5 Lift by a Linear Connection

**Proposition.** Let $\nabla$ be a linear connection on a fat manifold $\overline{M}$ and

$$\gamma : \mathbb{I} \to M$$

an embedding of a closed interval $\mathbb{I} = [a,b]$. Then there exists a fat curve
$$\overline{\gamma} : \overline{\mathbb{I}} \to \overline{M},$$
over $\gamma$ such that $\nabla$ is $\overline{\gamma}$-related with the trivial connection $D$ on $\overline{\mathbb{I}}$. Moreover, $\overline{\gamma}$ is unique up to an uniform fat identity map on the domain.

**Proof.** Let $\overline{\mathbb{I}}_0$ be the bundle induced by $\gamma$ from $\overline{M}$, $\overline{\gamma}_0$ the induced map and $\Delta$ the linear connection induced by $\overline{\gamma}_0$ from $\nabla$ (see Definition 2.2.4).

Consider the fat identity map
$$\overline{f} : \overline{\mathbb{I}} \to \overline{\mathbb{I}}_0,$$
$\overline{\mathbb{I}}$ being a fat interval, such that the trivial connection $D$ in $\overline{\mathbb{I}}$ is $\overline{f}$-related with $\Delta$ (see Proposition 2.2.2). It is easy to see that $D$ is related with $\nabla$ through the fat map
$$\overline{\gamma} \stackrel{\text{def}}{=} \overline{\gamma}_0 \circ \overline{f}.$$
Conversely, if
$$\overline{\gamma}' : \overline{\mathbb{I}}' \to \overline{M}$$
is a fat map over $\gamma$ that relates the trivial connection $D'$ on $\overline{\mathbb{I}}'$ with $\nabla$, then, by universal property of the induced bundle, there exists the unique fat identity map
$$\overline{f}' : \overline{\mathbb{I}}' \to \overline{\mathbb{I}}_0$$
such that
$$\overline{\gamma}' = \overline{\gamma}_0 \circ \overline{f}'.$$
By the uniqueness property of induced connections (Proposition 2.2.4), the linear connection induced by $\overline{f}'$ from $\Delta$ coincides with the linear connection induced by $\overline{\gamma}$ from $\nabla$, i.e., with $D'$. This shows that $D'$ is $\overline{f}'$-related with $\Delta$.

Thus the correspondence $\overline{f} \leftrightarrow \overline{\gamma}_0 \circ \overline{f}$ establishes a one-to-one correspondence between fat identity maps $\overline{f}$ that relate the trivial connection $D$ in $\overline{\mathbb{I}}$ with $\Delta$ and fat maps over $\gamma$ that relate $D$ with $\nabla$. Now the result follows directly from Proposition 2.2.2. □

**Definition.** Let $\nabla$ be a linear connection on a fat manifold $\overline{M}$ and
$$\gamma : \mathbb{I} \to M$$
be an embedding of a closed interval $\mathbb{I} = [a,b]$. A fat curve
$$\overline{\gamma} : \overline{\mathbb{I}} \to \overline{M}$$
over $\gamma$ such that $\nabla$ is $\overline{\gamma}$-related with the trivial connection $D$ on $\overline{\mathbb{I}}$ is called a *lift of $\gamma$ by $\nabla$*.

### 2.2.6 Parallel Translation

Since the lift $\overline{\gamma}$ is unique up to an uniform fat identity map, it defines, according to n. 1.6.12, a *parallel translation* of fat points.

In the sequel the term 'parallel translation' (along $\gamma$) will be reserved for the identification isomorphisms as follows.

**Definition.** Let $\nabla$ be a linear connection on a fat manifold $\overline{M}$,
$$\gamma : \mathbb{I} \to M$$
an embedding of a closed interval $\mathbb{I} = [a, b]$ and
$$m_0 = \gamma(t_0), \quad m_1 = \gamma(t_1), \quad t_0, t_1 \in \mathbb{I}.$$
The isomorphism
$$\mathbf{T}^\gamma_{m_0, m_1} : \overline{m_0} \to \overline{m_1}$$
corresponding (see n. 1.6.12) to a lift of $\gamma$ by $\nabla$ is called the *parallel translation of $\overline{m_0}$ to $\overline{m_1}$ along $\gamma$ by $\nabla$*.

The notions of lift and parallel translation are, in fact, naturally defined for all curves, *i.e.*, not necessarily embedded nor necessarily with endpoints. But the general case is not so immediate from the definition of a linear connection (we invite the reader to reflect for a while on this extension). The general definition will become immediate in the next chapter, after having introduced the notion of compatible connections.

If $\overline{\gamma}$ is a lift of $\gamma$ by $\nabla$, then the family of *parallel along $\gamma$ vectors* is the family of vectors $\mathbf{v}_t = \overline{\gamma}(t, v)$ for a fixed $v \in F$. In other words, the family of parallel vectors is the image of a constant section of $\overline{\mathbb{I}}$ by $\overline{\gamma}$. By definition, each vector of this family may be viewed as the parallel translation of any other.

It is worth noticing the transitivity of parallel translations:
$$\mathbf{T}^\gamma_{m_0, m_2} = \mathbf{T}^\gamma_{m_1, m_2} \circ \mathbf{T}^\gamma_{m_0, m_1}.$$

### 2.2.7 Constant Sections

Let $p \in \Gamma(\overline{M})$ be such that
$$\nabla_X(p) = 0, \quad \forall X \in \mathrm{D}(M).$$
Since $D$ and $\nabla$ are related through every lifted fat curve $\overline{\gamma}$, and since constant sections $s$ of $\overline{\mathbb{I}}$ are characterized by the condition
$$D_X(s) = 0, \quad X \in \mathrm{D}(\mathbb{I}),$$
we see that the vectors $p(m)$ of $\overline{M}$ are identified with each other via the parallel translations, independently of the connecting them curve $\gamma$.

**Definition.** Let $\nabla$ be a linear connection on a fat manifold $\overline{M}$. A section $p$ of $\overline{M}$ is called *constant with respect to* $\nabla$, if
$$\nabla_X(p) = 0$$
for all $X \in D(M)$.

### 2.2.8

The fact that a linear connection is defined by first order differential operators guarantees that the lifting is a differential operation of first order. One consequence of this fact is that if two curves are tangent at a point, then their lifts are 'fat tangent' at the corresponding fat point. The exact meaning of that is the following.

Let $\overline{d/dt}$, $\overline{d/dt'}$ be standard fat fields on fat intervals $\overline{\mathbb{I}}$, $\overline{\mathbb{I}'}$ over closed bounded intervals, and $\gamma : \mathbb{I} \to M$, $\gamma' : \mathbb{I}' \to M$ two embedded curves on the base of a fat manifold $\overline{M}$. Assume now that $\gamma$ and $\gamma'$ pass through the same point and are tangent at it, *i.e.*,

$$\gamma(t_0) = \gamma'(t_0') = m \quad \text{and} \quad d_{t_0}\gamma\left(\left.\frac{d}{dt}\right|_{t_0}\right) = d_{t_0'}\gamma'\left(\left.\frac{d}{dt'}\right|_{t_0'}\right)$$

where
$$\left.\frac{d}{dt}\right|_{t_0} = t_0 \circ \frac{d}{dt} \quad \text{and} \quad \left.\frac{d}{dt'}\right|_{t_0'} = t_0' \circ \frac{d}{dt'}.$$

**Exercise.** Let $\nabla$ be a linear connection on $\overline{M}$ and consider the lifts by $\nabla$, $\overline{\gamma} : \overline{\mathbb{I}} \to \overline{M}$ and $\overline{\gamma'} : \overline{\mathbb{I}'} \to \overline{M}$ of $\gamma$, $\gamma'$, respectively.

Show that

$$\overline{d}_{t_0}\overline{\gamma}\left(\left.\overline{\frac{d}{dt}}\right|_{t_0}\right) = \overline{d}_{t_0'}\overline{\gamma'}\left(\left.\overline{\frac{d}{dt}}\right|_{t_0'}\right)$$

(see Definition 1.3.7), where

$$\left.\overline{\frac{d}{dt}}\right|_{t_0} = \overline{h}_{t_0} \circ \overline{\frac{d}{dt}} \quad \text{and} \quad \left.\overline{\frac{d}{dt'}}\right|_{t_0'} = \overline{h}_{t_0'} \circ \overline{\frac{d}{dt'}}.$$

### 2.2.9 *Linear Connections and Non-projective Modules*

From a naive geometrical point of view a connection is a means that makes possible parallel translations of fibers one to other along paths connecting their base points. This induces to think that fibers of a pseudobundle associated with a module $P$ admitting a linear connection are isomorphic one to other assuming that the base manifold is connected and, moreover, that the module $P$ is projective. From the algebraic point of view developed in this book not only projective modules admit connections as is shown in the example that follows. This bears evidence that a due preference should be done to the algebraic approach.

**Example.** Let $A = C^\infty(\mathbb{R})$ and $\mu_0$ the ideal of functions vanishing at the origin $0 \in \mathbb{R}$. Recall that the vector space of $l$-jets of smooth function on $\mathbb{R}$ at 0 is defined as

$$J_0^l \mathbb{R} \stackrel{\text{def}}{=} \frac{A}{\mu_0^{l+1}}$$

and the vector space of *infinite jets* as

$$J_0^\infty \mathbb{R} \stackrel{\text{def}}{=} \frac{A}{\bigcap_{l \in \mathbb{N}} \mu_0^l}.$$

It is immediate to recognize in $J_0^\infty \mathbb{R}$ the inverse limit of spaces $J_0^l \mathbb{R}$, related one to other by natural projections (*cf.* [Nestruev (2003), 9.64]). Let

$$j_0^\infty : A \twoheadrightarrow J_0^\infty \mathbb{R}, \quad f \mapsto [f] = f + \bigcap_{l \in \mathbb{N}} \mu_0^l$$

be the canonical projection. The space $J_0^\infty \mathbb{R}$ has an $A$-module structure induced by $j_0^\infty$, *i.e.*,

$$f \cdot j_0^\infty(g) = j_0^\infty(fg), \quad f, g \in A.$$

Now, consider the linear connection $\nabla$ in $J_0^\infty \mathbb{R}$ determined by

$$\nabla_{\frac{d}{dt}}(j_0^\infty(f)) = j_0^\infty\left(\frac{df}{dt}\right).$$

In a sense, $\nabla$ is induced by $j_0^\infty$ from the trivial connection in $A$.

On the other hand, it is easily seen that all fibers of the pseudobundle associated with $P$ are zero except that over the origin. So, this pseudobundle over the connected manifold $\mathbb{R}$ is not a bundle and hence the (finitely generated) module $P$ is not projective (see Theorem 0.3.21).

### 2.2.10 *Linear Connections on a 'Singular Manifold'*

The algebraic definition of a linear connection works well also on manifolds with singularities. This is illustrated by an example below in which connections on the *cross* (*cf.* [Nestruev (2003), 2.11, Exercise 1]) are discussed.

Consider the cross
$$\mathbf{K} \stackrel{\text{def}}{=} \{(x_1, x_2) \in \mathbb{R}^2 : x_1 x_2 = 0\}$$
in $\mathbb{R}^2$, i.e., the union of the two coordinate axes
$$\mathbb{R}_1 \stackrel{\text{def}}{=} \{(x_1, 0) : x_1 \in \mathbb{R}\}, \quad \mathbb{R}_2 \stackrel{\text{def}}{=} \{(0, x_2) : x_2 \in \mathbb{R}\}.$$
The algebra of *smooth functions on* $\mathbf{K}$ is defined as
$$C^\infty(\mathbf{K}) \stackrel{\text{def}}{=} \{f|_\mathbf{K} : f \in C^\infty(\mathbb{R}^2)\}.$$
$(\mathbf{K}, C^\infty(\mathbf{K}))$ is not a smooth manifold, in spite of its 'regular' status of an algebraic subset, but only a *smooth set* (see n. 0.2.20). It can be viewed as a manifold with a singular point at the origin $O = (0,0) \in \mathbf{K} \subseteq \mathbb{R}^2$. As topological space $\mathbf{K}$ is homeomorphic to the dual space $|C^\infty(\mathbf{K})|$.

Consider now the algebra
$$A = \{(f_1, f_2) : f_1 \in C^\infty(\mathbb{R}_1), f_2 \in C^\infty(\mathbb{R}_2), f_1(O) = f_2(O)\}.$$
Restrictions to coordinate axes define an isomorphism
$$C^\infty(\mathbf{K}) \xrightarrow{\sim} A\,;$$
*cf.* [Nestruev (2003), 9.35]. In the sequel smooth functions on the cross will be often identified with the corresponding pairs of functions constituting the algebra $A$.

By definition *vector fields* on the cross $\mathbf{K}$ are derivations of the $\mathbb{R}$-algebra $C^\infty(\mathbf{K})$. They are naturally identified, see [Nestruev (2003), 9.45], with pairs
$$(X_1, X_2)$$
of vector fields on the axes $\mathbb{R}_1$, $\mathbb{R}_2$, respectively, that vanish at $O$, i.e.,
$$(X_1)_O = 0 \quad \text{and} \quad (X_2)_O = 0,$$
in conformity with the above identification of $\mathbb{R}$-algebras $C^\infty(\mathbf{K})$ and $A$:
$$(X_1, X_2)(f_1, f_2) = (X_1(f_1), X_2(f_2)).$$
The $C^\infty(\mathbf{K})$-module
$$P = C^\infty(\mathbf{K}) \otimes_\mathbb{R} F,$$
$F$ being the standard fiber, is free and hence projective. The associated pseudobundle is naturally interpreted as a standard trivial vector bundle $\overline{\mathbf{K}}$ over $\mathbf{K}$. Smooth sections of this bundle are identified naturally with pairs
$$(p_1, p_2) : p_1(O) = p_2(O),$$

where $p_1 \in C^\infty(\mathbb{R}_1) \otimes_\mathbb{R} F$, $p_2 \in C^\infty(\mathbb{R}_2) \otimes_\mathbb{R} F$ are sections of standard trivial bundles $\overline{\mathbb{R}_1}$, $\overline{\mathbb{R}_2}$, respectively. A linear connection in $P$ is interpreted geometrically to be a *linear connection on* $\overline{\mathbf{K}}$.

Observe that the bundle $\overline{\mathbf{K}}$ may be seen as the restriction of the standard trivial bundle $\overline{\mathbb{R}^2}$ over $\mathbb{R}^2$ with the standard fiber $F$ to $\mathbf{K}$. Denote by

$$\overline{f} : \overline{\mathbf{K}} \to \overline{\mathbb{R}^2}$$

the corresponding induced map. Proposition 2.2.4 can be easily extended to this case. Therefore, if $\nabla$ is a linear connection in $\overline{\mathbb{R}^2}$, then the *induced* linear connection $\nabla^{\overline{\mathbf{K}}}$ on $\overline{\mathbf{K}}$ is defined literally by the same way as as it was done for closed fat submanifolds (see Definition 2.2.4). Namely, we put

$$\nabla^{\overline{\mathbf{K}}}_{(X_1, X_2)}(p_1, p_2) = \nabla_X(p)|_{\mathbf{K}}$$

where $X$ is a vector field on $\mathbb{R}^2$ compatible with $(X_1, X_2)$ with respect to the inclusion map $f : \mathbf{K} \to \mathbb{R}^2$ and $p$ is a section of $\overline{\mathbb{R}}$ that restricts to $(p_1, p_2)$.

Denote the restrictions of $\nabla$ to closed fat submanifolds $\overline{\mathbb{R}_1}$, $\overline{\mathbb{R}_2}$ by $\nabla^1$ and $\nabla^2$, respectively. As a section over the cross $\nabla_X(p)|_{\mathbf{K}}$ is identified with the pair $(p_1', p_2')$ with $p_1'$, $p_2'$ being the restrictions of $\nabla_X(p)|_{\mathbf{K}}$ to the axes $\mathbb{R}_1$ and $\mathbb{R}_2$, respectively. Obviously, $p_1'$, $p_2'$ may be viewed as restrictions of $\nabla_X(p)$ to the axes as well.

Now again by definition of induced linear connection we have

$$p_1' = \nabla^1_{X_1}(p_1), \quad p_2' = \nabla^2_{X_2}(p_2) .$$

This leads to the following description of $\nabla^{\overline{\mathbf{K}}}$:

$$\nabla^{\overline{\mathbf{K}}}_{(X_1, X_2)}(p_1, p_2) = \left(\nabla^1_{X_1}(p_1), \nabla^2_{X_2}(p_2)\right) ,$$

or, shortly,

$$\nabla^{\overline{\mathbf{K}}} = (\nabla_1, \nabla_2) .$$

Note that there are vector fields on one of two axes that do not extend from it to the cross. By this reason, the proof of Proposition 2.2.4 does not literally extend to this case and, therefore, we could not directly deduce the above description by means of restrictions from $\mathbf{K}$ to the axes. By the same reason it is not a priori clear if *every* linear connection on $\overline{\mathbf{K}}$ admits a description as a pair of linear connections over the axes. Now we shall discuss this more subtle question.

So, let $\square$ be an arbitrary linear connection on $\overline{\mathbf{K}}$. Recall that the tangent space to $\mathbf{K}$ at the origin $O$ is two-dimensional (see [Nestruev (2003), 9.35]). On the other hand, all vector fields on $\overline{\mathbf{K}}$ vanish at $O$. By this reason, to define the lift of a nonzero tangent vector $\xi \in T_O \mathbf{K}$ by applying the approach of n. 2.1.10 is not possible. This may create a sensation that $\Box$ does not induce connections on axes on the whole. It is, however, not so as is shown below.

The module $D(\mathbf{K})$ of all vector fields is (not freely) generated by

$$Z_1 \stackrel{\text{def}}{=} \left( x_1 \frac{\partial}{\partial x_1}, 0 \right) \quad \text{and} \quad Z_2 \stackrel{\text{def}}{=} \left( 0, x_2 \frac{\partial}{\partial x_2} \right).$$

Let $p = (p_1, p_2) \in P$ and $(p'_1, p'_2) = \Box_{Z_1}(p)$. Then

$$(0, x_2 p'_2) = (0, x_2) \Box_{Z_1}(p) = \Box_{(0, x_2) Z_1}(p) = \Box_0(p) = (0, 0).$$

So, $p'_2 = 0$ and $p'_1(O) = p'_2(O) = 0$. According to Hadamard's lemma (see [Nestruev (2003), 2.8]), there is a unique section $(q'_1, 0)$ such that $(p'_1, 0) = (x_1, 0)(q'_1, 0)$. Define a linear connection $\Box_1$ on $\mathbb{R}_1$ by putting

$$(\Box_1)_{\frac{\partial}{\partial x_1}}(p_1) = q'_1.$$

With an obvious extension of the notion of related connections to the considered situation it is easy to see that $\Box_1$ is related with $\Box$ with respect to the embedding of the first axis into the cross. Similarly, a connection, say, $\Box_2$, is constructed on the second axis and we see that $\Box$ is completely determined by the pair of linear connections $(\Box_1, \Box_2)$.

Finally, it is easy to see that for all pairs of linear connections $(\nabla_1, \nabla_2)$ on the axes, there is a linear connection on $\overline{\mathbb{R}^2}$ that restricts to them. This shows that linear connections on $\overline{\mathbf{K}}$ are in one-to-one correspondence with pairs of linear connections on the axes and that these extend to linear connections on the whole plane.

## 2.3 Curvature

Consider a linear connection $\nabla$ on an $A$-module $P$. Covariant derivatives $\nabla_{X_1}$ and $\nabla_{X_2}$ are der-operators over $X_1$ and $X_2$, respectively. So, their commutator $[\nabla_{X_1}, \nabla_{X_2}]$ is a der-operator over $[X_1, X_2]$ (see Proposition 1.2.10) as well as $\nabla_{[X_1, X_2]}$. Now a natural question is: do these two der-operators coincide? In other words, is it true that the $A$-module of all covariant derivatives $\{\nabla_X\}_{X \in D(A)}$ is a Lie subalgebra of $\text{Der}(P)$? Examples shows that, generally, the answer is negative. One of them is based on the description of linear connections given in n. 2.1.4, and it is the following.

## 2.3.1

**Example.** Let $A = C^\infty(\mathbb{R}^2)$, $P = A^m$ and $\nabla$ be a linear connection in $P$. Then $\nabla_{X_i} = \underline{\partial_i} + \Gamma_i$ with $X_i = \partial_i = \partial/\partial x_i$, $i = 1, 2$ (*cf.* n. 1.2.8). Recall that

$$\underline{\partial_i} p = \begin{pmatrix} \partial_i p_1 \\ \vdots \\ \partial_i p_m \end{pmatrix},$$

for a column

$$p = \begin{pmatrix} p_1 \\ \vdots \\ p_m \end{pmatrix} \in A^m$$

and similarly for a matrix with entries in $A$. In this notation we have

$$[\nabla_{X_1}, \nabla_{X_2}] - \nabla_{[X_1, X_2]} = [\nabla_{\partial_1}, \nabla_{\partial_2}] - \nabla_{[\partial_1, \partial_2]} = [\underline{\partial_1} + \Gamma_1, \underline{\partial_2} + \Gamma_2]$$
$$= [\Gamma_1, \Gamma_2] + [\underline{\partial_1}, \Gamma_2] + [\Gamma_1, \underline{\partial_2}].$$

If $p \in P$, then, obviously,

$$\underline{\partial_i}(\Gamma_j(p)) = (\underline{\partial_i}\Gamma_j)p + \Gamma_j(\underline{\partial_i}p),$$

so that $[\underline{\partial_i}, \Gamma_j] = \underline{\partial_i}\Gamma_j$ and

$$[\nabla_{X_1}, \nabla_{X_2}] - \nabla_{[X_1, X_2]} = [\Gamma_1, \Gamma_2] + \underline{\partial_1}\Gamma_2 - \underline{\partial_2}\Gamma_1.$$

The matrix $[\Gamma_1, \Gamma_2] + \underline{\partial_1}\Gamma_2 - \underline{\partial_2}\Gamma_1$ is, generally, non-zero. For instance, it happens when $\Gamma_1$ and $\Gamma_2$ are noncommuting matrices with constant entries.

Though $[\nabla_{X_1}, \nabla_{X_2}]$ and $\nabla_{[X_1, X_2]}$ are not equal, they are der-operators over the same vector field $[X_1, X_2]$. Hence in view of n. 1.2.5 $[\nabla_{X_1}, \nabla_{X_2}] - \nabla_{[X_1, X_2]}$ is an $A$-module endomorphism of $P$ measuring the deviation of

$$\nabla : \mathrm{D}(A) \to \mathrm{Der}(A)$$

from being a Lie algebra homomorphism. This explains the importance of the following notion.

**Definition.** The *curvature tensor* of a linear connection $\nabla$ in an $A$-module $P$ is the function

$$R^\nabla : \mathrm{D}(A) \times \mathrm{D}(A) \to \mathrm{End}(P), \quad (X, Y) \mapsto [\nabla_X, \nabla_Y] - \nabla_{[X,Y]}.$$

## 2.3.2

**Proposition.** The curvature tensor $R$ of a linear connection $\nabla$ is $A$-bilinear and skew-symmetric.

**Proof.** Skew-symmetricity and additivity are obvious. Further, if $f \in A$ and $X, Y \in D(A)$, then

$$\begin{aligned}
R(fX, Y) &= [\nabla_{fX}, \nabla_Y] - \nabla_{[fX,Y]} \\
&= \nabla_{fX}\nabla_Y - \nabla_Y\nabla_{fX} - \nabla_{f[X,Y]} + \nabla_{Y(f)X} \\
&= f\nabla_X\nabla_Y - f\nabla_Y\nabla_X - Y(f)\nabla_X - f\nabla_{[X,Y]} + Y(f)\nabla_X \\
&= f\left([\nabla_X, \nabla_Y] - \nabla_{[X,Y]}\right) = fR(X, Y) . \qquad \square
\end{aligned}$$

Note that when $P = \Gamma(\overline{M})$, $R$ is not a tensor on $M$ in the strict sense of this term. However, $R$ is commonly called 'tensor' partially due to some historical reasons and partially to the $A$-linearity property. Being $A$-bilinear and skew-symmetric $R$ can be viewed as a differential form with values in the $A$-module of endomorphisms of $P$, i.e.,

$$R \in \Lambda^2\left(\operatorname{End}(P)\right)$$

(see Definition 0.5.1).

### 2.3.3 Flatness

**Definition.** A linear connection with zero curvature tensor is called *flat*.

Note that a linear connection $\nabla : D(A) \to \operatorname{Der}(A)$ is flat if and only if it is a Lie algebra homomorphism.

**Example.** The trivial linear connection and the connections of Examples 2.1.6, 2.1.7 and 2.1.8 are all flat. Every linear connection on a one-dimensional fat manifold is flat.

## 2.3.4

**Exercise.** Let $\nabla$ be a linear connection on a fat manifold $\overline{M}$ and $\bar{i} : \overline{N} \hookrightarrow \overline{M}$ be the fat embedding of a closed fat submanifold. Consider the restriction $\nabla|_{\overline{N}}$ of $\nabla$ on $\overline{N}$.

Prove that

$$R^{\nabla|_{\overline{N}}}(Y_1, Y_2)(p|_N) = R^{\nabla}(X_1, X_2)(p)\big|_N, \quad p \in \Gamma(\overline{M}),$$

when $X_1, X_2$ are chosen $i$-compatible with $Y_1, Y_2$, respectively.

## 2.4 Operations with Linear Connections

In this section is shown that connections are in a natural conformity with basic operations of linear algebra. Concrete situations discussed below are to be considered as patterns on the basis of which the reader can learn how to *assemble* new connections from given ones in many other contexts.

### 2.4.1 *Linear Connection in Modules of Homomorphisms*

Below $A$ stands for a commutative $k$-algebra, $k$ being a field.

**Definition.** Let $P$ and $Q$ be $A$-modules supplied with linear connections $\nabla$ and $\Delta$, respectively. The linear connection $\mathrm{Hom}\,(\nabla, \Delta)$ in $\mathrm{Hom}\,(P,Q)$ associated with $\nabla$ and $\Delta$ is defined by putting

$$\mathrm{Hom}\,(\nabla, \Delta)_X \stackrel{\mathrm{def}}{=} \mathrm{Hom}(\nabla_X, \Delta_X)$$

with $X \in \mathrm{D}\,(A)$ and $\mathrm{Hom}(\nabla_X, \Delta_X)$ given by Definition 1.5.1.

It is easily verified that the above construction gives in reality a linear connection in $\mathrm{Hom}\,(P,Q)$.

### 2.4.2 *Curvature of* $\mathrm{Hom}\,(\nabla, \Delta)$

**Proposition.** For the curvature tensor of the linear connection

$$\Box = \mathrm{Hom}\,(\nabla, \Delta)$$

the following relation holds:

$$R^\Box (X,Y)(\varphi) = R^\Delta (X,Y) \circ \varphi - \varphi \circ R^\nabla (X,Y)$$

with $X, Y \in \mathrm{D}\,(A)$, $\varphi \in \mathrm{Hom}\,(P,Q)$.

*Proof.* This is a direct computation:

$$\begin{aligned}
R^\Box (X,Y)(\varphi) &= \Box_X (\Box_Y (\varphi)) - \Box_Y (\Box_X (\varphi)) - \Box_{[X,Y]}(\varphi) \\
&= \Box_X (\Delta_Y \circ \varphi - \varphi \circ \nabla_Y) - \Box_Y (\Delta_X \circ \varphi - \varphi \circ \nabla_X) \\
&\quad - (\Delta_{[X,Y]} \circ \varphi - \varphi \circ \nabla_{[X,Y]}) \\
&= \Delta_X \circ (\Delta_Y \circ \varphi) - (\Delta_Y \circ \varphi) \circ \nabla_X - \Delta_X \circ (\varphi \circ \nabla_Y) + (\varphi \circ \nabla_Y) \circ \nabla_X \\
&\quad - \Delta_Y \circ (\Delta_X \circ \varphi) + (\Delta_X \circ \varphi) \circ \nabla_Y + \Delta_Y \circ (\varphi \circ \nabla_X) - (\varphi \circ \nabla_X) \circ \nabla_Y \\
&\quad - \Delta_{[X,Y]} \circ \varphi + \varphi \circ \nabla_{[X,Y]} = R^\Delta (X,Y) \circ \varphi - \varphi \circ R^\nabla (X,Y) \,. \quad \Box
\end{aligned}$$

### 2.4.3 Associated Linear Connections in Endomorphisms

In the particular case when $P = Q$ and $\nabla = \Delta$ the previous construction gives a connection in $\mathrm{End}(P)$ as follows.

**Definition.** The linear connection $\mathrm{Hom}\,(\nabla, \nabla)$ on $\mathrm{End}(P)$ is called *associated with* $\nabla$ and denoted by $\nabla^{\mathrm{End}}$.

Explicitly we have
$$\nabla_X^{\mathrm{End}} \varphi = [\nabla_X, \varphi]$$
with $X \in \mathrm{D}(A)$ and $\varphi \in \mathrm{End}\,(P)$.

According to Proposition 2.4.2 the curvature of $\nabla^{\mathrm{End}}$ reads as
$$R^{\nabla^{\mathrm{End}}}(X, Y)(\varphi) = \left[R^\nabla(X, Y), \varphi\right].$$

### 2.4.4 Dual Linear Connection

In the particular case when $Q = A$ and $\Delta$ is the trivial connection $D$ the construction of n. 2.4.1 gives a linear connection in the dual module $P^\vee$ of $P$.

**Definition.** Let $\nabla$ be a linear connection in an $A$-module $P$ and $D$ the trivial connection in $A$. The linear connection $\mathrm{Hom}\,(\nabla, D)$ in the dual module $P^\vee$, denoted by $\nabla^\vee$, is called *dual to* $\nabla$.

The explicit formula is
$$(\nabla^\vee)_X (\psi) = X \circ \psi - \psi \circ \nabla_X$$
with $\psi \in P^\vee$, $p \in P$ and $X \in \mathrm{D}\,(A)$. It shows that $(\nabla^\vee)_X$ is nothing else but the dual to $\nabla_X$ der-operator and there will not be any ambiguity in the notation
$$\nabla_X^\vee.$$

Since $D$ is flat, Proposition 2.4.2, gives the following explicit formula for the curvature of $\nabla^\vee$:
$$R^{\nabla^\vee}(X, Y)(\psi) = -\psi \circ R^\nabla(X, Y)$$
with $\psi \in P^\vee$ and $X, Y \in \mathrm{D}\,(A)$.

## 2.4.5 Associated Linear Connections in Tensor Products

**Definition.** Let $P$ and $Q$ be $A$-modules supplied with linear connections $\nabla$ and $\Delta$, respectively. The linear connection $\nabla \boxtimes \Delta$ in $P \otimes_A Q$ associated with $\nabla$ and $\Delta$ is defined by putting

$$(\nabla \boxtimes \Delta)_X \stackrel{\text{def}}{=} \nabla_X \boxtimes \Delta_X ,$$

with $X \in \mathrm{D}(A)$ and $\nabla_X \boxtimes \Delta_X$ given by Definition 1.5.2.

## 2.4.6 Curvature of $\nabla \boxtimes \Delta$

**Proposition.** The following explicit formula for the curvature of $\nabla \boxtimes \Delta$ takes place:

$$R^{\nabla \boxtimes \Delta}(X,Y) = R^\nabla(X,Y) \otimes \mathrm{id}_Q + \mathrm{id}_P \otimes R^\Delta(X,Y), \quad X,Y \in \mathrm{D}(A) .$$

**Proof.** Let $X, Y \in \mathrm{D}(A)$, $p \in P$, $q \in Q$ and $\square = \nabla \otimes \Delta$. We have

$$(\square_X \circ \square_Y)(p \otimes q) = \square_X (\nabla_Y(p) \otimes q + p \otimes \nabla_Y(q))$$
$$= \nabla_X(\nabla_Y(p)) \otimes q + \nabla_Y(p) \otimes \Delta_X(q)$$
$$+ \nabla_X(p) \otimes \Delta_Y(q) + p \otimes \Delta_X(\Delta_Y(q)) .$$

This implies

$$[\square_X, \square_Y](p \otimes q) = [\nabla_X, \nabla_Y](p) \otimes q + p \otimes [\Delta_X, \Delta_Y](q) .$$

Hence

$$R^\square(X,Y)(p \otimes q) = [\square_X, \square_Y](p \otimes q) - \square_{[X,Y]}(p \otimes q)$$
$$= [\nabla_X, \nabla_Y](p) \otimes q + p \otimes [\Delta_X, \Delta_Y](q) - \nabla_{[X,Y]}(p) \otimes q - p \otimes \Delta_{[X,Y]}(q)$$
$$= ([\nabla_X, \nabla_Y] - \nabla_{[X,Y]})(p) \otimes q + p \otimes ([\Delta_X, \Delta_Y] - \Delta_{[X,Y]})(q)$$
$$= R^\nabla(X,Y)(p) \otimes q + p \otimes R^\Delta(X,Y)(q) . \quad \square$$

## 2.4.7 Associated Linear Connection in Bilinear Forms

Let $P$ be an $A$-module, $\nabla$ a linear connection in $P$ and $\mathrm{Bil}(P)$ the module of $A$-bilinear forms on $P$. Denote by $(\nabla \boxtimes \nabla)^\vee$ the linear connection in $(P \otimes P)^\vee$ that is dual to the connection $\nabla \boxtimes \nabla$ in $P \otimes P$. In view of the natural isomorphism

$$\mathrm{Bil}(P) \cong (P \otimes P)^\vee ,$$

we have a corresponding linear connection in $\mathrm{Bil}(P)$, which will be denoted by $\nabla^{\mathrm{Bil}}$. Recall that a bilinear form $b \in \mathrm{Bil}(P)$ and the corresponding to

it via the above mentioned isomorphism linear function $\beta$ on $P \otimes P$ are related by the formula
$$b(p_1, p_2) = \beta(p_1 \otimes p_2), \quad p_1, p_2 \in P.$$
Having this in mind and putting $\square = \nabla \boxtimes \nabla$ we find (for all $X \in D(A)$):
$$\begin{aligned}\nabla_X^{\mathrm{Bil}}(b)(p_1,p_2) &= \square_X^{\vee}(\beta)(p_1 \otimes p_2) = X(\beta(p_1 \otimes p_2)) - \beta(\square_X(p_1 \otimes p_2)) \\ &= X(\beta(p_1 \otimes p_2)) - \beta(\nabla_X(p_1) \otimes p_2 + p_1 \otimes \nabla_X(p_2)) \\ &= X(\beta(p_1 \otimes p_2)) - \beta(\nabla_X(p_1) \otimes p_2) - \beta(p_1 \otimes \nabla_X(p_2)) \\ &= X(b(p_1,p_2)) - b(\nabla_X(p_1), p_2) - b(p_1, \nabla_X(p_2)) .\end{aligned}$$
This leads to the following explicit description of $\nabla^{\mathrm{Bil}}$:
$$\left(\nabla_X^{\mathrm{Bil}} b\right)(p_1, p_2) = X(b(p_1, p_2)) - b(\nabla_X(p_1), p_2) - b(p_1, \nabla_X(p_2)) .$$

### 2.4.8 Curvature of the Associated Connection in Bilinear Forms

Denote by $R^{\mathrm{Bil}}$ the curvature tensor of $\nabla^{\mathrm{Bil}}$. Then
$$\begin{aligned}\left(R^{\mathrm{Bil}}(X,Y)(b)\right)(p_1,p_2) &= (R^{(\nabla \boxtimes \nabla)^{\vee}}(X,Y)(\beta))(p_1 \otimes p_2) \\ &= -\beta\left(R^{\nabla \boxtimes \nabla}(X,Y)(p_1 \otimes p_2)\right) \\ &= -\beta\left(R^{\nabla}(X,Y)(p_1) \otimes p_2 + p_1 \otimes R^{\nabla}(X,Y)(p_2)\right) \\ &= -\beta\left(R^{\nabla}(X,Y)(p_1) \otimes p_2\right) - \beta(p_1 \otimes R^{\nabla}(X,Y)(p_2)) \\ &= -b\left(R^{\nabla}(X,Y)(p_1), p_2\right) - b(p_1, R^{\nabla}(X,Y)(p_2)) ,\end{aligned}$$
and, therefore,
$$\left(R^{\mathrm{Bil}}(X,Y)(b)\right)(p_1, p_2) = -b\left(R^{\nabla}(X,Y)(p_1), p_2\right) - b(p_1, R^{\nabla}(X,Y)(p_2)) .$$

### 2.4.9

Let $\nabla$ be a linear connection on a fat manifold $\overline{M}$. Assume that the module $P = \Gamma(\overline{M})$ is free, i.e., $\overline{M}$ is a trivial bundle, as well as the module $D(M)$. Let $(e_1, \ldots, e_r)$ be a basis of $P$ and $(\partial_1, \ldots, \partial_n)$ be a basis of $D(M)$. The corresponding basis
$$\{b_{ij}\}, \quad 1 \leq i, j \leq r ,$$
in the module $\mathrm{Bil}(P)$ of $A$-bilinear forms on $P$ (which is, therefore, free as well) is defined by putting
$$b_{ij}(e_{i'}, e_{j'}) = \delta_i^{i'} \delta_j^{j'}$$

(here the $\delta$'s are the Kronecker symbols).

Denote by $\Gamma_{ij}^k$ the Christoffel symbols of $\nabla$ in bases $(e_1, \ldots, e_r)$ and $(\partial_1, \ldots, \partial_n)$ and let us introduce the slightly unusual notation $\widetilde{\Gamma}_{k(ij)}^{(lm)}$ for the Christoffel symbols of $\nabla^{\text{Bil}}$ in bases $\{b_{ij}\}$ and $(\partial_1, \ldots, \partial_n)$, with its obvious meaning.

**Exercise.** Express the Christoffel symbols $\widetilde{\Gamma}$ in terms of the Christoffel symbols $\Gamma$.

The result of the above exercise,
$$\widetilde{\Gamma}_{k(ij)}^{(lm)} = -\Gamma_{kl}^i \delta_m^j - \delta_l^i \Gamma_{km}^j ,$$
is valid locally for any fat manifold (*cf.* n. 2.1.5).

### 2.4.10 *Localization of Linear Connections*

Any connection can be restricted, or, better to say, localized, to an open submanifold. Essentially, this possibility is due to the fact that differential operators are *local*, *i.e.*, localizable to open subsets. Below it is shown how covariant derivatives are localized. This is the only we need for our purposes.

**Lemma.** Let $\bar{i} : \overline{N} \to \overline{M}$ be a fat embedding of an open fat submanifold, $X$ and $Y$ vector fields on $M$ and $\nabla$ a linear connection on $\overline{M}$. Then
$$X|_N = Y|_N \Longrightarrow \nabla_X|_N = \nabla_Y|_N .$$

***Proof.*** Let $Z = X - Y$, $s \in \Gamma(\overline{M})$ and $n \in N$. Consider a function $f \in C^\infty(M)$ such that
$$f(n) = 0 \text{ and } f|_{M-N} = 1$$
(see [Nestruev (2003), 4.17 (ii)]). Then $Z = fZ$ because $Z|_N = X|_N - Y|_N = 0$ and
$$\nabla_Z(s)(n) = \nabla_{fZ}(s)(n) = (f\nabla_Z(s))(n) = f(n)(\nabla_Z(s))(n) = 0 .$$
So, $\nabla_Z(s)|_N = 0$ for all $s \in P$, and, therefore, $\nabla_Z|_N = 0$.

Hence
$$\nabla_X|_N - \nabla_Y|_N = (\nabla_X - \nabla_Y)|_N = \nabla_{X-Y}|_N = \nabla_Z|_N = 0$$
and
$$\nabla_X|_N = \nabla_Y|_N . \qquad \square$$

**Proposition.** Let $\nabla$ be a linear connection on a fat manifold $\overline{M}$ and $\overline{N}$ an open fat submanifold of $\overline{M}$. Then there exists a unique linear connection $\nabla|_{\overline{N}}$ on $\overline{N}$ that is related with $\nabla$ through the fat embedding of $\overline{N}$ in $\overline{M}$.

**Proof.** Let $X \in \mathrm{D}(N)$ and for each $n \in N$ consider a vector field $X_n \in \mathrm{D}(M)$ such that $X|_{U_n} = X_n|_{U_n}$, with $U_n \subseteq N$ being a neighborhood of $n$. In view of the above lemma and locality of fat fields (see n. 1.3.6), $(\nabla|_{\overline{N}})_X$ can be defined as the fat field obtained by gluing together $\{\nabla_{X_n}|_{\overline{U_n}}\}_{n \in N}$ (see Definition 1.3.11). The easy verification that

$$X \mapsto (\nabla|_{\overline{N}})_X$$

gives the required connection is left to the reader. □

**Definition.** The linear connection $\nabla|_{\overline{N}}$ defined above is called the *localization of* $\nabla$ *to* $\overline{N}$, or also the *restriction of* $\nabla$ *to* $\overline{N}$.

### 2.4.11 Gluing Linear Connections

It is possible now to describe how to glue together a family of linear connections.

**Exercise.** Let $\overline{M}$ be a fat manifold and $\{U_i\}_{i \in \mathcal{I}}$ be an open covering of $M$. Let also $\nabla_i$ be a linear connection in the open fat submanifold $\overline{U_i}$ over $U_i$, $i \in \mathcal{I}$, and suppose that for all $i, j \in \mathcal{I}$ the restrictions of $\nabla_i$ and $\nabla_j$ to the open fat submanifold $\overline{U_i \cap U_j}$ over $U_i \cap U_j$ coincide. Show that there exists a unique linear connection $\nabla$ on $\overline{M}$ such that its restriction to $\overline{U_i}$ is $\nabla_i$ for all $i \in \mathcal{I}$.

### 2.4.12

**Exercise.** Show that every fat manifold possess linear connections.

**Hint.** Take into account n. 2.1.9 and use local triviality and a partition of unity.

## 2.5 Linear Connections and Inner Structures

When inner structures are considered, a natural notion of *structure-preserving* connections arises. Here we illustrate this fact by means of the complex and the orthogonal guiding examples (*cf.* Sect. 1.7).

## 2.5.1 Complex Linear Connections

Intuitively, a connection is structure-preserving if any parallel transport of one fat point to another is an identification of the corresponding structures they are supplied with. For instance, if such a structure is an endomorphism of the bundle $\overline{M} \to M$, say, $\varphi$, then the endomorphism $\varphi_{m_1}$ is identified with $\varphi_{m_2}$ each time when the fat point $\overline{m_1}$ is identified with $\overline{m_2}$ via a parallel transport isomorphism. This geometrically clear idea is, however, not very practical because it is not possible to check what happens along each single parallel transport. But it means, as it is not difficult to see, that the endomorphism $\varphi$ is 'constant' with respect to the considered connection $\nabla$, *i.e.*, that

$$\nabla_X^{\text{End}}(\varphi) = 0, \quad \forall X \in \mathrm{D}(M).$$

In such a case we say that $\nabla$ *preserves* $\varphi$.

A general algebraic paraphrase of the above said is the following.

**Definition.** Let $\nabla$ be a linear connection in an $A$-module $P$ and $\varphi \in$ End $P$. Then $\nabla$ is said to *preserve* $\varphi$ if $\nabla_X^{\text{End}}(\varphi) = 0$ for all $X \in \mathrm{D}(M)$.

In particular, this definition applies to a linear connection and a complex structure $J$ on a fat manifold $\overline{M}$. In this case the condition can be characterized in terms of symmetry group/algebra as follows.

**Proposition.** The following conditions are equivalent:

(1) $\nabla_X^{\text{End}}(J) = 0$ for all $X \in \mathrm{D}(M)$;
(2) $\overline{f} \in \mathrm{GL}(\overline{M}, J), X \in \mathrm{D}(M) \Rightarrow \nabla_X^{\text{End}}(\overline{f}^*) \in \mathfrak{gl}(\overline{M}, J)$;
(3) $\varphi \in \mathfrak{gl}(\overline{M}, J), X \in \mathrm{D}(M) \Rightarrow \nabla_X^{\text{End}}(\varphi) \in \mathfrak{gl}(\overline{M}, J)$.

**Proof.** The implications (1)$\Rightarrow$(3)$\Rightarrow$(2) are immediate. Assume now that (2) holds. The Jacobi identity for operators $\nabla_X, J$ and $\overline{f}^*$ shows that $\nabla_X^{\text{End}}(J)$ commutes with $\overline{f}^*$ for all $\overline{f} \in \mathrm{GL}(\overline{M}, J)$. Now the result of Exercise 1.7.5 implies that $\nabla_X^{\text{End}}(J) = a\,\mathrm{id} + bJ$. On the other hand, it follows from $J^2 = -\,\mathrm{id}$ that

$$\nabla_X^{\text{End}}(J) \circ J + J \circ \nabla_X^{\text{End}}(J) = 0,$$

which immediately leads to $a = b = 0$. $\square$

Note that $\nabla$ preserves $J$ if $\nabla_X$ is $\mathbb{C}$-linear for all $X$. This leads to an alternative interpretation of the condition in terms of the extended algebra $\hat{A}$, with $A = \mathrm{C}^\infty(M)$ (see n. 1.7.2). Namely, it is easy to see that derivations of this algebra are of the form $\hat{X}+\mathrm{i}\hat{Y}$ with $\hat{X},\hat{Y}$ being the extensions of $X,Y \in \mathrm{D}(M)$. Hence they form an $\hat{A}$-module $\mathrm{D}(M)_\mathbb{C}$ obtained from $\mathrm{D}(M)$ by the extension of scalars $A \to \hat{A}$, i.e., by *complexification*. This shows that

$$\hat{\nabla} \,:\, \hat{X}+\mathrm{i}\hat{Y} \mapsto \hat{\nabla}_{\hat{X}+\mathrm{i}\hat{Y}} \stackrel{\text{def}}{=} \nabla_X + J \circ \nabla_Y$$

is a connection in the $\hat{A}$-module $\Gamma(\overline{M})$ if and only if $\nabla$ preserves $J$.

### 2.5.2  Orthogonal Linear Connections

The general idea from the previous subsection applied to a bilinear form $b$ on an $A$-module $P$ is realized as follows.

**Definition.** A linear connection $\nabla$ in an $A$-module $P$ is said to *preserve* a bilinear form $b$ on $P$ if $\nabla^{\mathrm{Bil}}_X(b) = 0$ for all $X \in \mathrm{D}(M)$.

In more explicit terms, according to the definition of $\nabla^{\mathrm{Bil}}$, $\nabla$ preserves $b$ if

$$X\big(b(p_1,p_2)\big) = b\big(\nabla_X(p_1),p_2\big) + b\big(p_1,\nabla_X(p_2)\big), \quad X \in \mathrm{D}(A),\ p_1,p_2 \in P. \tag{2.1}$$

The notion of a linear connection on a fat manifold $\overline{M}$ that preserves an inner pseudo-metric $g$ on it is the particular case of this definition applied to the $C^\infty(M)$–module $\Gamma(\overline{M})$. It is not difficult to see that a linear connection that preserves $g$ is characterized by the fact that any parallel translation of a fat point $\overline{m}_1$ to a fat point $\overline{m}_2$ is a pseudo-orthogonal transformation of the pseudo-scalar product $g_{m_1}$ to $g_{m_2}$.

**Exercise.** Let $\nabla$ be a linear connection on an $A$-module $P$ that preserves a bilinear form $b$ on $P$. Moreover, let $\Delta$ be a linear connection on $P$ and set $\rho = \Delta - \nabla \in \Lambda^1(\mathrm{End}\,P)$ (cf. n. 2.1.9). Show that

(1) $\Delta$ preserves $b \iff \rho(X) \in \mathrm{o}(P,b),\ \forall X \in \mathrm{D}(A)$;
(2) $\varphi \in \mathrm{o}(P,b), X \in \mathrm{D}(A) \Rightarrow \nabla^{\mathrm{End}}_X(\varphi) \in \mathrm{o}(P,b)$;
(3) $R^\nabla(X,Y) \in \mathrm{o}(P,b),\ \forall X,Y \in \mathrm{D}(A)$;

**Hint.** For (2), calculate

$$X\big(g(\varphi(p_1),p_2) + g(p_1,\varphi(p_2))\big)$$
$$- X\big(g(\varphi(\nabla_X(p_1)),p_2) + g(\nabla_X(p_1),\varphi(p_2))\big)$$
$$- X\big(g(\varphi(p_1),\nabla_X(p_2)) + g(p_1,\varphi(\nabla_X(p_2)))\big)$$

by means of (2.1) and note that it vanish because $\varphi \in o(P, b)$. For (3), calculate $[X, Y](b(p_1, p_2))$, $X(Y(b(p_1, p_2)))$ and $Y(X(b(p_1, p_2)))$ by means of (2.1), and compare them.

Let $g$ be an inner pseudo-metric on a fat manifold $\overline{M}$. Now we are going to clarify relations between a linear connection $\nabla$ that preserves $g$ and inner symmetries of $g$. If $\overline{f}$ is a such one, then, by definition,

$$g(p_1, p_2) = g\left(\overline{f}^*(p_1), \overline{f}^*(p_2)\right), \quad p_1, p_2 \in \Gamma(\overline{M}), \quad (2.2)$$

and hence

$$X(g(p_1, p_2)) = X\left(g\left(\overline{f}^*(p_1), \overline{f}^*(p_2)\right)\right), \quad X \in D(M). \quad (2.3)$$

Recall that $\nabla$ preserves $g$ when

$$X(g(p_1, p_2)) = g(\nabla_X(p_1), p_2) + g(p_1, \nabla_X(p_2)) \quad (2.4)$$

for all $X \in D(M)$ and $p_1, p_2 \in \Gamma(\overline{M})$. It follows from (2.2), (2.3) and (2.4) that

$$g\left(\overline{f}^*(\nabla_X(p_1)), \overline{f}^*(p_2)\right) + g\left(\overline{f}^*(p_1), \overline{f}^*(\nabla_X(p_2))\right)$$
$$= g\left(\nabla_X(\overline{f}^*(p_1)), \overline{f}^*(p_2)\right) + g\left(\overline{f}^*(p_1), \nabla_X(\overline{f}^*(p_2))\right),$$

which gives

$$g\left(\nabla_X^{\text{End}}\left(\overline{f}^*\right)(p_1), \overline{f}^*(p_2)\right) + g\left(\overline{f}^*(p_1), \nabla_X^{\text{End}}\left(\overline{f}^*\right)(p_2)\right) = 0. \quad (2.5)$$

In view of (2.2) equality (2.5) tells that

$$\overline{f}^{*-1} \circ \nabla_X^{\text{End}}(\overline{f}^*) \in o(\overline{M}, g). \quad (2.6)$$

Thus

$$\overline{f} \in O(\overline{M}, g), X \in D(M) \Rightarrow \overline{f}^{*-1} \circ \nabla_X^{\text{End}}(\overline{f}^*) \in o(\overline{M}, g).$$

**Proposition.** Let $\nabla$ be a linear connection and $g$ an inner pseudo-metric on a fat manifold $\overline{M}$. Then the following conditions are equivalent:

(1) $\nabla_X^{\text{Bil}}(g) = \alpha g$, $\forall X \in D(M)$ with $\alpha \in C^\infty(M)$ depending on $X$;
(2) $\overline{f} \in O(\overline{M}, g), X \in D(M) \Rightarrow \overline{f}^{*-1} \circ \nabla_X^{\text{End}}(\overline{f}^*) \in o(\overline{M}, g)$;
(3) $\varphi \in o(\overline{M}, g), X \in D(M) \Rightarrow \nabla_X^{\text{End}}(\varphi) \in o(\overline{M}, g)$.

**Proof.** Assume that $\overline{f} \in \mathrm{O}(\overline{M}, g)$. Then (2.2) and (2.3) hold and (2.6) $\iff$ (2.5). Independently of validity of (2.4), (2.5) means that

$$X\big(g\,(p_1, p_2)\big) - g\big(\nabla_X (p_1), p_2\big) - g\big(p_1, \nabla_X (p_2)\big)$$
$$= X\left(g\big(\overline{f}^* (p_1), \overline{f}^* (p_2)\big)\right) - g\left(\nabla_X \big(\overline{f}^* (p_1)\big), \overline{f}^* (p_2)\right)$$
$$- g\left(\overline{f}^* (p_1), \nabla_X \big(\overline{f}^* (p_2)\big)\right),$$

or, equivalently, that

$$\nabla^{\mathrm{Bil}}_X(g)(p_1, p_2) = \nabla^{\mathrm{Bil}}_X(g)\big(\overline{f}^* (p_1), \overline{f}^* (p_2)\big), \quad X \in \mathrm{D}(M), p_1, p_2 \in \Gamma(\overline{M}).$$

This means that $\overline{f}^*$ is an inner symmetry of $\nabla^{\mathrm{Bil}}_X(g)$. Now, by applying Proposition 1.7.5 to $\nabla^{\mathrm{Bil}}_X(g)$ we conclude that (1) $\iff$ (2).

Similar computations show that if $\varphi$ is an infinitesimal symmetry of $g$, then $\nabla^{\mathrm{End}}_X(\varphi)$ is an infinitesimal symmetry of $g$ if and only if $\varphi$ is an infinitesimal symmetry of $\nabla^{\mathrm{Bil}}_X(g)$. According to n. 1.7.6, this implies that (1) $\iff$ (3). $\square$

### 2.5.3 Lie Connections

The question how to characterize gauge Lie algebras among $A$-Lie algebras, $A$ being a $k$-algebra, becomes natural in the light of the discussion in n. 1.7.8. It is rather obvious that a direct reference to principal bundles is not very practical. With the following notion this question can be resolved in a much simpler 'infinitesimal' manner.

**Definition.** Let $\mathfrak{g}$ be an $A$-Lie algebra, whose product is denoted by $\langle \cdot, \cdot \rangle$. A linear connection $\nabla$ in $\mathfrak{g}$ is a *Lie connection* if for all $X \in \mathrm{D}(A)$, $\nabla_X$ is a derivation of $\mathfrak{g}$, that is,

$$\nabla_X (\langle g_1, g_2 \rangle) = \langle \nabla_X (g_1), g_2 \rangle + \langle g_1, \nabla_X (g_2) \rangle.$$

It will be shown later (see n. 3.6.1) that a $C^\infty(M)$-Lie algebra $\mathfrak{g} = \Gamma(\overline{M})$, $M$ being a connected fat manifold, is a gauge algebra if it admits a Lie connection. This is a consequence of the fact that the parallel translation from $m_1$ to $m_2$, $m_1, m_2 \in M$, along a curve is a Lie algebra isomorphism of Lie algebra structures in $\overline{m_1}$ and $\overline{m_2}$.

**Exercise.** Show without using this fact that the $A$-Lie algebra of Example 1.7.9 does not admit a Lie connection.

### 2.5.4 Levi-Civita Connection

**Definition.** Let $M$ be a manifold. A linear connection in $D(M)$ is *torsion-free* if
$$\nabla_X(Y) - \nabla_Y(X) = [X, Y], \quad X, Y \in D(M).$$

**Proposition.** Let $(M, g)$ be a pseudo-Riemannian manifold. There exists exactly one torsion-free linear connection on $D(M)$ that preserves $g$.

*Proof.* If $\nabla$ is a linear connection that preserves $g$, for all $X, Y, Z \in D(M)$ we have
$$g(\nabla_X(Y), Z) + g(\nabla_X(Z), Y) = X(g(Y, Z)), \quad (2.7)$$
$$g(\nabla_Y(Z), X) + g(\nabla_Y(X), Z) = Y(g(Z, X)), \quad (2.8)$$
$$g(\nabla_Z(X), Y) + g(\nabla_Z(Y), X) = Z(g(X, Y)). \quad (2.9)$$
Moreover, if $\nabla$ is torsion-free we have
$$g(\nabla_X(Y), Z) - g(\nabla_Y(X), Z) = g([X, Y], Z), \quad (2.10)$$
$$g(\nabla_Y(Z), X) - g(\nabla_Z(Y), X) = g([Y, Z], X), \quad (2.11)$$
$$g(\nabla_Z(X), Y) - g(\nabla_X(Z), Y) = g([Z, X], Y). \quad (2.12)$$
If we sum (2.7), (2.8), (2.10), (2.12) and subtract (2.9), (2.11), we get
$$2g(\nabla_X(Y), Z)$$
$$= X(g(Y, Z)) + Y(g(Z, X)) - Z(g(X, Y))$$
$$+ g([X, Y], Z) - g([Y, Z], X) + g([Z, X], Y)$$
Now set
$$\omega_{X,Y} : D(M) \to \mathbb{R}, \quad Z \mapsto X(g(Y, Z)) + Y(g(Z, X)) - Z(g(X, Y))$$
$$+ g([X, Y], Z) - g([Y, Z], X) + g([Z, X], Y)$$
so that $\omega_{X,Y} \in \Lambda^1(M)$, and recall that, by definition of a nondegenerate form, the homomorphism $\varphi : D(M) \to \Lambda^1(M)$ induced by $g$ (that is, $\varphi(W)(Z) \overset{\text{def}}{=} g(W, Z)$, $W, Z \in D(M)$) is an isomorphism. This shows that if $\nabla$ is a torsion-free linear connection preserves $g$, then for all $X \in D(M)$, $\nabla_X$ must coincide with the map
$$\Box_X : D(M) \to D(M), \quad Y \mapsto \varphi^{-1}\left(\frac{1}{2}\omega_{X,Y}\right).$$
Therefore there is at most one such connection. On the other hand, it is straightforward to check that $\Box_X$ is a der-operator over $X$ and that $X \mapsto \Box_X$ is linear, which proves the existence. $\square$

The above proposition is often called fundamental theorem of Riemannian Geometry and the connection in the statement is called the *Levi-Civita connection* of the pseudo-Riemannian manifold $(M, g)$. Much of linear connection theory owe its motivations to this topic.

## Chapter 3

# Covariant Differential

The covariant differential of a connection is the determined by it universal covariant derivative. This predetermines importance of this notions. Moreover, it is, conceptually, very convenient to take it as definition of the original connection. According to this point of view, a connection in an $A$-module $P$ may be understood as a (graded) der-operator over the (graded) 'vector field' d in $\Lambda(A)$ in $\Lambda(A)$–module of $P$-valued differential forms. A gradual passage from the previous, geometrically transparent point of view to this new conceptual one is done in this chapter.

This change of the view-point facilitates construction of further, more delicate elements in the theory of connections. Two of them, namely, morphisms of connections, formalized in terms of compatibility, and gauge/fat structures, are discussed in this chapter with a special attention.

## 3.1 Fat de Rham Complexes

In this section some de Rham-like complexes naturally associated with a module/fat manifold are introduced. In doing that we imitate the standard descriptive definition of the exterior differential which is sufficient for our goals here. A conceptual approach requires more 'spacetime' and would deviate us from the main lines of this book.

### 3.1.1 *Fat forms*

Let $k$ be a field, $A$ a commutative $k$-algebra, $P$ an $A$-module and $s$ a nonnegative integer.

**Definition.** A *fat (differential) $s$-form on $P$* is an alternating $s$-linear over

A function
$$\mathrm{Der}(P) \times \cdots \times \mathrm{Der}(P) \longrightarrow P.$$

A *fat s-form* on a fat manifold $\overline{M}$ is a fat $s$-form on $\Gamma(\overline{M})$.

So, a fat $s$-form on $\overline{M}$ is an alternating $s$-linear over $\mathrm{C}^\infty(M)$ function
$$\overline{\mathrm{D}}\left(\overline{M}\right) \times \cdots \times \overline{\mathrm{D}}\left(\overline{M}\right) \longrightarrow \Gamma\left(\overline{M}\right)$$

The $A$-module of all fat $s$-forms on $P$ will be denoted by
$$\overline{\Lambda}^s(P)$$
and the graded $A$-module
$$\bigoplus_s \overline{\Lambda}^s(P)$$
by $\overline{\Lambda}^\bullet(P)$. For a fat manifold, *i.e.*, for $P = \Gamma(\overline{M})$, we shall use symbols
$$\overline{\Lambda}^s\left(\overline{M}\right) \quad \text{and} \quad \overline{\Lambda}^\bullet\left(\overline{M}\right),$$
respectively.

### 3.1.2  Semi-fat Forms

**Definition.** A *semi-fat (differential) s-form* on $P$ is an alternating $s$-linear over $A$ function
$$\mathrm{Der}(P) \times \cdots \times \mathrm{Der}(P) \longrightarrow A.$$

A *semi-fat s-form* on a fat manifold $\overline{M}$ is a semi-fat $s$-form on $\Gamma(\overline{M})$.

In other words, a semi-fat $s$-form on $\overline{M}$ is an alternating $s$-linear over $\mathrm{C}^\infty(M)$ function
$$\overline{\mathrm{D}}(\overline{M}) \times \cdots \times \overline{\mathrm{D}}(\overline{M}) \longrightarrow \mathrm{C}^\infty(M).$$

The module of semi-fat $s$-forms on $P$ will be denoted by $\underline{\overline{\Lambda}}^s(P; A)$. The module of semi-fat $s$-forms on a fat manifold $\overline{M}$ will be denoted simply by $\underline{\overline{\Lambda}}^s(M)$.

### 3.1.3

If $P$ is faithful, then every der-operator is over a unique derivation (see n. 1.2.5). In such a case the $A$-module homomorphism

$$\mathrm{Der}(P) \to \mathrm{D}(A), \quad \overline{X} \to X ,$$

is well-defined and in its turn induces an $A$-module homomorphism

$$\Lambda^s(A) \to \overline{\Lambda}^s(P;A) .$$

In other words, ordinary forms can be also interpreted as semi-fat forms. Namely, if $\omega \in \Lambda^s(A)$, then its semi-fat interpretation $\overline{\omega}$ is defined as

$$\overline{\omega}(\overline{X_1},\ldots,\overline{X_s}) = \omega(X_1,\ldots,X_s), \quad \overline{X_i} \in \mathrm{Der}(P).$$

For a fat manifold every vector field can be lifted to a fat vector field, *i.e.*, the above map of der-operators is surjective. By this reason the homomorphism

$$\Lambda^s(M) \to \overline{\Lambda}^s(M)$$

is, in fact, a monomorphism.

### 3.1.4 *Thickened Forms*

In order to distinguish between various types of differential forms we shall use the following terminology in the case the context requires that.

**Definition.** A $P$-valued $s$-form over $A$ (see Definition 0.5.1) will be called, synonymously, *thickened (differential) $s$-form on $P$*. A *thickened form* on a fat manifold $\overline{M}$ is a thickened form on $\Gamma(\overline{M})$.

So, a thickened $s$-form on $\overline{M}$ is an alternating $s$-linear over $\mathrm{C}^\infty(M)$ function

$$\mathrm{D}(M) \times \cdots \times \mathrm{D}(M) \longrightarrow \Gamma(\overline{M}) .$$

Recall that the $A$-module of all thickened $s$-forms is denoted by $\Lambda^s(P)$ and $\Lambda^\bullet(P) = \bigoplus_s \Lambda^s(P)$ (see Definition 0.5.1). For a fat manifold

$$\Lambda^s\left(\overline{M}\right) \quad \text{and} \quad \Lambda^\bullet\left(\overline{M}\right)$$

will be generally used instead of $\Lambda^s\left(\Gamma(\overline{M})\right)$ and $\Lambda^\bullet\left(\Gamma(\overline{M})\right)$, respectively.

## 3.1.5

By the same reason as in n. 3.1.3 thickened forms on a faithful module $P$ are interpreted naturally as fat forms on it, *i.e.*, a natural $A$-module homomorphism

$$\Lambda^s(P) \to \overline{\Lambda}^s(P)$$

is well-defined. In the case of fat manifolds this homomorphism, *i.e.*,

$$\Lambda^s(\overline{M}) \to \overline{\Lambda}^s(\overline{M}),$$

is injective.

### 3.1.6  Geometric Description

Differential forms introduced above may be seen as fields of suitable *geometric quantities* or, equivalently, as sections of suitable vector bundles. For fat forms it looks as follows.

Fix a fat point $\overline{m} \in \overline{M}$. According to Exercise 1.3.12 a fat $s$-form

$$\overline{\omega} : \overline{\mathrm{D}}(\overline{M}) \times \cdots \times \overline{\mathrm{D}}(\overline{M}) \longrightarrow \Gamma(\overline{M})$$

defines an alternating $s$-linear function of vector spaces

$$\overline{\omega}_m : \overline{T}_m\overline{M} \times \cdots \times \overline{T}_m\overline{M} \longrightarrow \overline{m}$$

such that

$$\overline{\omega}_m(\overline{X_{1m}}, \ldots, \overline{X_{sm}}) = \overline{\omega}(\overline{X_1}, \ldots, \overline{X_s})(m), \qquad \overline{X_1}, \ldots, \overline{X_s} \in \overline{\mathrm{D}}(\overline{M})$$

(see n. 0.1.5 (3)). The family $\{\overline{\omega}_m\}_{m \in M}$ determines, obviously, the form $\overline{\omega}$ which, by this reason, may be seen as a (smooth) field of such functions.

Similarly, a semi-fat $s$-form is geometrically described by a field of alternating $s$-linear functions of the form

$$\overline{T}_m\overline{M} \times \cdots \times \overline{T}_m\overline{M} \longrightarrow \mathbb{R}$$

and a thickened $s$-form by a field of alternating $s$-linear functions

$$T_mM \times \cdots \times T_mM \longrightarrow \overline{m}.$$

### 3.1.7 Wedge Products

Let $S_{r+s}$ be the permutations group of $\{1,\ldots,r+s\}$, with $r$, $s$ being nonnegative integers. Define $S_{r,s} \subseteq S_{r+s}$ by

$$S_{r,s} = \{\sigma \in S_{r+s} : \sigma(1) < \cdots < \sigma(r) \text{ and } \sigma(r+1) < \cdots < \sigma(r+s)\}.$$

Recall that the *wedge product* of an $r$-form $\omega \in \Lambda^r(A)$ and an $s$-form $\varkappa \in \Lambda^s(A)$ is defined by

$$(\omega \wedge \varkappa)(X_1,\ldots,X_{r+s})$$
$$= \sum_{\sigma \in S_{r,s}} (-1)^{|\sigma|} \omega\left(X_{\sigma(1)},\ldots,X_{\sigma(r)}\right) \varkappa\left(X_{\sigma(r+1)},\ldots,X_{\sigma(r+s)}\right),$$

$$X_1,\ldots,X_{r+s} \in \mathrm{D}(A)$$

(see n. 0.5.13).

If $\varkappa$ is a thickened $s$-form, then this formula defines a thickened $s+r$-form, called the *(external) wedge product* of $\omega \in \Lambda^r(A)$ and $\varkappa \in \Lambda^s(P)$, provided that the products in its right-hand side means the $A$-module multiplication in $P$. With respect to the so-defined wedge product $\Lambda^\bullet(P)$ becomes a graded $\Lambda^\bullet(A)$–module.

The above formula in which the arguments $X_i$'s from $\mathrm{D}(A)$ are replaced by der-operators from $\mathrm{Der}(P)$ defines two other wedge products, one between semi-fat forms, *i.e.*,

$$\Lambda^r(P;A) \times \Lambda^s(P;A) \longrightarrow \Lambda^{r+s}(P;A)$$

and another between semi-fat and fat forms, *i.e.*,

$$\Lambda^r(P;A) \times \overline{\Lambda}^s(P) \longrightarrow \overline{\Lambda}^{r+s}(P).$$

With these products $\overline{\Lambda}^\bullet(P;A)$ becomes a graded algebra while $\overline{\Lambda}^\bullet(P)$ a graded $\overline{\Lambda}^\bullet(P;A)$–module.

In the case when $P$ is faithful (as in the case of fat manifolds) further wedge products arise. First, the 'semi-fat' interpretation of ordinary differential forms given in n. 3.1.3, combined with already defined products leads to wedge products between ordinary forms and any other type of forms considered earlier. Schematically they are

$$\Lambda^r(A) \times \Lambda^s(P;A) \longrightarrow \Lambda^{r+s}(P;A), \quad \Lambda^r(A) \times \overline{\Lambda}^s(P) \longrightarrow \overline{\Lambda}^{r+s}(P).$$

Similarly, the 'fat' interpretation of thickened form given in n. 3.1.5 allows to define wedge products of the following type:

$$\Lambda^r(P;A) \times \Lambda^s(P) \longrightarrow \overline{\Lambda}^{r+s}(P).$$

We shall keep the notation $\wedge$ for all these wedge products.

The Leibnitz rule of Proposition 0.5.19 generalizes, in an obvious manner, to
$$\bar{\mathrm{i}}_X(\omega_r \wedge \overline{\varkappa}) = \mathrm{i}_X(\omega_r) \wedge \overline{\varkappa} + (-1)^r \omega_r \wedge \bar{\mathrm{i}}_X(\overline{\varkappa})$$
with $\omega_r \in \Lambda^r(A)$ and $\overline{\varkappa} \in \overline{\Lambda}^\bullet(P)$.

### 3.1.8 Exterior Fat and Semi-fat Differential

Recall that the degree $s$ component
$$\mathrm{d}_s : \Lambda^s(A) \to \Lambda^{s+1}(A)$$
of the *de Rham differential* is given by the formula
$$\mathrm{d}_s(\omega)(X_1, \ldots, X_{s+1}) = \sum_i (-1)^{i+1} X_i\left(\omega\left(X_1, \ldots, \widehat{X_i}, \ldots, X_{s+1}\right)\right)$$
$$+ \sum_{i<j} (-1)^{i+j} \omega\left([X_i, X_j], X_1, \ldots, \widehat{X_i}, \ldots, \widehat{X_j}, \ldots, X_{s+1}\right),$$
$$\omega \in \Lambda^s(A), X_1, \ldots, X_{s+1} \in \mathrm{D}(A)$$
(see Definition 0.5.11).

This formula still makes sense when $\omega$ in it is a fat $s$-form on $P$ and arguments $X_i$'s are replaced by der-operators in $P$. So interpreted it defines the $s$-th component of the *fat (exterior) differential*
$$\overline{\mathrm{d}}_s : \overline{\Lambda}^s(P) \to \overline{\Lambda}^{s+1}(P) \ .$$
Putting them together we obtain the exterior fat differential
$$\overline{\mathrm{d}} : \overline{\Lambda}^\bullet(P) \to \overline{\Lambda}^\bullet(P)$$
on $P$ (or *on* $\overline{M}$ in the case when $P = \Gamma(\overline{M})$).

Since the formulas defining both ordinary and fat differentials are formally identical, the same computation as for ordinary differential shows that
$$\overline{\mathrm{d}}^2 = 0$$
(*cf.* Proposition 0.5.12) and hence transforms $\overline{\Lambda}^\bullet(P)$ into a cochain complex called the *der-complex of $P$* (or *of $\overline{M}$* if $P = \Gamma(\overline{M})$).

Note that the zeroth degree component of $\overline{\mathrm{d}}$ acts in a particularly simple manner:
$$p \longmapsto \overline{\mathrm{d}}p : \overline{X} \mapsto \overline{X}(p)$$

and is called the *(basic) fat differential* of $P$.

Suppose now that $P$ is faithful. In this case a slight modification of the above formula allows to define the *semi-fat exterior differential* and the *semi-fat basic differential*:

$$d_s(\omega)(\overline{X_1}, \ldots, \overline{X_{s+1}}) = \sum_i (-1)^{i+1} X_i\left(\omega\left(\overline{X_1}, \ldots, \widehat{\overline{X_i}}, \ldots, \overline{X_{s+1}}\right)\right)$$
$$+ \sum_{i<j} (-1)^{i+j} \omega\left([\overline{X_i}, \overline{X_j}], \overline{X_1}, \ldots, \widehat{\overline{X_i}}, \ldots, \widehat{\overline{X_j}}, \ldots, \overline{X_{s+1}}\right)$$

with $\omega \in \overline{\Lambda}^s(P;A)$, $\overline{X_1}......\overline{X_{s+1}} \in \mathrm{Der}(P)$ and $X_i$ being the unique derivation ($P$ is faithful!) corresponding to the der-operator $\overline{X_i}$. As before one easily observes that the same formal computation as for ordinary differential d proves that the square of the semi-fat differential vanishes. This way $\Lambda^\bullet(P;A)$ becomes a complex called *semi-fat*. It is easy to see that the homomorphism of n. 3.1.3 sending ordinary differential forms to semi-fat ones is a cochain map.

In this context thickened forms make an exception, because there is no canonical way to supply them with a differential. In particular, the basic formula for the standard differential which was used so far cannot be adapted to thickened forms. The arising problem concerns terms in the first summation of it: an action of $\mathrm{D}(A)$ on $P$ is necessary in order to give a sense to them. Obviously, there is no natural action of this kind for a general module $P$. However, such an action can be constructed by associating with a derivation $X \in \mathrm{D}(A)$ a der-operator in $P$. In other words, it can be done with help of a connection in $P$. These considerations formalized properly lead to discover the concept of *covariant differential* associated to a linear connection as it is done in the next section. It appears to be an analogue of the differential in the above discussed complexes.

### 3.1.9 *Leibnitz Rule for Fat and Semi-fat Forms*

Both fat and semi-fat differentials introduced in the previous subsection are first order differential operators in both graded and ordinary senses. This is a consequence of the fact that they are subject to the following Leibnitz rules:

$$d(\omega_r \wedge \varkappa) = d\omega_r \wedge \varkappa + (-1)^r \omega_r \wedge d\varkappa, \quad \omega_r \in \Lambda^r(P;A), \varkappa \in \Lambda^\bullet(P;A),$$

$$\overline{d}(\omega_r \wedge \overline{\varkappa}) = d\omega_r \wedge \overline{\varkappa} + (-1)^r \omega_r \wedge \overline{d\varkappa}, \quad \omega_r \in \Lambda^r(P;A), \overline{\varkappa} \in \overline{\Lambda}^\bullet(P).$$

The cohomology of fat and semi-fat complexes are not widely known in spite of their importance. Essentially they were described in [Rubtsov (1980)].

In the rest of this chapter we shall focus on thickened forms, rather than on fat and semi-fat ones, in view of their natural relation with connections.

## 3.2 Covariant Differential

Let $A$ be as before a commutative $k$-algebra, with $k$ being a field.

### 3.2.1

With a linear connection $\nabla$ in an $A$-module $P$ the operator
$$\mathrm{d}_\nabla : P \to \Lambda^1(P), \qquad p \longmapsto \mathrm{d}_\nabla(p) : X \mapsto \nabla_X(p)$$
is naturally associated.

In the case when $P = A$ and $\nabla$ is the trivial connection $D$ the covariant differential $\mathrm{d}_D : A \to \Lambda^1(A)$ coincides with the standard d. This analogy suggests that the operator $\mathrm{d}_\nabla$ could be characterized by a suitable property. In fact, it is the following Leibnitz-like rule:

$$\mathrm{d}_\nabla(ap) = \mathrm{d}\,a \wedge p + a\,\mathrm{d}_\nabla(p), \quad a \in A, p \in P\,. \tag{3.1}$$

Indeed, if $X \in \mathrm{D}(A)$, then

$$\mathrm{d}_\nabla(ap)(X) = \nabla_X(ap) = X(a)p + a\nabla_X(p) = ((\mathrm{d}\,a)(X))\,(p) + a\,\mathrm{d}_\nabla(p)(X)$$
$$= (\mathrm{d}\,a \wedge p + a\,\mathrm{d}_\nabla(p))\,(X)\,.$$

If $\mathrm{D}(A)$ is finitely generated and projective, as in the case of fat manifolds, then
$$\Lambda^1(P) = \Lambda^1(A) \otimes P\,,$$
and the Leibnitz rule can be rewritten in the form
$$\mathrm{d}_\nabla(ap) = \mathrm{d}\,a \otimes p + a\,\mathrm{d}_\nabla(p), \quad a \in A, p \in P\,.$$

It follows from (3.1) that (in notation of n. 0.1.2) the commutator
$$[\mathrm{d}_\nabla, a]$$
is the $A$-module homomorphism $P \to \Lambda^1(P)$ that sends $p$ to $\mathrm{d}\,a \wedge p$. Therefore, $\mathrm{d}_\nabla$ is a differential operator of first order.

The operator $\mathrm{d}_\nabla$ is called the *covariant differential of* $\nabla$.

### 3.2.2

Let $\nabla$ be a linear connection in an $A$-module $P$, $\varphi : P \to P$ an endomorphism and $b : P \times P \to A$ a bilinear form.

**Example.** The connection $\nabla$ preserves $\varphi$ if and only if $d_{\nabla\text{End}}(\varphi) = 0$. The connection $\nabla$ preserves $b$ if and only if $d_{\nabla\text{Bil}}(b) = 0$.

### 3.2.3

It is important to note that a $k$-linear operator $\Delta : P \to \Lambda^1(P)$ satisfying (3.1) determines a unique linear connection $\nabla$ by associating with a derivation $X \in D(A)$ the operator $\nabla_X : P \to P$ defined by

$$p \mapsto \Delta(p)(X) \,.$$

Indeed, the linearity with respect to $X$ is obvious and the Leibnitz rule for $\Delta$ implies that the so-defined operator $\nabla_X$ is actually a der-operator over $X$.

Thus a linear connection might be defined as a $k$-linear operator

$$\Delta : P \to \Lambda^1(P)$$

satisfying the Leibnitz rule

$$\Delta(ap) = d\,a \wedge p + a\Delta(p), \quad a \in A, p \in P \,. \tag{3.2}$$

This point of view is less intuitive, but more preferable in many situations. An instance of that we shall see in connection with the notion of compatibility of linear connections.

### 3.2.4 *Splitting of First Order Jets*

In this subsection another useful characterization of the covariant differential is given.

Let $\overline{M}$ be a fat manifold, $A = C^\infty(M)$ and $P = \Gamma(\overline{M})$. Recall that the projective module $\mathcal{J}^1(P)$ of *first order jets*, together with the 1-jet operator $j_1 : P \to \mathcal{J}^1(P)$, represents the functor

$$Q \mapsto \text{Diff}_1(P, Q)$$

in the category of *geometric* $A$-modules (see [Nestruev (2003), 11.59–11.64]). We denote it also by $\mathcal{J}^1(\overline{M})$. This means that for each $\Delta \in \text{Diff}_1(P, Q)$ there exists a unique homomorphism $h_\Delta : \mathcal{J}^1(P) \to P$ such that $h_\Delta \circ j_1 = \Delta$.

According to n. 1.2.1, the symbol of a fist order differential operator $\Box : P \to Q$ connecting two arbitrary modules over a commutative algebra $A$ is identified with a derivation $A \to \mathrm{Hom}(P,Q)$ so that the exact sequence of $A$-modules

$$0 \to \mathrm{Diff}_0(P,Q) = \mathrm{Hom}(P,Q) \to \mathrm{Diff}_1(P,Q) \to \mathrm{D}(\mathrm{Hom}(P,Q))$$

holds. Note that in the considered situation all these modules are geometric. This means that the sequence of natural transformations of functors

$$0 \to \mathrm{Hom}(P,\cdot) \to \mathrm{Diff}_1(P,\cdot) \to \mathrm{D}(\mathrm{Hom}(P,\cdot))$$

is exact. Passing to the representing objects in the category of geometric modules and taking into account Proposition 0.5.10, one gets the exact sequence of $A$-homomorphisms

$$\Lambda^1(P) \stackrel{\mathrm{csmbl}_P}{\longrightarrow} \mathcal{J}^1(P) \stackrel{\pi_{1,0}}{\to} P \to 0 \,. \tag{3.3}$$

Here $\pi_{1,0} = h_{\mathrm{id}_P} : aj_1(p) \mapsto ap$ is the restriction of order of jets homomorphism and $\mathrm{csmbl}_P$ is called the *cosymbol map*. Explicitly,

$$(h_\Delta \circ \mathrm{csmbl}_P)(\mathrm{d}\, a \otimes p) = [\Delta](ap) = \Delta(ap) - a\Delta(p),$$
$$\Delta \in \mathrm{Diff}_1(P,Q), a \in A, p \in P \quad (3.4)$$

(with geometric $Q$) and hence

$$\mathrm{csmbl}_P(\mathrm{d}\, a \otimes p) = [j_1](a)(p) = j_1(ap) - aj_1(p), \quad a \in A, p \in P \,.$$

Note that for $P = A$

$$h_{\mathrm{d}} \circ \mathrm{csmbl}_A = \mathrm{id}_{\Lambda^1(A)} \,,$$

so that a natural splitting $\mathcal{J}^1(A) = \mathrm{Ker}\, h_{\mathrm{d}} \oplus \Lambda^1(A)$ is defined.

On the contrary, for a generic module $P$ there is no natural choice of a *splitting* homomorphism $\mathcal{J}^1(P) \to \Lambda^1(P)$, i.e., such that the composition

$$\Lambda^1(P) \stackrel{\mathrm{csmbl}}{\to} \mathcal{J}^1(P) \to \Lambda^1(P)$$

is the identity homomorphism. In view of (3.4), a choice of splitting homomorphism is equivalent to a choice of operator $\Delta : P \to \Lambda^1(P)$ that satisfies (3.2), i.e., $\Delta = \mathrm{d}_\nabla$ for a linear connection $\nabla$. In other words, a linear connection in $\overline{M}$ may be alternatively defined as a splitting of $\mathcal{J}^1(\overline{M})$ by means of the cosymbol map. Note that the natural splitting of $\mathcal{J}^1(A)$ considered above corresponds to the trivial connection in $A$.

Since sequence (3.3) is exact the second direct summand of the above splitting is isomorphic to $P$. More exactly, put

$$i_\nabla \stackrel{\mathrm{def}}{=} j_1 - \mathrm{csmbl}_P \circ \mathrm{d}_\nabla : P \to \mathcal{J}^1(P) \,.$$

Then $i_\nabla$ is an $A$-module homomorphism. Indeed, by definition of $\mathrm{csmbl}_P$, we have

$$i_\nabla(ap) = j_1(ap) - \mathrm{csmbl}_P(\mathrm{d}_\nabla(ap)) = j_1(ap) - \mathrm{csmbl}_P(\mathrm{d}\, a \otimes p + a\, \mathrm{d}_\nabla(p))$$
$$= a j_1(p) - a\, \mathrm{csmbl}_P(\mathrm{d}_\nabla(p)) = a i_\nabla(p)\,.$$

On the other hand, $\pi_{1,0} \circ i_\nabla = \mathrm{id}_P$, because $\pi_{1,0} \circ j_1 = \mathrm{id}_P$ and $\pi_{1,0} \circ \mathrm{csmbl}_P = 0$.

Thus

$$\mathcal{J}^1(P) = \mathrm{im}(i_\nabla) \oplus \mathrm{im}(\mathrm{csmbl}_P)$$

with $\mathrm{Im}(i_\nabla)$ and $\mathrm{Im}(\mathrm{csmbl}_P)$ isomorphic to $P$ and $\Lambda^1(P)$, respectively, as required.

Moreover, from Exercise 2.4.12 it follows immediately that the natural (connection-independent) sequence

$$0 \to \Lambda^1(\overline{M}) \xrightarrow{\mathrm{csmbl}} \mathcal{J}^1(\overline{M}) \xrightarrow{\pi_{1,0}} \Gamma(\overline{M}) \to 0$$

is exact. However, since $P$ is projective, this fact is easily deduced from the exactness of the similar sequence for $P = A$.

The above said leads to an alternative definition of a linear connection in $\overline{M}$, i.e., as a homomorphism $i_\nabla : \Gamma(\overline{M}) \to \mathcal{J}^1(\overline{M})$ such that $\pi_{1,0} \circ i_\nabla = \mathrm{id}_P$.

It is worth mentioning that the above considerations remain valid for modules $P$ over arbitrary commutative algebra, admitting a linear connection, if $\mathcal{J}^1(P)$ is the representing object of the functor $Q \mapsto \mathrm{Diff}_1(P, Q)$ in a 'good' (=*differentially closed*) category of $A$-modules to which $P$ belongs.

### 3.2.5 Dioles

By comparing Leibnitz rule (3.1) for covariant derivative and that of a der-operator

$$\square\,(ap) = X(a)\, p + a\square\,(p)$$

(see (1.4), p. 100) one can see that they are essentially the same. Namely, the first of them is obtained from the second by replacing in it the derivation $X : A \to A$ by the derivation $\mathrm{d} : A \to \Lambda^1(A)$. This suggests to interpret the covariant differential as a der-operator over d. In other words, the covariant differential may be seen as a *fat derivation* over the ordinary derivation d. This and other similar facts show that it is both natural and useful to generalize in a due manner the notion of a der-operator. This is done below.

Let $A$ be a commutative $k$-algebra, $k$ being a field. A *diole* over an $A$-module $P$ is a pair $(Q, R)$ of $A$-modules together with an $A$-module homomorphism
$$\varphi : Q \otimes P \to R .$$
Given a diole $(Q, R, \varphi)$ a *der-operator*
$$\square : P \to R$$
over a derivation
$$D : A \to Q$$
is a $k$-linear operator such that
$$\square(ap) = \varphi(D(a) \otimes p) + a\square(p) .$$
For instance, the pair $(\Lambda^1(A), \Lambda^1(P))$ together with the homomorphism
$$\omega \otimes p \mapsto \omega \wedge p ,$$
forms a diole and this way the covariant differential $d_\nabla$ appears to be a der-operator over d.

Note that the diole related to ordinary der-operators is the pair $(A, P)$ together with the identity homomorphism $\mathrm{id}_P : P = A \otimes P \to P$. More generally, let $\varphi : A \to B$ be a homomorphism of commutative algebras and $Q$ be a $B$-module considered also as an $A$-module via $\varphi$. The diole associated with a homomorphism of $A$-modules $\overline{\varphi} : P \to Q$ is composed of the pair $(B, Q)$ together with the $B$-module homomorphism $B \otimes_A P \to Q$ corresponding to $\overline{\varphi}$ according to the universal property of scalar extensions. This way we recover der-operators along $\overline{\varphi}$ (see n. 1.2.11).

### 3.2.6

Similarly to the ordinary differential, the covariant differential extends naturally to all thickened forms:
$$(d_\nabla)_s : \Lambda^s(P) \to \Lambda^{s+1}(P)$$
is defined by
$$(d_\nabla)_s(\overline{\omega})(X_1, \ldots, X_{s+1}) = \sum_i (-1)^{i+1} \nabla_{X_i}(\overline{\omega}(X_1, \ldots, \widehat{X_i}, \ldots, X_{s+1}))$$
$$+ \sum_{i<j} (-1)^{i+j} \overline{\omega}([X_i, X_j], X_1, \ldots, \widehat{X_i}, \ldots, \widehat{X_j}, \ldots, X_{s+1}),$$
$$X_1, \ldots, X_{s+1} \in \mathrm{D}(A), \ \overline{\omega} \in \Lambda^s(P) .$$

**Definition.** The *covariant differential associated with* $\nabla$ is the first degree graded $k$-module endomorphism

$$d_\nabla : \Lambda^\bullet(P) \to \Lambda^\bullet(P)$$

whose component of degree $s$ is $(d_\nabla)_s$.

It is easy to see that $(d_\nabla)_s$ is well-defined, *i.e.*, that $(d_\nabla)_s(\overline{\omega})$ is really a thickened $s+1$–form.

In the sequel a homogeneous component $(d_\nabla)_s$ will be often denoted simply by $d_\nabla$. The component of degree zero is sometimes called the *basic covariant differential*.

### 3.2.7

Let $X \in D(A)$ and

$$\bar{i}_X : \Lambda^1(P) \to P$$

be the degree 0 component of the insertion operator of $X$ into $\Lambda^\bullet(P)$, *i.e.*, the map given by

$$\overline{\omega} \longmapsto \overline{\omega}(X).$$

Then, obviously,

$$\bar{i}_X \circ d_\nabla = \nabla_X.$$

Also it is worth noticing that

$$p \text{ is constant with respect to } \nabla \iff d_\nabla(p) = 0.$$

### 3.2.8

Let us now consider the sequence

$$P \xrightarrow{d_\nabla} \Lambda^1(P) \xrightarrow{d_\nabla} \Lambda^2(P) \xrightarrow{d_\nabla} \ldots \xrightarrow{d_\nabla} \Lambda^s(P) \xrightarrow{d_\nabla} \ldots .$$

It is natural to ask if it is a complex, *i.e.*, if $d_\nabla^2 = 0$. In the case when $\nabla$ is a trivial connection in $P = A$, $d_\nabla$ coincides with the standard differential and, therefore, the answer is positive (see Proposition 0.5.12). However, in the general case the answer is negative as one can see by computing directly $d_\nabla^2$. This computation is practically identical to that of $d^2$ and is obtained formally from the latter by replacing the terms $X(\cdot)$ in it by $\nabla_X(\cdot)$. The difference between these two cases is that at the end expressions $\left[\nabla_{X_i}, \nabla_{X_j}\right] - \nabla_{[X_i,X_j]} = R^{\nabla}(X_i, X_j)$ emerge in place of $[X_i, X_j] - [X_i, X_j] = 0$. The final result is:

$$d_{\nabla}^2(\overline{\omega})(X_1, \ldots, X_s)$$
$$= \sum_{i<j}(-1)^{i+j-1} R^{\nabla}(X_i, X_j)\left[\overline{\omega}\left(X_1, \ldots, \widehat{X_i}, \ldots, \widehat{X_j}, \ldots, X_s\right)\right].$$

The above formula can be expressed in a much more expressive way by means of a further useful generalization of the wedge product introduced below.

### 3.2.9 Multiplication by End(P)–valued Forms

**Definition.** Let $P$ be an $A$-module and

$$R : D(A) \times \cdots \times D(A) \to \text{End}(P)$$

a differential $r$-form over $A$ with values in End $(P)$. The *wedge product*

$$R \wedge \overline{\omega} \in \Lambda^{r+s}(P)$$

of $R$ and a form $\overline{\omega} \in \Lambda^s(P)$ is defined by the formula

$$(R \wedge \overline{\omega})(X_1, \ldots, X_{r+s})$$
$$= \sum_{\sigma \in S_{r,s}} (-1)^{|\sigma|} R\left(X_{\sigma(1)}, \ldots, X_{\sigma(r)}\right)\left(\overline{\omega}\left(X_{\sigma(r+1)}, \ldots, X_{\sigma(r+s)}\right)\right),$$

$$X_1, \ldots, X_{r+s} \in D(A).$$

Here $S_{r,s}$ stands for the set of permutations of $\{1, \ldots, r+s\}$ such that
$S_{r,s} = \{\sigma \in S_{r+s} : \sigma(1) < \cdots < \sigma(r) \text{ and } \sigma(r+1) < \cdots < \sigma(r+s)\}$.
Similarly, the wedge product

$$\Lambda^r(\text{End}(P)) \times \Lambda^s(\text{End}(P)) \longrightarrow \Lambda^{r+s}(\text{End}(P))$$

is defined. We shall keep the notation $\wedge$ for these wedge products too.

Because of the similarity of these definitions with that of the wedge product in $\Lambda^{\bullet}(A)$, the usual properties of it have some counterparts in this context (with some caution about the commutation rules), and the proofs are basically the same. For instance, $\Lambda^{\bullet}(\text{End}(P))$ is a graded associative algebra and $\Lambda^{\bullet}(P)$ becomes a graded $\Lambda^{\bullet}(\text{End}(P))$–module. Moreover, take into account that the homomorphism $A \to \text{End}(P)$, that with each $a \in A$ assigns the multiplication by $a$ operator, obviously induces a homomorphism $\Lambda^{\bullet}(A) \to \Lambda^{\bullet}(\text{End}(P))$. Then the $\Lambda^{\bullet}(A)$–module structure on $\Lambda^{\bullet}(P)$

obtained from the $\Lambda^\bullet\,(\mathrm{End}(P))$–module structure coincides with the usual one. Finally, if $R \in \Lambda^r\,(\mathrm{End}(P))$ and $\omega$ is the image in $\Lambda^s\,(\mathrm{End}(P))$ of some form in $\Lambda^s(A)$, then

$$R \wedge \omega = (-1)^{rs} \omega \wedge R\,.$$

### 3.2.10  The Square of the Covariant Differential

According to Proposition 2.3.2 the curvature $R$ of a linear connection $\nabla$ in $P$ is a 2-form of $A$ with values in the $A$-module $\mathrm{End}\,(P)$, i.e.,

$$R \in \Lambda^2(\mathrm{End}\,(P))$$

and its wedge product with a form $\overline{\omega} \in \Lambda^{s-2}\,(P)$ (see Definition 3.2.9) is

$$(R \wedge \overline{\omega})\,(X_1,\ldots,X_s)$$
$$= \sum_{i<j} (-1)^{i+j-1} R\,(X_i,X_j)\left[\overline{\omega}\left(X_1,\ldots,\widehat{X_i},\ldots,\widehat{X_j},\ldots,X_s\right)\right],$$
$$X_1,\ldots,X_s \in \mathrm{D}(A)\,.$$

**Proposition.** Let $\nabla$ be a linear connection in an $A$-module $P$. Then

$$\mathrm{d}_\nabla^2\,(\overline{\omega}) = R^\nabla \wedge \overline{\omega}, \qquad \overline{\omega} \in \Lambda^\bullet(P)\,.$$

**Proof.**  Immediately from the above said and n. 3.2.8.  $\square$

**Corollary.** The sequence

$$P \xrightarrow{\mathrm{d}_\nabla} \Lambda^1\,(P) \xrightarrow{\mathrm{d}_\nabla} \cdots \xrightarrow{\mathrm{d}_\nabla} \Lambda^s\,(P) \xrightarrow{\mathrm{d}_\nabla} \cdots$$

is a cochain complex if and only if $\nabla$ is flat.

### 3.2.11  Leibnitz Rule for the Covariant Differential

**Proposition.** Let $\nabla$ be a linear connection in an $A$-module $P$. Then

$$\mathrm{d}_\nabla\,(\omega_r \wedge \overline{\varkappa}) = \mathrm{d}\,\omega_r \wedge \overline{\varkappa} + (-1)^r\,\omega_r \wedge \mathrm{d}_\nabla \overline{\varkappa}, \qquad \omega_r \in \Lambda^r\,(A),\ \overline{\varkappa} \in \Lambda^\bullet\,(P)\,.$$

**Proof.** A direct computation, practically identical to that for the standard wedge product (see Proposition 0.5.14).  $\square$

This Leibnitz rule shows the covariant differential $\mathrm{d}_\nabla$ to be a graded der-operator over the graded derivation $\mathrm{d}$.

**Corollary.** Each component of the covariant differential $d_\nabla$ is a first order differential operator as well as $d_\nabla$ itself.

**Proof.** In view of the above proposition the commutator $[d_\nabla, a]$, $a \in A$, is the (wedge) multiplication by $d\, a$ operator and, therefore, is a homomorphism of $A$-modules, *i.e.*, a zeroth order differential operator. □

### 3.2.12 *Generalization to* $\mathrm{End}(P)$*-valued Forms*

**Proposition.** Let $\nabla$ be a linear connection in an $A$-module $P$ and $T \in \Lambda^r(\mathrm{End}(P))$, $\overline{\omega} \in \Lambda^s(P)$. Then
$$d_\nabla(T \wedge \overline{\omega}) = d_{\nabla^{\mathrm{End}}}(T) \wedge \overline{\omega} + (-1)^r T \wedge d_\nabla(\overline{\omega}) .$$

**Proof.** It follows from the definition of $\nabla^{\mathrm{End}}$ (see Definition 2.4.3) that
$$\nabla_Z \circ T(X_1, \ldots, X_r) = \nabla_Z^{\mathrm{End}}(T(X_1, \ldots, X_r)) + T(X_1, \ldots, X_r) \circ \nabla_Z ,$$
with $X_1, \ldots, X_r \in \mathrm{D}(A)$. Hence
$$\nabla_Z(T(X_1, \ldots, X_r)(\overline{\omega}(Y_1, \ldots, Y_s)))$$
$$= \nabla_Z^{\mathrm{End}}(T(X_1, \ldots, X_r))(\overline{\omega}(Y_1, \ldots, Y_s))$$
$$+ T(X_1, \ldots, X_r)(\nabla_Z(\overline{\omega}(Y_1, \ldots, Y_s))) ,$$
with $X_1, \ldots, X_r, Y_1, \ldots, Y_s, Z \in \mathrm{D}(A)$. The last formula is an exact analogue of
$$\nabla_Z(\omega(X_1, \ldots, X_r)\overline{\varkappa}(Y_1, \ldots, Y_s))$$
$$= Z(\omega(X_1, \ldots, X_r))\overline{\varkappa}(Y_1, \ldots, Y_s)$$
$$+ \omega(X_1, \ldots, X_r)\nabla_Z(\overline{\varkappa}(Y_1, \ldots, Y_s)) ,$$
for $\omega \in \Lambda^r(A)$ and $\overline{\varkappa} \in \Lambda^s(P)$. This analogy shows that literally the same computation as in Proposition 3.2.11 (or in Proposition 0.5.14) proves the formula. □

### 3.2.13

**Proposition.** If $\omega \in \Lambda^\bullet(P)$, then
$$d_\nabla(R \wedge \omega) = R \wedge d_\nabla(\omega)$$
with $R = R^\nabla$.

**Proof.** In view of Proposition 3.2.10 we have
$$d_\nabla(R \wedge \omega) = d_\nabla(d_\nabla^2(\omega)) = d_\nabla^2(d_\nabla(\omega)) = R \wedge d_\nabla(\omega) . \qquad \square$$

### 3.2.14 Bianchi Identity

The above Proposition seems to contradict Proposition 3.2.12. But the fact is that

$$d_{\nabla^{\mathrm{End}}}(R) \wedge \overline{\omega} = 0, \qquad \omega \in \Lambda^{\bullet}(P).$$

This formula leads to the following fundamental result, called *Bianchi identity*.

**Proposition.** Let $\nabla$ be a linear connection in an $A$-module $P$. Then

$$d_{\nabla^{\mathrm{End}}}(R^{\nabla}) = 0.$$

**Proof.** Observe that for $\overline{\omega} = p \in P = \Lambda^0(P)$ and $S \in \Lambda^3(\mathrm{End}(P))$

$$(S \wedge p)(X, Y, Z) = S(X, Y, Z)(p)$$

with $X, Y, Z \in \mathrm{D}(A)$. □

## 3.3 Compatible Linear Connections

Many areas in the modern Differential Geometry are organized into categories naturally related with the basic category of all manifolds. For instance, one of them is formed of fat manifolds of a given type. So, it is natural to expect the same in what concerns the category of fat manifolds and one of the first questions is whether linear connections could be organized in a reasonable category. What are objects of such a category is absolutely clear. These should be pairs of the form $(\overline{M}, \nabla)$ with $\nabla$ being a connection in $\overline{M}$. But the question what are morphisms in this suspected category is not very easy. For instance, it could seem natural to define a morphism

$$(\overline{N}, \Delta) \to (\overline{M}, \nabla)$$

as a fat map $\overline{f}$ such that $\Delta$ and $\nabla$ are $\overline{f}$-related. But it turns out that the composition of two such 'morphisms' fails, generally, to be a 'morphism'.

Another way to approach this problem is to pass to the algebraic counterparts $(\Lambda^{\bullet}(\overline{M}), d_{\nabla})$ and $(\Lambda^{\bullet}(\overline{N}), d_{\Delta})$ and then look for morphisms of this new objects.

A category in which these new objects, called *cd-modules*, live is discussed below. As a result we shall discover the fundamental concept of *compatible linear connections*.

### 3.3.1 cd-Algebras and cd-Modules

We start with the necessary definitions. In this section $k$ stands for a field, $\mathcal{A}$ for a graded commutative $k$-algebra (see n. 0.1.1) the product in which is denoted by $\wedge$.

**Definition.** A graded commutative algebra $\mathcal{A}$ supplied with a graded derivation $\mathrm{d} : \mathcal{A} \to \mathcal{A}$ is called a *cd-algebra*.

### 3.3.2

We need also graded $\mathcal{A}$-modules. A typical one will be denoted by $\mathcal{P}$ and we shall continue to use '$\wedge$' for the module multiplication in it.

**Definition.** A graded $\mathcal{A}$-module $\mathcal{P}$ supplied with a graded $k$-module endomorphism

$$\overline{\mathrm{d}} : \mathcal{P} \to \mathcal{P}$$

of the same degree as $\mathrm{d} : \mathcal{A} \to \mathcal{A}$, say $l$, such that

$$\overline{\mathrm{d}}\left(\omega_r \wedge \overline{\varkappa}\right) = \mathrm{d}\,\omega_r \wedge \overline{\varkappa} + (-1)^{lr} \omega_r \wedge \overline{\mathrm{d}}\overline{\varkappa}, \quad \omega_r \in \mathcal{A}_r, \overline{\varkappa} \in \mathcal{P},$$

is called a *cd-module over* $(\mathcal{A}, \mathrm{d})$ (or simply *over $\mathcal{A}$*). A cd-module is called *flat* if $\overline{\mathrm{d}}^2 = 0$.

### 3.3.3

Let now $(\mathcal{P}, \overline{\mathrm{d}})$ and $\left(\mathcal{P}', \overline{\mathrm{d}}'\right)$ be cd-modules over the same cd-algebra $\mathcal{A}$ and let $l$ be the degree of the differentials.

**Definition.** A degree $r$ *cd-module homomorphism*

$$\varphi : \mathcal{P} \to \mathcal{P}'$$

is a degree $r$ $\mathcal{A}$-module homomorphism such that

$$\overline{\mathrm{d}}' \circ \varphi = (-1)^{lr} \varphi \circ \overline{\mathrm{d}}.$$

### 3.3.4

Let $(\mathcal{P}, \overline{\mathrm{d}})$ and $(\mathcal{P}', \overline{\mathrm{d}}')$ be cd-modules over a cd-algebra $\mathcal{A}$ and suppose that both $\overline{\mathrm{d}}$ and $\overline{\mathrm{d}}'$ are of degree 1. Then a zero degree $\mathcal{A}$-module homomorphism

$\varphi : \mathcal{P} \to \mathcal{P}'$ is a cd-module homomorphism if and only if the diagram

$$\begin{array}{ccccccccc} P_0 & \xrightarrow{\overline{d}} & P_1 & \xrightarrow{\overline{d}} & P_2 & \xrightarrow{\overline{d}} & \cdots & \xrightarrow{\overline{d}} & P_s & \xrightarrow{\overline{d}} & \cdots \\ \varphi \downarrow & & \varphi \downarrow & & \varphi \downarrow & & & & \varphi \downarrow & & \\ Q_0 & \xrightarrow{\overline{d}'} & Q_1 & \xrightarrow{\overline{d}'} & Q_2 & \xrightarrow{\overline{d}'} & \cdots & \xrightarrow{\overline{d}'} & Q_s & \xrightarrow{\overline{d}'} & \cdots \end{array}$$

is commutative.

In the case when the cd-modules are flat, a cd-homomorphism is, in particular, a cochain map.

### 3.3.5

**Example.** The algebra $\Lambda^\bullet(A)$ together with the exterior differential is a cd-algebra (see Proposition 0.5.14). If $\nabla$ is a linear connection on an $A$-module $P$, then $(\Lambda^\bullet(P), d_\nabla)$ is a cd-module over $\Lambda^\bullet(A)$ (by n. 3.1.7 and Proposition 3.2.11).

### 3.3.6

**Example.** The algebra $(\Lambda^\bullet(P;A), d)$ of semi-fat forms over a faithful module $P$ (see n. 3.1.7) supplied with the exterior semi-fat differential (see n. 3.1.8) is a cd-algebra and the module $\left(\overline{\Lambda}^\bullet(P), \overline{d}\right)$ of fat forms supplied with the exterior fat differential is a cd-module over it.

### 3.3.7 Linear Connections and cd-Module Structures on Thickened Forms

**Proposition.** Let $P$ be an $A$-module. If $\Lambda^\bullet(P)$ is generated as $\Lambda^\bullet(A)$-module by $P$, its zeroth degree component, then

$$\nabla \leftrightarrow (\Lambda^\bullet(P), d_\nabla)$$

is a bijection between linear connections in $P$ and cd-module structures $(\Lambda^\bullet(P), \Delta)$ over $(\Lambda^\bullet(A), d)$ with $\Delta$ of degree 1.

*Proof.* The only fact that requires a proof is that for a given cd-module $(\Lambda^\bullet(P), \Delta)$ over $(\Lambda^\bullet(A), d)$ with $\Delta$ being of degree 1 there exists a unique linear connection $\nabla$ in $P$ such that $\Delta = d_\nabla$. To this end, first, note that by definition of cd-modules the zeroth component of $\Delta$ satisfies the Leibnitz rule (3.1), p. 178. By this reason (see n. 3.2.3) there exists a unique linear connection $\nabla$ in $P$ such that the zero degree component of $d_\nabla$ coincides with that of $\Delta$. Now it remains to prove that all other components of $\Delta$ and $d_\nabla$ coincide as well.

Since $\Lambda^\bullet(P)$ is generated as $\Lambda^\bullet(A)$–module by $P$, it suffices to verify that on elements of the form $\omega \wedge p$, $\omega \in \Lambda^\bullet(A)$ homogeneous, $p \in P$ only. But this is obvious:

$$d_\nabla(\omega \wedge p) = d(\omega) \wedge p + (-1)^s \omega \wedge d_\nabla(p) = d(\omega) \wedge p + (-1)^s \omega \wedge \Delta(p) = \Delta(\omega \wedge p)$$

with $s = \deg \omega$. $\square$

The hypothesis of the above propositions is satisfied if $D(A)$ is projective and finitely generated as, in particular, in the case of fat manifolds. In fact, in this case

$$\Lambda^\bullet(P) = \Lambda^\bullet(A) \otimes P$$

(see Proposition 0.5.10) and, therefore, $\Lambda^\bullet(P)$ is manifestly generated as $\Lambda^\bullet(A)$–module by $P$.

Thus linear connections on a fat manifold $\overline{M}$ are identified with (degree 1) cd-module structures in the $\Lambda^\bullet(M)$–module of thickened forms $\Lambda^\bullet(\overline{M})$.

### 3.3.8 *Homomorphisms of Thickened Forms Induced by Fat Maps*

Consider a fat map

$$\overline{f} : \overline{N} \to \overline{M}$$

and set

$$A = C^\infty(M), \quad B = C^\infty(N), \quad P = \Gamma(\overline{M}), \quad Q = \Gamma(\overline{N}).$$

Recall that $\overline{f}$ generates the homomorphism of $A$-modules

$$\overline{f}^* : P \to Q$$

and a natural identification

$$Q = B \otimes_A P$$

(see n. 1.1.6) takes place.

The identifications

$$\Lambda^\bullet(P) = \Lambda^\bullet(A) \otimes_A P \quad \text{and} \quad \Lambda^\bullet(Q) = \Lambda^\bullet(B) \otimes_B Q.$$

are guaranteed by Proposition 0.5.10 while the identification

$$\Lambda^\bullet(B) \otimes_B Q = \Lambda^\bullet(B) \otimes_A P$$

comes from elementary properties of scalar extensions.

Hence we have
$$\Lambda^\bullet(Q) = \Lambda^\bullet(B) \otimes_A P.$$

The induced by $f$ homomorphism of the de Rham complexes (see n. 0.5.16)
$$\Lambda^\bullet f^* : \Lambda^\bullet(A) \to \Lambda^\bullet(B)$$
extends naturally to thickened forms
$$\Lambda^\bullet \overline{f}^* \stackrel{\text{def}}{=} \Lambda^\bullet f^* \otimes_A \mathrm{id}_P : \Lambda^\bullet(A) \otimes_A P \to \Lambda^\bullet(B) \otimes_A P$$
$$\| \qquad\qquad \|$$
$$\Lambda^\bullet(\overline{M}) \qquad \Lambda^\bullet(\overline{N})$$

It is easy to see that $\Lambda^\bullet \overline{f}^*$ is a zero degree homomorphism of $\Lambda^\bullet(M)$–modules with respect of the (left) $\Lambda^\bullet(M)$–module structure in $\Lambda^\bullet(\overline{N})$ induced by the restriction of scalars via $\Lambda^\bullet f^*$.

**Definition.** The homomorphism
$$\Lambda^\bullet \overline{f}^* : \Lambda^\bullet(\overline{M}) \to \Lambda^\bullet(\overline{N})$$
of the (graded) $\Lambda^\bullet(M)$–modules will be called *induced by* $\overline{f}$. Its $s$-th degree component
$$\Lambda^s(\overline{M}) \to \Lambda^s(\overline{N})$$
will be denoted by $\Lambda^s \overline{f}^*$.

### 3.3.9

Let $\overline{f} : \overline{M} \to \overline{M}$ be a fat identity map of $\overline{M}$ into itself, *i.e.*, a gauge transformation. Since $f = \mathrm{id}_M$, $\Lambda^\bullet f^* = \mathrm{id}_{\Lambda^\bullet(M)}$. But the homomorphism
$$\Lambda^\bullet \overline{f}^* = \mathrm{id}_{\Lambda^\bullet(M)} \otimes \mathrm{id}_{\Gamma(\overline{M})} : \Lambda^\bullet(M) \otimes \Gamma(\overline{M}) \to \Lambda^\bullet(M) \otimes \Gamma(\overline{M})$$
$$\| \qquad\qquad \|$$
$$\Lambda^\bullet(\overline{M}) \qquad \Lambda^\bullet(\overline{M})$$
in spite of its appearance is *not* generally the identity of $\Lambda^\bullet(\overline{M})$. The point is that the notation used is ambiguous. Namely, $\Lambda^\bullet(\overline{M})$ is considered here as a tensor product $\Lambda^\bullet(M) \otimes \Gamma(\overline{M})$ in two different ways. First, on the domain of $\Lambda^\bullet \overline{f}^*$ it stands for the universal bilinear function
$$(\omega, p) \mapsto \omega \wedge p$$
while on the range it stands for
$$(\omega, p) \mapsto \omega \wedge \overline{f}^*(p).$$
So, the symbol $\otimes$ in $\Lambda^\bullet \overline{f}^* = \mathrm{id}_{\Lambda^\bullet(M)} \otimes \mathrm{id}_{\Gamma(M)}$ changes its meaning when passing from the domain and the range and $\Lambda^\bullet \overline{f}^*$ is the automorphism
$$\omega \wedge p \mapsto \omega \wedge \overline{f}^*(p).$$

### 3.3.10 Geometrical Interpretation of the Induced Homomorphism of Thickened Forms

Recall that a thickened $s$-form $\overline{\omega} \in \Lambda^s(\overline{M})$ may be interpreted as the field of maps
$$\overline{\omega}_m : T_m M \times \cdots \times T_m M \longrightarrow \overline{m}$$
(see n. 3.1.6). Such field for the form $\Lambda^\bullet(\overline{f}^*)(\overline{\omega})$ is described by the formula
$$(\Lambda^\bullet(\overline{f}^*)(\overline{\omega}))_n(\xi_1, \ldots, \xi_s) = \overline{f}_n^{-1}\left(\overline{\omega}_{f(n)}\left(\mathrm{d}_n f(\xi_1), \ldots, \mathrm{d}_n f(\xi_s)\right)\right),$$
$$\overline{\omega} \in \Lambda^s(\overline{M}),\ n \in N,\ \xi_1, \ldots, \xi_s \in T_n N$$
which is deduced easily from the geometric description of $\Lambda^\bullet(f^*)$ (see n. 0.5.17) and the definition of $\Lambda^\bullet \overline{f}^*$.

### 3.3.11

**Proposition.** Let $\overline{f} : \overline{N} \to \overline{M}$ be a fat map and $X, Y$ vector fields on $M$ and $N$, respectively. If $X$ and $Y$ are $f$-compatible, then
$$\bar{\mathrm{i}}_Y \circ \Lambda^\bullet \overline{f}^* = \Lambda^\bullet \overline{f}^* \circ \bar{\mathrm{i}}_X$$
(see n. 0.5.18).

**Proof.** Directly from the pointwise description of $\Lambda^\bullet \overline{f}^*$ given in the previous n. 3.3.10. □

The following fact is an immediate consequence of the proposition (*cf.* Corollary 0.5.21).

**Corollary.** Let $\overline{f} : \overline{N} \to \overline{M}$ be a fat map, $X_1, \ldots, X_s$ vector fields on $M$ and $Y_1, \ldots, Y_s$ be vector fields on $N$ such that for all $i = 1, \ldots, s$ vector fields $X_i$ and $Y_i$ are $f$-compatible. Then for all $\overline{\omega} \in \Lambda^s(\overline{M})$
$$\Lambda^\bullet \overline{f}^*(\overline{\omega})(Y_1, \ldots, Y_s) = \overline{f}^*(\overline{\omega}(X_1, \ldots, X_s)).$$

### 3.3.12 Compatible Linear Connections

Let $\overline{f} : \overline{N} \to \overline{M}$ be a fat map, $\nabla$ a linear connection on $\overline{M}$ and $\square$ a linear connection on $\overline{N}$.

By Example 3.3.5 $(\Lambda^\bullet(\overline{N}), \mathrm{d}_\square)$ is a cd-module over $\Lambda^\bullet(N)$. On the other hand, $\Lambda^\bullet(\overline{N})$ is also a $\Lambda^\bullet(M)$–module via $\Lambda^\bullet f^*$. This way $(\Lambda^\bullet(\overline{N}), \mathrm{d}_\square)$ acquires a structure of cd-module over $\Lambda^\bullet(M)$, in view of the fact that $\Lambda^\bullet f^*$ commutes with the exterior differentials on $M$ and $N$.

**Definition.** Linear connections $\nabla$ and $\square$ are called *compatible with respect to* $\overline{f}$ or, shortly, $\overline{f}$-*compatible* if $\Lambda^\bullet \overline{f}^*$ is a homomorphism of cd-modules over $\Lambda^\bullet(M)$, that is

$$d_\square \circ \Lambda^\bullet \overline{f}^* = \Lambda^\bullet \overline{f}^* \circ d_\nabla \;.$$

Compatibility is, obviously, a *transitive* property, that is, if $\nabla_1$ and $\nabla_2$ are $\overline{f}_1$-compatible and $\nabla_2$, $\nabla_3$ are $\overline{f}_2$-compatible, then $\nabla_1$ and $\nabla_3$ are $(\overline{f}_2 \circ \overline{f}_1)$-compatible.

This allows to organize connections in fat manifolds of a given type into a category. Objects of this category are pairs of the form $(\overline{M}, \nabla)$ and morphisms are fat maps along which the corresponding connections are compatible.

### 3.3.13

It follows from commutativity of the diagram

$$\begin{array}{ccc} \Gamma(\overline{M}) & \xrightarrow{d_\nabla} & \Lambda^1(\overline{M}) \\ \overline{f}^* \downarrow & & \downarrow \Lambda^1 \overline{f}^* \\ \Gamma(\overline{N}) & \xrightarrow{d_\square} & \Lambda^1(\overline{N}) \end{array}$$

that $\Lambda^\bullet \overline{f}^*$ is a cd-module homomorphism and, obviously, vice versa. In fact, if $\omega \in \Lambda^s(M)$ and $p \in \Gamma(\overline{M})$, then

$$d_\square \left( \Lambda^\bullet \overline{f}^*(\omega \otimes p) \right) = d_\square \left( \Lambda^\bullet f^*(\omega) \wedge \Lambda^\bullet \overline{f}^*(p) \right)$$
$$= d\left( \Lambda^\bullet f^*(\omega) \right) \wedge \Lambda^\bullet \overline{f}^*(p) + (-1)^s \Lambda^\bullet f^*(\omega) \wedge d_\square \left( \Lambda^\bullet \overline{f}^*(p) \right)$$
$$= \Lambda^\bullet f^*(d\omega) \wedge \Lambda^\bullet \overline{f}^*(p) + (-1)^s \Lambda^\bullet f^*(\omega) \wedge \Lambda^\bullet \overline{f}^*(d_\nabla(p))$$
$$= \Lambda^\bullet \overline{f}^* \left( d\omega \wedge p + (-1)^s \omega \wedge d_\nabla(p) \right)$$
$$= \Lambda^\bullet \overline{f}^* \left( d_\nabla(\omega \otimes p) \right) \;.$$

In other words, $\square$ and $\nabla$ are $\overline{f}$-compatible if and only if the above diagram is commutative.

### 3.3.14

**Proposition.** If linear connections $\nabla$ and $\square$ are $\overline{f}$-compatible, then they are $\overline{f}$-related.

**Proof.** Let $X$, $Y$ be $f$-compatible vector fields on $M$ and $N$, respectively. In order to show that $\nabla_X$ and $\square_Y$ are $\overline{f}$-compatible consider the commutative diagram

$$\begin{array}{ccc} \Lambda^1\left(\overline{M}\right) & \xrightarrow{\Lambda^1 \overline{f}^*} & \Lambda^1\left(\overline{N}\right) \\ {\scriptstyle \overline{i}_X}\downarrow & & \downarrow{\scriptstyle \overline{i}_Y} \\ \Gamma\left(\overline{M}\right) & \xrightarrow{\overline{f}^*} & \Gamma\left(\overline{N}\right) \end{array},$$

(Proposition 3.3.11). The following commutative diagram expresses the fact that $\Lambda^\bullet \overline{f}^*$ is a cd-homomorphism of $\Lambda^\bullet(M)$–modules

$$\begin{array}{ccc} \Gamma\left(\overline{M}\right) & \xrightarrow{\overline{f}^*} & \Gamma\left(\overline{N}\right) \\ {\scriptstyle d_\nabla}\downarrow & & \downarrow{\scriptstyle d_\square} \\ \Lambda^1\left(\overline{M}\right) & \xrightarrow{\Lambda^1 \overline{f}^*} & \Lambda^1\left(\overline{N}\right) \end{array}.$$

By combining these two commutative diagrams we get the the third one

$$\begin{array}{ccc} \Gamma\left(\overline{M}\right) & \xrightarrow{\overline{f}^*} & \Gamma\left(\overline{N}\right) \\ {\scriptstyle \overline{i}_X \circ d_\nabla}\downarrow & & \downarrow{\scriptstyle \overline{i}_Y \circ d_\square} \\ \Gamma\left(\overline{M}\right) & \xrightarrow{\overline{f}^*} & \Gamma\left(\overline{N}\right) \end{array}. \qquad (3.5)$$

Since (see n. 3.2.7)

$$\overline{i}_X \circ d_\nabla = \nabla_X \quad \text{and} \quad \overline{i}_Y \circ d_\square = \square_Y,$$

diagram (3.5) tells exactly that $\nabla_X$ and $\square_Y$ are $\overline{f}$-compatible. $\square$

### 3.3.15

The inverse to Proposition 3.3.14 is false, *i.e.*, $\overline{f}$-related linear connections are not necessarily $\overline{f}$-compatible. An example of that is as follows.

Let $\overline{M}$ be a fat manifold over $M = \mathbb{R}^2$ with $\Gamma\left(\overline{M}\right) = C^\infty(\mathbb{R}^2)$ and $\overline{N}$ a fat manifold over $N = \mathbb{R}$ with $\Gamma\left(\overline{N}\right) = C^\infty(\mathbb{R})$. Fix a (certainly existing) bijective map $\sigma : \mathbb{N} \to \mathbb{Q}_0^+$ between naturals and nonnegative rational numbers with $\sigma(1) = 0$, fix a function $a \in C^\infty(\mathbb{R})$ that vanishes on $(-\infty, 0]$

and is constantly 1 on $[1, +\infty)$ (see [Nestruev (2003), Corollary 2.5]) and define
$$b(t) = \begin{cases} (a(t) - 1)t, & t < 1 ; \\ \sigma(i) + a(t - i) \cdot (\sigma(i + 1) - \sigma(i)), & i \in \mathbb{N}, t \in [i, i + 1) . \end{cases}$$
The function $b$ is smooth because all derivatives of $a$ vanish at $t = 0$, $t = 1$.

Let
$$f : N \to M, \quad t \mapsto \bigl(b(t), \sin(\pi t a(t))\bigr) .$$
It is easily verified that

- $f(t) = f(-\sigma(t)) = (\sigma(t), 0), \quad t \in \mathbb{N}$;
- $(d_{t_0} f) (d / d t|_{t_0})$ and $(d_{-\sigma(t_0)} f) (d / d t|_{-\sigma(t_0)})$ are linearly independent in $T_{(\sigma(t_0), 0)} M$ for all $t_0 \in \mathbb{N}$.

By these properties, if a vector field $Y$ on $N$ is $f$-compatible with some vector field $X$ on $M$, then for every negative rational $t$, the tangent vector $Y_t$ must vanish. Hence, by continuity, $Y_t = 0$ for $t \leq 0$.

Let $\overline{f}$ be the fat map with $\overline{f}^* = f^*$, fix a nonzero $c \in C^\infty(\mathbb{R})$ that vanishes on $[0, \infty)$ (e.g., $c(t) = a(-t)$), define a linear connection $\nabla$ on $\overline{N}$ by
$$\nabla_{g \frac{\partial}{\partial t}}(q) = g \frac{\partial q}{\partial t} + gcq, \quad g \in C^\infty(\mathbb{R}), \quad q \in \Gamma(\overline{N}) = C^\infty(\mathbb{R})$$
and consider the trivial connection $D$ on $\overline{M}$.

With $Y = g \partial / \partial t$ being $f$-compatible with some vector field $X$ on $M$, we have that $g$ vanishes for $t \leq 0$, and therefore $gc = 0$. Hence
$$\nabla_Y(\overline{f}^*(p)) = g \frac{\partial \overline{f}^*(p)}{\partial t} = Y(f^*(p)) = f^*(X(p)) = \overline{f}^*(D_X(p)), \quad p \in \Gamma(\overline{M}),$$
i.e., $\nabla_Y$ and $D_X$ are $\overline{f}$-compatible. This shows that $\nabla$ and $D$ are $\overline{f}$-related.

Finally, with $x$ being the first coordinate function on $M = \mathbb{R}^2$ we have
$$d_\nabla \left(\overline{f}^*(x)\right) \left(\frac{\partial}{\partial t}\right) = \nabla_{\frac{\partial}{\partial t}}(b) = \frac{\partial b}{\partial t} + cb$$
while
$$\Lambda^1 \overline{f}^* (d_D(x)) \left(\frac{\partial}{\partial t}\right) = \Lambda^1 \overline{f}^* (d x) \left(\frac{\partial}{\partial t}\right) = d b \left(\frac{\partial}{\partial t}\right) = \frac{\partial b}{\partial t} .$$
Since $cb$ is nonzero ($c(t) \neq 0$ for some $t < 0$ and $b(t) \neq 0$ for all $t < 0$), this shows that $\nabla$ and $D$ are not $\overline{f}$-compatible.

### 3.3.16 Induced Linear Connections

**Proposition.** Let $\bar{f} : \overline{N} \to \overline{M}$ be a fat map and $\nabla$ a linear connection on $\overline{M}$. Then there exists a unique linear connection $\square$ on $\overline{N}$ $\bar{f}$-compatible with $\nabla$.

**Proof.** Set
$$A = C^\infty(M), \quad B = C^\infty(N), \quad P = \Gamma\left(\overline{M}\right), \quad Q = \Gamma\left(\overline{N}\right).$$
Then, according to n. 3.3.8,
$$\Lambda^\bullet\left(\overline{N}\right) = \Lambda^\bullet(B) \otimes_A P,$$
and
$$\omega \wedge \Lambda^\bullet(\overline{f}^*)(p) = \omega \otimes p, \quad \omega \in \Lambda^\bullet(B),\ p \in P.$$

It follows from n. 3.3.13 and Proposition 3.2.11 that a linear connection $\square$ is $\bar{f}$-compatible with $\nabla$ if and only if its covariant differential $d_\square$ satisfies the following condition:
$$d_\square(b \otimes p) = db \otimes p + b\,\Lambda^\bullet(\overline{f}^*)(d_\nabla(p)), \quad b \in B,\ p \in P.$$
On the other hand the formula
$$\Delta(b \otimes p) = db \otimes p + b\,\Lambda^\bullet(\overline{f}^*)(d_\nabla(p)), \quad b \in B,\ p \in P$$
correctly determines an additive operator
$$\Delta : Q \to \Lambda^1(Q)$$
(see n. 0.1.4) that satisfies the characteristic Leibniz rule (3.1), p. 178:
$$\Delta(bq) = db \otimes q + b\Delta(q).$$
In fact, it suffices to check it on $q = b' \otimes p$, with $b' \in B$ and $p \in P$:
$$\Delta(bb' \otimes p) = d(bb') \otimes p + bb'\,\Lambda^\bullet(\overline{f}^*)(d_\nabla(p))$$
$$= db \otimes (b' \otimes p) + b\left(db' \otimes p + b'\,\Lambda^\bullet(\overline{f}^*)(d_\nabla(p))\right).$$
Hence $\Delta$ is actually the (basic) covariant differential of a unique linear connection $\square$, as required. $\square$

**Example.** Let $\nabla$ be a linear connection on a fat manifold $\overline{M}$ and $\bar{i} : \overline{N} \to \overline{M}$ be a fat map over a closed embedding $i : N \to M$. In n. 2.2.4 the induced linear connection $\nabla|_{\overline{N}}$ on $\overline{N}$ was defined as the unique one that is $\bar{i}$-related with $\nabla$. In view of Propositions 3.3.14 and 3.3.16 one can see now that the unique linear connection $\bar{i}$-compatible with $\nabla$ must coincide with $\nabla|_{\overline{N}}$.

By the same reason the unique linear connection on an open fat submanifold of $\overline{M}$ that is compatible with $\nabla$ with respect to the underlying fat embedding is the localization of $\nabla$ (see Definition 2.4.10).

**Definition.** Let $\overline{f} : \overline{N} \to \overline{M}$ be a fat map and $\nabla$ a linear connection on $\overline{M}$. The unique linear connection $\square$ on $\overline{N}$ that is $\overline{f}$-compatible with $\nabla$ is called *induced by $\overline{f}$ from $\nabla$*.

### 3.3.17 Lift and Parallel Translation for Arbitrary Smooth Curves

Now we are able to define lift and parallel translation along arbitrary (possibly singular) curves as it was promised in Sect. 2.2.6.

**Proposition.** Let $\nabla$ be a linear connection on a fat manifold $\overline{M}$ and
$$\gamma : \mathbb{I} \to M$$
be a curve. Then there exists a fat curve
$$\overline{\gamma} : \overline{\mathbb{I}} \to \overline{M},$$
over $\gamma$ such that $\nabla$ is $\overline{\gamma}$-compatible with the trivial connection $D$ on $\overline{\mathbb{I}}$. Moreover, $\overline{\gamma}$ is unique up to an uniform fat identity map on the domain.

***Proof.*** Note that with respect to fat diffeomorphisms linear connections are compatible if and only if they are related. Therefore, Proposition 2.2.2 in which 'related' is replaced by 'compatible' holds. After this modification the proof repeats literally that of Proposition 2.2.5. In fact, it is even simpler because the compatibility is transitive with respect to compositions property. $\square$

**Definition.** Let $\nabla$ be a linear connection on a fat manifold $\overline{M}$ and
$$\gamma : \mathbb{I} \to M$$
be a curve. A fat curve
$$\overline{\gamma} : \overline{\mathbb{I}} \to \overline{M}$$
over $\gamma$ such that $\nabla$ is $\overline{\gamma}$-compatible with the trivial connection $D$ on $\overline{\mathbb{I}}$ is called a *lift of $\gamma$ by $\nabla$*, or simply a *$\nabla$-lift*. The isomorphism
$$\mathbf{T}^{\gamma}_{t_0,t_1} = \overline{\gamma}_{t_1} \circ \left(\overline{\gamma}_{t_0}\right)^{-1} : \overline{m_0} = \overline{\gamma}\left(\overline{t_0}\right) \longrightarrow \overline{m_1} = \overline{\gamma}\left(\overline{t_1}\right)$$
(here we used the identification of fat points of $\overline{\mathbb{I}}$ with the standard fiber $F$; *cf.* n. 1.6.12) is called the *parallel translation from $m_0 = \gamma(t_0)$ to $m_1 = \gamma(t_1)$ along $\gamma$ via $\nabla$*.

The above said illustrates naturalness and advantages of the concept of compatibility of connections which in its turn is based on that of covariant differential.

### 3.3.18  *Holonomy*

Denote by $\mathrm{GL}(\overline{m})$ the group of all automorphisms of the vector space $\overline{m}$. Recall that a *loop* at $m$ is a curve $\gamma : [a,b] \to M$ such that $\gamma(a) = \gamma(b) = m$ and put $\mathbf{T}_m^\gamma = \mathbf{T}_{a,b}^\gamma : \overline{m} \to \overline{m}$ (with respect to a given linear connection $\nabla$), $\mathbf{T}_m^\gamma \in \mathrm{GL}(\overline{m})$.

**Definition.** The subgroup of $\mathrm{GL}(\overline{m})$ that is the closure of the subgroup generated by all automorphisms of the form $\mathbf{T}_m^\gamma$ is called the *holonomy group of $\nabla$ at $m$* and is denoted by $\mathrm{Hol}(\nabla, m)$.

Being a closed subgroup of $\mathrm{GL}(\overline{m})$, $\mathrm{Hol}(\nabla, m)$ is a Lie subgroup according to a well-known result (see [Lee (2003), Theorem 20.10 (p. 526)]).

If $\mathbf{T} : \overline{m_0} \to \overline{m_1}$ is a parallel translation along a curve $\sigma$, then

$$\mathrm{Hol}(\nabla, m_0) \ni g \mapsto \mathbf{T} \circ g \circ \mathbf{T}^{-1} \in \mathrm{Hol}(\nabla, m_1) \tag{3.6}$$

is an isomorphism depending on $\sigma$. If $M$ is connected, then it is not difficult to construct the *holonomy bundle over $M$* whose fiber over $m$ is $\mathrm{Hol}(\nabla, m)$. In spite of the fact that fibers of this bundle are isomorphic Lie groups, it is not a principal bundle. It possesses a natural nonlinear connection with respect to which parallel translations are of the form (3.6). We do not use this fact in the sequel and, so, skip the details.

Informally speaking, the holonomy group $\mathrm{Hol}(\nabla, m)$ gives a cumulative estimate, as viewed by an observer at $m$, of how much $\nabla$ is 'twisted'.

**Exercise.** Let $\nabla$ be a flat linear connection in $\overline{M}$. Is $\mathrm{Hol}(\nabla, m)$ a discrete subgroup of $\mathrm{GL}(\overline{m})$ if $M$ is compact?

### 3.3.19  *Another Description of Compatibility*

To be complete, we shall discuss below a geometric approach to compatibility, induced connections, *etc.*, à la Grothendieck. According to it, a construction, proof, *etc.*, are to be done in two steps, one for injection (embedding) and another for surjection (projection). After that the result is obtained by passing to the graph of the map in consideration. In the case of connections the first step, namely, the restriction to closed submanifolds is rather simple and is already done (see n. 2.2.4). For the second one a fat cylinder $\overline{N_M}$ over $\overline{M}$ (see n. 1.6.13) is necessary, and also a geometrical description of the induced by the fat projection linear connection.

Recall that vector fields on $M \times N$ decompose naturally into 'horizontal' and 'vertical' parts, *i.e.*,

$$\mathrm{D}(M \times N) = \mathrm{D}_N(M \times N) \oplus \mathrm{D}_M(M \times N)$$

(see n. 0.4.19; *cf.* also Proposition 0.4.18 and n. 0.4.20). A horizontal vector field $X \in \mathrm{D}_N(M \times N)$ is a vector field on $M \times N$ that is compatible with the zero vector field with respect to the projection onto $N$. Similarly are defined vertical vector fields $Y \in \mathrm{D}_M(M \times N)$.

In view of Proposition 0.5.9 and n. 0.4.19 the following canonical identifications takes place:

$$\mathrm{D}_N(M \times N) = \mathrm{C}^\infty(M \times N) \otimes_{\mathrm{C}^\infty(M)} \mathrm{D}(M)$$

and

$$\mathrm{D}_M(M \times N) = \mathrm{C}^\infty(M \times N) \otimes_{\mathrm{C}^\infty(N)} \mathrm{D}(N).$$

According to them the vector field $1 \otimes X \in \mathrm{D}_N(M \times N)$ with $X \in \mathrm{D}(M)$ is the unique horizontal vector field compatible with $X$ with respect to the projection onto $M$. Similarly are characterized vector fields $1 \otimes Y \in \mathrm{D}_M(M \times N)$ with $Y \in \mathrm{D}(N)$. These vector fields will be denoted by $\tilde{X}$, $\tilde{Y}$, respectively.

Thus

$$\mathrm{D}(M \times N) = \left(\mathrm{C}^\infty(M \times N) \otimes_{\mathrm{C}^\infty(M)} \mathrm{D}(M)\right) \oplus \left(\mathrm{C}^\infty(M \times N) \otimes_{\mathrm{C}^\infty(N)} \mathrm{D}(N)\right).$$

Let $\nabla$ be a linear connection on $\overline{M}$ and $\square$ be a $\overline{\pi}$-related with $\nabla$ linear connection on the fat cylinder $\overline{N_M}$. Since $\tilde{X}$ is $\pi$-compatible with $X \in \mathrm{D}_M$, the fat vector field $\square_{\tilde{X}}$ over $\tilde{X}$ is induced by $\nabla_X$ (see n. 1.5.3) and hence is uniquely determined by $X$ and $\nabla$. In explicit terms this means that

$$\square_{\tilde{X}}(f \otimes p) = \tilde{X}(f) \otimes p + f \otimes \nabla_X(p), \quad f \otimes p \in \mathrm{C}^\infty(M \times N) \otimes \Gamma(\overline{M}) = \Gamma(\overline{N_M}).$$

Similarly, if $Y \in \mathrm{D}_N$, then $\tilde{Y}$ is $\pi$-compatible with the zero field on $M$. By this reason $\square_{\tilde{Y}}$ is a fat vector field over $\tilde{Y}$ induced by the zero fat vector field on $\overline{M}$. As such it is uniquely determined by $\tilde{Y}$, *i.e.*,

$$\square_{\tilde{Y}}(f \otimes p) = \tilde{Y}(f) \otimes p, \quad f \otimes p \in \mathrm{C}^\infty(M \times N) \otimes \Gamma(\overline{M}) = \Gamma(\overline{N_M}).$$

In other words, $\square_{\tilde{Y}} = \tilde{Y} \otimes \mathrm{id}_{\Gamma(\overline{M})}$ because $\tilde{Y}$ is $\mathrm{C}^\infty(M)$-linear.

According to the canonical decomposition of $\mathrm{D}(M \times N)$, a generic vector field $Z$ on $M \times N$ has the form $Z = f\tilde{X} + g\tilde{Y}$, $f, g \in \mathrm{C}^\infty(M \times N)$. So,

$\Box_Z = f\Box_{\tilde{X}} + g\Box_{\tilde{Y}}$. Taking into account the above formulas for $\Box_{\tilde{X}}$ and $\Box_{\tilde{Y}}$ we get the following explicit formula for the connection $\Box$:

$$\Box_{f\otimes X + g\otimes Y}(h\otimes p) = f\tilde{X}(h)\otimes p + fh\otimes \nabla_X(p) + g\tilde{Y}(h)\otimes p,$$
$$f, g, h \in C^\infty(M\times N), X\in D(M), Y\in D(N), p\in \Gamma(\overline{M}).$$

So, this formula proves the uniqueness of the connection $\overline{\pi}$-related with $\nabla$. With an easy check of correctness it also proves the existence of such a connection. Moreover, Proposition 3.3.14 shows that $\Box$ is nothing else but the induced by $\overline{\pi}$ from $\nabla$ linear connection.

Thus the induced by the canonical projection connection on the fat cylinder can be constructed geometrically without appealing to covariant differential. In its turn this allows to construct the induced by an *arbitrary* fat map $\overline{f}: \overline{N} \to \overline{M}$ linear connection by means of the Grothendieck trick. Namely, consider the graph map of $f$:

$$(f, \mathrm{id}_N): N \to M \times N.$$

In view of Proposition 0.3.14 there exists a unique over this graph fat map $\overline{g}: \overline{N} \to \overline{N_M}$ such that $\overline{f} = \overline{\pi}\circ \overline{g}$. The graph is a closed embedding and, therefore, the connection $\Box$ can be restricted to it (see n. 2.2.4). Obviously, this restriction coincides with the induced by $\overline{g}$ from $\Box$ connection and by this reason is precisely the linear connection induced from $\nabla$ by $\overline{f}$.

Thus this is a 'naive' geometric way to construct the induced connection for arbitrary fat maps. Of course, it is rather cumbersome and much less manageable with respect to the conceptual algebraic one. But the Grothendieck trick could give a useful initial impulse in a situation which is conceptually unclear.

## 3.4 Linear Connections Along Fat Maps

If in the definition of linear connection (see Definition 2.1.1) one replaces vector fields, both ordinary and fat, by vector fields, respectively, ordinary and fat, *along a fat map*, say $\overline{f}$, then he will get the notion of a *linear connection along* $\overline{f}$. This generalization is very natural and useful. In particular, it is closely related with the notion of compatibility. Fundamentals of the corresponding theory are discussed in this section.

### 3.4.1 Thickened Forms Along Fat Maps

**Definition.** A *thickened form along a fat map* $\overline{f} : \overline{N} \to \overline{M}$ is a $C^\infty(M)$-module homomorphism
$$D(M) \to \Gamma(\overline{N}).$$

In other words, a thickened form along $\overline{f}$ is a thickened form of the $C^\infty(M)$-module obtained from $C^\infty(N)$-module $\Gamma(\overline{N})$ by the restriction of scalars via $f^*$. The $C^\infty(N)$-module of all thickened forms along $\overline{f}$ will be denoted by $\Lambda^\bullet(\overline{M})_{\overline{f}}$, and its degree $s$ component by $\Lambda^s(\overline{M})_{\overline{f}}$.

The following 'thickened version' of Proposition 0.5.24 holds.

**Proposition.** The module $\Lambda^s(\overline{M})_{\overline{f}}$ of thickened $s$-forms along a fat map $\overline{f} : \overline{N} \to \overline{M}$, together with the map
$$\nu : \Lambda^s\left(\overline{M}\right) \to \Lambda^s\left(\overline{M}\right)_{\overline{f}}, \quad \overline{\omega} \mapsto \overline{f}^* \circ \overline{\omega},$$
possesses the universal property of extension via $f^*$ of the $C^\infty(M)$-module $\Lambda^s\left(\overline{M}\right)$.

In other words,
$$\Lambda^s(\overline{M})_{\overline{f}} = C^\infty(N) \otimes_{C^\infty(M)} \Lambda^s\left(\overline{M}\right)$$
and
$$1 \otimes \overline{\omega} = \overline{f}^* \circ \overline{\omega}, \quad \overline{\omega} \in \Lambda^s\left(\overline{M}\right).$$

*Proof.* Set as usual
$$A = C^\infty(M), \quad B = C^\infty(N), \quad P = \Gamma\left(\overline{M}\right), \quad Q = \Gamma\left(\overline{N}\right).$$
By Proposition 0.5.10 we have
$$\Lambda^s(\overline{M}) = \Lambda^s(M) \otimes P, \quad \Lambda^s(\overline{M})_{\overline{f}} = \Lambda^s(M) \otimes Q.$$
Also $Q = B \otimes P$ according to n. 1.1.6. Therefore,
$$\Lambda^s(\overline{M})_{\overline{f}} = \Lambda^s(M) \otimes B \otimes P = B \otimes \Lambda^s(M) \otimes P = B \otimes \Lambda^s(\overline{M})$$
as required.

Concerning the second assertion in the proposition it suffices to prove it for forms of the type
$$\overline{\omega} = \omega \otimes p \in \Lambda^s(\overline{M}) = \Lambda^s(M) \otimes P$$
only. But in this case $1 \otimes \overline{\omega} = 1 \otimes \omega \otimes p = \omega \otimes \overline{f}^*(p)$ and Proposition 0.5.10 shows again that
$$(1 \otimes \overline{\omega})(X_1, \ldots, X_s) = \omega(X_1, \ldots, X_s)\overline{f}^*(p) = \overline{f}^*(\omega(X_1, \ldots, X_s)p)$$
$$= \overline{f}^*(\overline{\omega}(X_1, \ldots, X_s)) = (\overline{f}^* \circ \overline{\omega})(X_1, \ldots, X_s),$$
i.e.,
$$1 \otimes \overline{\omega} = \overline{f}^* \circ \overline{\omega}$$
as required. □

### 3.4.2 Insertion of $D(M)_f$ into $\Lambda^\bullet(\overline{M})_{\overline{f}}$

According to Definition 0.5.18, the insertion operator

$$\Lambda^\bullet(\overline{M})_{\overline{f}} \to \Lambda^\bullet(\overline{M})_{\overline{f}}, \quad \overline{\omega} \mapsto X \lrcorner \overline{\omega}$$

is associated with a vector field $X$ on $M$. Exactly as in the case of differential forms along smooth maps this construction can be, in fact, extended to vector fields along $f$ (cf. Definition 0.5.28). Indeed, recall that vector fields along $f$ constitute a module $D(M)_f$ and the identification

$$D(M)_f = C^\infty(N) \otimes_{C^\infty(M)} D(M)$$

such that

$$1 \otimes X = f^* \circ X, \quad X \in D(M)$$

holds (see Proposition 0.5.9).

This allows to identify a thickened $s$-form $\overline{\omega}$ along $\overline{f}$ with an alternating $s$-linear function of $C^\infty(N)$–modules

$$D(M)_f \times \cdots \times D(M)_f \to \Gamma(\overline{N}),$$

according to n. 0.1.5, (3). Up to this identification, if $Y$ is a vector field along $f$, the *insertion of $Y$ operator* into $\Lambda^\bullet(\overline{M})_{\overline{f}}$ is defined by setting

$$\bar{i}_Y(\overline{\omega})(X_1, \ldots, X_{s-1}) \stackrel{\text{def}}{=} \overline{\omega}(Y, f^* \circ X_1, \ldots, f^* \circ X_{s-1}),$$

$$X_1, \ldots, X_{s-1} \in D(M)$$

(if $s = 0$ set $\bar{i}_Y(\overline{\omega}) = 0$).

The notation

$$Y \lrcorner \overline{\omega}$$

or

$$\overline{\omega}_Y$$

will be also used.

### 3.4.3 Linear Connections Along Fat Maps

**Definition.** Let $\varphi : A \to B$ be a commutative $k$-algebra homomorphism, $k$ being a field, and $\overline{\varphi} : P \to Q$ a homomorphism of modules over $\varphi$ (cf. n. 1.2.11). A *linear connection along* $\overline{\varphi}$ is an $A$-module homomorphism

$$\nabla : D(A)_\varphi \to \text{Der}(P)_{\overline{\varphi}}$$

such that $\nabla(X)$ is a der-operator (along $\overline{\varphi}$) over $X \in D(A)_\varphi$.

A *linear connection along a fat map* $\overline{f}$ is a linear connection along the associated module homomorphism $\overline{f}^*$.

As in the classical case we shall write $\nabla_X$ instead of $\nabla(X)$, $X \in \mathrm{D}(A)_\varphi$.
We repeat in exact terms that a linear connection along a fat map $\overline{f} : \overline{N} \to \overline{M}$ is a $C^\infty(N)$–module homomorphism

$$\nabla : \mathrm{D}(M)_f \to \overline{\mathrm{D}}\left(\overline{M}\right)_{\overline{f}}$$

(here $\overline{\mathrm{D}}\left(\overline{M}\right)_{\overline{f}} \stackrel{\text{def}}{=} \overline{\mathrm{D}}\left(\overline{M}\right)_{\overline{f}^*}$) such that $\nabla_X$ is a fat along $\overline{f}$ field over $X$.

### 3.4.4 Geometric Description of Linear Connections Along Fat Maps

Recall that a fat vector field $\overline{Y}$ along $\overline{f} : \overline{N} \to \overline{M}$ is geometrically described as a field that associates with a fat point $\overline{n}$ of $\overline{N}$ a fat tangent vector $\overline{Y}_n$ to $\overline{M}$ at $\overline{f}(\overline{n})$ (see n. 1.3.13).

Let now $\nabla$ be a linear connection along $\overline{f}$. Fix a point $n \in N$ and a tangent vector $\xi$ to $M$ at $f(n)$. Consider then a vector field $Y$ along $f$ such that $Y_n = \xi$ and the fat tangent vector $\overline{\xi}$ at $f(n)$ that the fat field $\nabla_Y$ associates with $\overline{n}$. As in n. 2.1.10 it is not difficult to prove that $\overline{\xi}$ is independent of the choice of $Y$ and, therefore, is determined completely by $n$ and $\xi$.

**Definition.** The so-defined fat tangent vector $\overline{\xi}$ is called the $\nabla$-*lift of* $\xi$ *at* $n$.

Thus, from the geometric viewpoint, a linear connection along $\overline{f}$ associates with every point $n \in N$ a lifting map $T_{f(n)}M \to \overline{T}_{f(n)}\overline{M}$.

### 3.4.5 Covariant Differential Along a Fat Map

Let $\nabla$ be a linear connection along a fat map $\overline{f} : \overline{N} \to \overline{M}$. With a $p \in \Gamma(\overline{M})$ is naturally associated the $C^\infty(N)$–module homomorphism

$$\mathrm{D}(M)_f \to \Gamma(\overline{N}), \quad Y \mapsto \nabla_Y(p) .$$

Since $\mathrm{D}(M)_f$ is obtained from $\mathrm{D}(M)$ by extension of scalars, the above operator corresponds to the differential form along $f$

$$X \mapsto \nabla_{f^* \circ X}(p), \ X \in \mathrm{D}(M) .$$

Thus $\nabla$ gives rise to a map

$$d_\nabla : \Gamma(\overline{M}) \to \Lambda^1(\overline{M})_{\overline{f}} ,$$

called the *(basic) covariant differential* associated with $\nabla$.

It immediately follows from the definitions that
$$Y \lrcorner\, d_\nabla(p) = \nabla_Y(p), \quad Y \in D(M)_f, p \in \Gamma(\overline{M}). \tag{3.7}$$
As in the ordinary case this construction extends naturally to higher degrees.

**Definition.** The *covariant differential* associated with a linear connection $\nabla$ along $\overline{f}$ is the graded $C^\infty(M)$-module homomorphism
$$d_\nabla : \Lambda^\bullet(\overline{M}) \to \Lambda^\bullet(\overline{M})_{\overline{f}}$$
whose degree $s$ components is defined by
$$(d_\nabla)_s(\overline{\omega})(X_1, \ldots, X_{s+1}) = \sum_i (-1)^{i+1} \nabla_{f^* \circ X_i}(\overline{\omega}(X_1, \ldots, \widehat{X_i}, \ldots, X_{s+1}))$$
$$+ \sum_{i<j} (-1)^{i+j} \overline{f}^*(\overline{\omega}([X_i, X_j], X_1, \ldots, \widehat{X_i}, \ldots, \widehat{X_j}, \ldots, X_{s+1})),$$
$$X_1, \ldots, X_{s+1} \in D(M), \overline{\omega} \in \Lambda^s(\overline{M}).$$

### 3.4.6 Linear Connections Along Maps Associated with Ordinary Linear Connections

Let $\overline{f} : \overline{N} \to \overline{M}$ be a fat map and
$$\nabla : D(M) \to \overline{D(M)}$$
a linear connection on $\overline{M}$.

Consider the $C^\infty(M)$-module homomorphism
$$\rho : \overline{D(M)} \to \overline{D(M)}_{\overline{f}}, \quad \square \mapsto \overline{f}^* \circ \square$$
coupled with the $C^\infty(M)$-module homomorphism
$$\nu : D(M) \to D(M)_f, \quad X \mapsto f^* \circ X$$
and recall that $D(M)_f$ as a $C^\infty(N)$-module is obtained from $D(M)$ by extension of scalars via $f^*$. By the universal property of scalar extensions there exists a unique linear connection $\nabla_{\overline{f}}$ along $\overline{f}$ characterized by the property $\nabla_{\overline{f}} \circ \nu = \rho \circ \nabla$, that is
$$\left(\nabla_{\overline{f}}\right)_{f^* \circ X} = \overline{f}^* \circ \nabla_X, \quad X \in D(M).$$
Indeed, if $X \in D(M)$, then
$$\nabla_{\overline{f}}(\nu(X))(ap) = \rho(\nabla_X)(ap) = \overline{f}^*(X(a)p + a\nabla_X(p))$$
$$= f^*(X(a))\overline{f}^*(p) + f^*(a)\overline{f}^*(\nabla_X(p)) = (\nu(X)(a))\overline{f}^*(p) + f^*(a)\rho(\nabla_X)(p)$$
$$= (\nu(X)(a))\overline{f}^*(p) + f^*(a)\nabla_{\overline{f}}(\nu(X))(p), \quad a \in C^\infty(M), p \in \Gamma(\overline{M}).$$

Since $D(M)_f$ as $B$-module is generated by the image of $\nu$, the above formula leads to

$$\nabla_{\overline{f}}(Y)(ap) = Y(a)\overline{f}^*(p) + f^*(a)\nabla_{\overline{f}}(Y)(p), \quad Y \in D(M)_f,$$

i.e., $\nabla_{\overline{f}}(Y)$ is a der-operator along $\overline{f}^*$ over $Y$.

**Definition.** *The so-constructed linear connection $\nabla_{\overline{f}}$ along $\overline{f}$ is called associated with $\nabla$.*

Given a field $k$ and a commutative $k$-algebra homomorphism $A \to B$, the associated connection along an $A$-module homomorphism of an $A$-module into a $B$-module can be defined exactly in the same way, provided that $D(A)$ is projective and finitely generated.

On the geometric side, it is easy to see that for a given tangent vector $\xi \in T_m M$ and for any point $n \in f^{-1}(m)$ the $\nabla_{\overline{f}}$-lift of $\xi$ at $n$ coincides with the $\nabla$-lift of $\xi$.

To describe the basic covariant differential $d_{\nabla_{\overline{f}}}$ note that for all $X \in D(M)$ and $p \in \Gamma(\overline{M})$ we have

$$d_{\nabla_{\overline{f}}}(p)(X) = (\nabla_{\overline{f}})_{f^* \circ X}(p) = \overline{f}^*(\nabla_X(p))$$
$$= \overline{f}^*(d_\nabla(p)(X)) = \left(\overline{f}^* \circ d_\nabla(p)\right)(X),$$

that is,

$$d_{\nabla_{\overline{f}}}(p) = \overline{f}^* \circ d_\nabla(p). \tag{3.8}$$

In other words, the (basic) covariant differential of $\nabla_{\overline{f}}$ is the composition

$$\Gamma(\overline{M}) \xrightarrow{d_\nabla} \Lambda^1(\overline{M}) \longmapsto \Lambda^1(\overline{M})_{\overline{f}},$$

(see Proposition 3.4.1).

**Exercise.** Let $\overline{g} : \overline{L} \to \overline{N}$ be another fat map. Show that for all $Y \in D(M)_f$ we have

$$\left(\nabla_{\overline{f} \circ \overline{g}}\right)_{g^* \circ Y} = \overline{g}^* \circ \left(\nabla_{\overline{f}}\right)_Y.$$

### 3.4.7

The notion of connection along a fat map leads to a new characterization of compatibility. To this end some simple facts are to be fixed.

First, note that
$$\Lambda^\bullet(\overline{M})_{\overline{f}} = \Lambda^\bullet(M) \otimes \Gamma(\overline{N}) = \Lambda^\bullet(M) \otimes C^\infty(N) \otimes \Gamma(\overline{M}) = \Lambda^\bullet(M)_f \otimes \Gamma(\overline{M}).$$
Second, it is not difficult to see that if $\mathrm{i}_Y$ is the insertion operator into $\Lambda^\bullet(M)_f$ (see Definition 0.5.28), then
$$\overline{\mathrm{i}}_Y = \mathrm{i}_Y \otimes \mathrm{id}_{\Gamma(\overline{M})}.$$
In other words,
$$Y \lrcorner (\omega \otimes p) = (Y \lrcorner \omega) \otimes p, \quad \omega \in \Lambda^\bullet(M)_f, Y \in \mathrm{D}(M)_f, p \in \Gamma(\overline{M}). \quad (3.9)$$
Third, it follows easily from the proof of Proposition 3.4.1 that
$$(f^* \circ \omega) \otimes p = \overline{f}^* \circ (\omega \otimes p), \quad \omega \in \Lambda^\bullet(M)_f, p \in \Gamma(\overline{M}). \quad (3.10)$$

### 3.4.8

The following 'thickened version' of Proposition 0.5.32 holds.

**Proposition.** Let $\overline{f} : \overline{N} \to \overline{M}$ be a fat map. Then
$$(\Lambda^1(\overline{f}^*)(\overline{\omega}))(Z) = (Z \circ f^*) \lrcorner (\overline{f}^* \circ \overline{\omega}), \quad \overline{\omega} \in \Lambda^1(\overline{M}), Z \in \mathrm{D}(N).$$

**Proof.** It suffices to verify the assertion for $\overline{\omega} = \omega \otimes p$, $\omega \in \Lambda^1(M)$ and $p \in \Gamma(\overline{M})$. Taking into account that $\Gamma(\overline{N}) = C^\infty(N) \otimes \Gamma(\overline{M})$, $\Lambda^1(\overline{M}) = \Lambda^1(M) \otimes \Gamma(\overline{M})$ and $\Lambda^1(\overline{M})_{\overline{f}} = \Lambda^1(M)_f \otimes \Gamma(\overline{M})$ this runs as follows:

$$\Lambda^1(\overline{f}^*)(\omega \otimes p))(Z) = \left(\Lambda^1(f^*)(\omega) \otimes p\right)(Z) = (\Lambda^1(f^*)(\omega))(Z) \otimes p$$
$$\stackrel{\mathrm{Prop.\ 0.5.32}}{=} ((Z \circ f^*) \lrcorner (f^* \circ \omega)) \otimes p$$
$$\stackrel{(3.9)}{=} (Z \circ f^*) \lrcorner ((f^* \circ \omega) \otimes p) \stackrel{(3.10)}{=} (Z \circ f^*) \lrcorner (\overline{f}^* \circ (\omega \otimes p)),$$

as required. $\square$

### 3.4.9 Characterization of Compatibility by Means of Associated Linear Connections

**Proposition.** Linear connections $\nabla$ and $\square$ on $\overline{M}$ and $\overline{N}$, respectively, are compatible with respect to $\overline{f} : \overline{N} \to \overline{M}$ if and only if
$$\square_Z \circ \overline{f}^* = \left(\nabla_{\overline{f}}\right)_{Z \circ f^*}, \quad \forall Z \in \mathrm{D}(N).$$

**Proof.** $\overline{f}$-compatibility of $\nabla$ and $\square$ is equivalent to commutativity of the diagram

$$\begin{array}{ccc} \Gamma(\overline{M}) & \xrightarrow{d_\nabla} & \Lambda^1(\overline{M}) \\ \overline{f}^* \downarrow & & \downarrow \Lambda^1 \overline{f}^* \\ \Gamma(\overline{N}) & \xrightarrow{d_\square} & \Lambda^1(\overline{N}) \end{array}$$

(see n. 3.3.13). But for all $p \in \Gamma(\overline{M})$ and $Z \in D(N)$ we have

$$((d_\square \circ \overline{f}^*)(p))(Z) = \square_Z(\overline{f}^*(p))$$

and

$$(\Lambda^1 \overline{f}^* \circ d_\nabla)(p)(Z) = \Lambda^1 \overline{f}^*(d_\nabla(p))(Z)$$
$$\stackrel{\text{Prop. 3.4.8}}{=} (Z \circ f^*) \lrcorner (\overline{f}^* \circ d_\nabla(p)) \stackrel{(3.8)}{=} (Z \circ f^*) \lrcorner d_{\nabla_{\overline{f}}}(p) \stackrel{(3.7)}{=} \left(\nabla_{\overline{f}}\right)_{Z \circ f^*}(p) .$$

Therefore, $\overline{f}$-compatibility is equivalent to

$$\square_Z(\overline{f}^*(p)) = \left(\nabla_{\overline{f}}\right)_{Z \circ f^*}(p), \quad p \in \Gamma(\overline{M}), Z \in D(N) ,$$

i.e.,

$$\square_Z \circ \overline{f}^* = \left(\nabla_{\overline{f}}\right)_{Z \circ f^*}, \quad \forall Z \in D(N) ,$$

as required. $\square$

### 3.4.10 Geometric Characterization of Compatibility

A geometric characterization of compatibility can be given in terms of the lifting procedure for tangent vectors. Let $\overline{f} : \overline{N} \to \overline{M}$ be a fat map and $\nabla$, $\square$ linear connections on $\overline{M}$ and $\overline{N}$, respectively. For a vector $\xi \in T_n N$ put $\overline{\xi} = \square_\xi$ and $\overline{d_n f(\xi)} = \nabla_{d_n f(\xi)}$ (see Definition 2.1.10).

**Proposition.** The linear connections $\nabla$ and $\square$ are $\overline{f}$-compatible if and only if

$$\overline{d_n f(\xi)} = \overline{d}_n \overline{f}(\overline{\xi}), \quad n \in N, \xi \in T_n N .$$

**Proof.** Let $Z$ be a vector field on $N$ and suppose that $\xi = Z_n$, $n \in N$. Since

$$d_n f(\xi) = \xi \circ f^* = Z_n \circ f^* = n \circ Z \circ f^* ,$$

it follows from n. 3.4.4 that
$$\overline{d_n f(\xi)} = \left(\left(\nabla_{\overline{f}}\right)_{Z \circ f^*}\right)_n$$
or, equivalently,
$$\overline{d_n f(\xi)} = \overline{f}_n \circ \overline{h}_n \circ \left(\nabla_{\overline{f}}\right)_{Z \circ f^*},$$
with
$$\overline{h}_n : \Gamma(\overline{M}) \to \overline{n}$$
being the evaluation map (see n. 1.3.13). On the other hand,
$$\overline{d_n \overline{f}}(\overline{\xi}) = \overline{d_n \overline{f}}\left((\square_Z)\overline{n}\right) = \overline{d_n \overline{f}}\left(\overline{h}_n \circ \square_Z\right) = \overline{f}_n \circ \overline{h}_n \circ \square_Z \circ \overline{f}^*.$$
Now the result comes immediately from Proposition 3.4.9. $\square$

### 3.4.11 Equivalent Conditions for Compatibility

Below we recapitulate the characterizations of $\overline{f}$-compatibility of two connections found previously:

(1) $\Lambda^{\bullet} \overline{f}^*$ is a homomorphism of cd-modules over $\Lambda^{\bullet}(M)$.
(2) The diagram

$$\begin{array}{ccc} \Gamma(\overline{M}) & \xrightarrow{d_{\nabla}} & \Lambda^1(\overline{M}) \\ \overline{f}^* \downarrow & & \downarrow \Lambda^1 \overline{f}^* \\ \Gamma(\overline{N}) & \xrightarrow{d_{\square}} & \Lambda^1(\overline{N}) \end{array}$$

commutes.
(3) $\square_Z \circ \overline{f}^* = \left(\nabla_{\overline{f}}\right)_{Z \circ f^*} \quad \forall Z \in D(N)$.
(4) $\overline{d_n \overline{f}}(\square_\xi) = \nabla_{d_n f(\xi)}, \quad n \in N, \xi \in T_n N$.

## 3.5 Covariant Lie Derivative

The *covariant Lie derivative* is another fundamental concept in the theory of linear connections. It is absolutely analogous to the ordinary one. Basically, the covariant Lie derivative is the velocity of the action of one-parameter group of fat diffeomorphisms generated by $\nabla_X$ on fat objects in consideration. The exact meaning of that is given below together with some basic properties.

### 3.5.1 Smooth Families of Fat Maps

First, we shall formalize the intuitively clear notion of a (smooth) family of fat maps (*cf.* Definition 0.6.8).

Let $\overline{T_N}$ be the $T$-cylinder over a fat manifold $\overline{N}$ with fat projection $\overline{\pi} : \overline{T_N} \to \overline{N}$ (see Definition 1.6.13). A fat map

$$\overline{G} : \overline{T_N} \to \overline{M}$$

gives rise to a family

$$\{\overline{G_t}\}_{t \in T} : \quad \overline{G_t} = \overline{G} \circ \overline{i_t}, \quad t \in T,$$

of fat maps $\overline{N} \to \overline{M}$ with $\overline{i_t} : \overline{N} \to \overline{T_N}$ being the fat embedding at $t$.

More generally, let $\overline{U}$ be an open fat submanifold of $\overline{T_N}$. For all $t \in T$ set $N_t = i_t^{-1}(U)$, denote again by $i_t$ the embedding $N_t \to U$ and consider the induced from $\overline{U}$ fat manifold $\overline{N_t}$ with the induced fat map $\overline{i_t} : \overline{N_t} \to \overline{U}$. A fat map

$$\overline{H} : \overline{U} \to \overline{M}$$

gives rise to a family

$$\{\overline{H_t}\}_{t \in T} : \quad \overline{H_t} = \overline{H} \circ \overline{i_t}, \quad t \in T.$$

**Definition.** The family $\{\overline{G_t}\}_{t \in T}$ corresponding to $\overline{G}$ is called a *smooth family of fat maps* $\overline{M} \to \overline{N}$. If $\overline{M} = \overline{N}$ and $\overline{G_t}$ is a fat diffeomorphism for each $t \in T$, then $\{\overline{G_t}\}_{t \in T}$ is called a *smooth family of fat diffeomorphisms* of $\overline{M}$.

The family $\{\overline{H_t}\}_{t \in T}$ corresponding to $\overline{H}$ is called a *smooth family of local fat maps of $\overline{N}$ into $\overline{M}$*.

If $T$ is a nonempty interval not reduced to a singleton and $\overline{U}$ is a relative fat interval, a smooth family of (possibly local) fat maps is called a *one-parameter family*.

### 3.5.2 Lift of One-parameter Families

Recall that the family of embeddings

$$j_n : \mathbb{I}_n \to \mathbb{I}_N, \quad n \in N$$

is associated with a relative interval $\mathbb{I}_N$.

**Definition.** Let $\nabla$ be a linear connection on $\overline{M}$ and $\{\overline{G_t}\}_{t \in \mathbb{I}}$ a one-parameter family corresponding to $\overline{G} : \overline{\mathbb{I}_N} \to \overline{M}$. The one-parameter family $\{\overline{G_t}\}_{t \in \mathbb{I}}$ is called a *lift of* $\{G_t\}_{t \in \mathbb{I}}$ *by* $\nabla$ (or simply $\nabla$-*lift*) if

$$\overline{G_{t_1}}(\mathbf{v}) = \mathbf{T}_{t_0,t_1}^{\nabla,\gamma_n}(\overline{G_{t_0}}(\mathbf{v})), \quad t_0, t_1 \in \mathbb{I}, \ \mathbf{v} \in \overline{n}, \ n \in N \,,$$

where $\mathbf{T}_{t_0,t_1}^{\nabla,\gamma_n}$ is the parallel translation from $t_0$ to $t_1$ along the curve

$$\gamma_n \stackrel{\text{def}}{=} G \circ j_n : \mathbb{I}_n \to M, \quad t \mapsto G_t(n) \,.$$

By analogy with the lift of a curve, it is natural to expect that the fat map $\overline{G}$ defining a lift of a one-parameter family could be characterized by a kind of compatibility condition with respect to $\nabla$. To this end, a notion of the 'trivial connection on $\overline{\mathbb{I}_N}$ over $\overline{N}$' is necessary. It is not difficult to do that, but it is not our goal here.

The following result assures the existence of a lift under the hypothesis that the map $G_0$ is global, *i.e.*, $N_0 = N$ at least. We emphasize that the only thing to be proved is smoothness of the family $\{\overline{G_t}\}_{t \in \mathbb{I}}$ from the above definition.

**Proposition.** Let $\nabla$ be a linear connection on a fat manifold $\overline{M}$, $\{G_t\}_{t \in \mathbb{I}}$ a one-parameter family and suppose that $\overline{g} : \overline{N} \to \overline{M}$ is a fat map over $g = G_0$. Then there exists a unique lift $\{\overline{G_t}\}_{t \in \mathbb{I}}$ of $\{G_t\}_{t \in \mathbb{I}}$ such that $\overline{G_0} = \overline{g}$.

***Proof.*** Let $\overline{\mathbb{I}_N}$ be the induced by $G : \mathbb{I}_N \to M$ from $\overline{M}$ fat manifold, $\overline{G}' : \overline{\mathbb{I}_N} \to \overline{M}$ the induced fat map and $\square$ the induced from $\nabla$ linear connection on $\overline{\mathbb{I}_N}$. Note also that, according to the universal property of induced bundles, there exists a unique fat map $\overline{i_0} : \overline{N} \to \overline{\mathbb{I}_N}$ over the embedding $N \to \mathbb{I}_N$ at $t = 0$ such that $\overline{g} = \overline{G}' \circ \overline{i_0}$.

Let $\overline{X} = \square_{\partial/\partial t}$, $\partial/\partial t$ being the standard vector field on the relative interval $\mathbb{I}_N$. Then, according to Theorem 1.6.15 ([1]), $\overline{i_0}$ is extended to a fat map

$$\overline{\Phi_{\overline{i_0}}} : \overline{\mathbb{I}_N}' \to \overline{\mathbb{I}_N}$$

by means of $\overline{X}$. Note that $\overline{\Phi_{\overline{i_0}}}$ is a fat identity map.

Now it is easy to see that the family $\{\overline{G_t}\}_{t \in \mathbb{I}}$ corresponding to

$$\overline{G} = \overline{G}' \circ \overline{\Phi_{\overline{i_0}}} \,,$$

possesses the required properties. The uniqueness is immediate. $\square$

---

[1] In this situation, it is quite easy to extend Theorem 1.6.15 to the case when $\overline{\mathbb{I}_N}$ has nonempty boundary. Details are left to the reader (see n. 1.6.10).

The above result shows that every one-parameter family $\{G_t\}_{t\in\mathbb{I}}$ with $G_0$ globally defined, admits many lifts. Indeed, if the fat manifold $\overline{N}$ is induced from $\overline{M}$ by $G_0$, then all other lifts are parametrized by fat identity maps $\overline{f}: \overline{N}' \to \overline{N}$: the corresponding family is $\{\overline{f} \circ \overline{G_t}\}_{t\in\mathbb{I}}$.

By substituting $G_0$ with a suitable local section of $\mathbb{I}_N \to N$, it is not difficult to prove that *every* one-parameter family admits lifts. However, we have no need in this generalization.

### 3.5.3

If $\{G_t\}_{t\in\mathbb{I}}$ is a one-parameter family of diffeomorphisms of $M$, a lift $\{\overline{G_t}\}_{t\in\mathbb{I}}$ is a one-parameter family of (possibly local) fat diffeomorphisms of $\overline{M}'$ into $\overline{M}$ with $\overline{M}'$ being a suitable fat manifold over $M$ which is, generally, different from $\overline{M}$. $\{\overline{G_t}\}_{t\in\mathbb{I}}$ is a one-parameter family of fat diffeomorphisms of $\overline{M}$, i.e., $\overline{M}' = \overline{M}$, if $G_0$ is lifted to a fat diffeomorphism $\overline{G_0}: \overline{M} \to \overline{M}$.

### 3.5.4 Standard Fat Field on a Relative Fat Interval

Since the standard vector field $\partial/\partial t$ on a relative interval $\mathbb{I}_M$ projects onto the zero vector field on $M$ (see n. 0.6.3), it is $C^\infty(M)$-linear as a map of $C^\infty(\mathbb{I}_M)$ into itself, assuming that $C^\infty(\mathbb{I}_M)$ is supplied with the $C^\infty(M)$-module structure via the projection $\mathbb{I}_M \to M$.

Consider $\overline{\mathbb{I}_M}$, a relative fat interval with base $\mathbb{I}_M$, so that

$$\Gamma(\overline{\mathbb{I}_M}) = C^\infty(\mathbb{I}_M) \otimes_{C^\infty(M)} \Gamma(\overline{M}),$$

and the fat field

$$\overline{\frac{\partial}{\partial t}} \stackrel{\text{def}}{=} \frac{\partial}{\partial t} \otimes \mathrm{id}_{\Gamma(\overline{M})}$$

on $\overline{\mathbb{I}_M}$. Obviously, $\overline{\partial/\partial t}$ is the fat field over $\partial/\partial t$ projecting to the zero fat field on $\overline{M}$.

**Definition.** The fat field $\overline{\partial/\partial t}$ on $\overline{\mathbb{I}_M}$ is called *standard*.

The fat field $\overline{\partial/\partial t}$ is $\overline{j_m}$-compatible with the standard fat field $\overline{\mathrm{d}/\mathrm{d}t}$ on $\overline{\mathbb{I}_m}$.

**Exercise.** Let $\{G_t\}_{t\in\mathbb{I}}$ be a one-parameter family and $\{\overline{G_t}\}_{t\in\mathbb{I}}$ be a lift by means of a linear connection $\nabla$ on $\overline{M}$. Denote by $\square$ the linear connection induced by $\overline{G} : \overline{\mathbb{I}_N} \to \overline{M}$ from $\nabla$. Show that

$$\Box_{\frac{\partial}{\partial t}} = \overline{\frac{\partial}{\partial t}}.$$

### 3.5.5

Let $\{G_t\}_{t \in \mathbb{I}}$ be a one-parameter family of local smooth maps of a manifold $N$ into a manifold $M$ corresponding to the map $G : \mathbb{I}_N \to M$. Recall that the operator

$$i_{t_0}^* \circ \frac{\partial}{\partial t} \circ G^* : C^\infty(M) \to C^\infty(N_{t_0}),$$

with $i_{t_0} : N_{t_0} \to \mathbb{I}_N$ being the embedding at $t_0 \in \mathbb{I}$, is sometimes denoted by

$$\left. \frac{\mathrm{d}}{\mathrm{d}t} \right|_{t=t_0} G_t^*$$

(*cf.* nn. 0.6.7, 0.6.12 and 0.6.19).

Similarly, if $\{\overline{G_t}\}_{t \in \mathbb{I}}$ is a one-parameter family of local fat maps of a fat manifold $\overline{N}$ into a fat manifold $\overline{M}$ corresponding to the fat map $\overline{G} : \overline{\mathbb{I}_N} \to \overline{M}$, the operator

$$\overline{i_{t_0}}^* \circ \overline{\frac{\partial}{\partial t}} \circ \overline{G}^* : \Gamma(\overline{M}) \to \Gamma(\overline{N_{t_0}}),$$

with $\overline{i_{t_0}} : \overline{N_{t_0}} \to \overline{\mathbb{I}_N}$ being the fat embedding at $t_0 \in \mathbb{I}$, will be sometimes denoted by

$$\left. \frac{\mathrm{d}}{\mathrm{d}t} \right|_{t=t_0} \overline{G_t}^*.$$

### 3.5.6

If $\{\overline{G_t}\}_{t \in \mathbb{I}}$ is the one-parameter group generated by a fat field $\overline{X}$, then

$$\overline{\frac{\partial}{\partial t}} \circ \overline{G}^* = \overline{G}^* \circ \overline{X}.$$

By applying on the left $\overline{i_{t_0}}^*$ to this formula, $\overline{i_{t_0}} : \overline{M_{t_0}} \to \overline{\mathbb{I}_M}$ being the fat embedding at $t_0$, we get

$$\overline{i_{t_0}}^* \circ \overline{\frac{\partial}{\partial t}} \circ \overline{G}^* = \overline{G_{t_0}}^* \circ \overline{X},$$

or, equivalently,

$$\left.\frac{d}{dt}\right|_{t=t_0} \overline{G_t}^* = \overline{G_{t_0}}^* \circ \overline{X} . \qquad (3.11)$$

For $t_0 = 0$ we have

$$\left.\frac{d}{dt}\right|_{t=0} \overline{G_t}^* = \overline{X} . \qquad (3.12)$$

### 3.5.7  *Geometric Interpretation of the Covariant Derivative*

Consider now a linear connection $\nabla$ on $\overline{M}$, the one-parameter group $\{G_t\}_{t\in\mathbb{R}}$ generated by a vector field $X$ on $M$ and its $\nabla$-lift $\{\overline{G_t}\}_{t\in\mathbb{R}}$ corresponding to the initial data $\overline{G_0} = \mathrm{id}_{\overline{M}}$ (see Proposition 3.5.2). From the result of Exercise 3.5.4 and Proposition 3.3.14 it follows that $\{\overline{G_t}\}_{t\in\mathbb{R}}$ is the one-parameter group generated by $\nabla_X$. Therefore, according to (3.11), we have

$$\left.\frac{d}{dt}\right|_{t=t_0} \overline{G_t}^* = \overline{G_{t_0}}^* \circ \nabla_X . \qquad (3.13)$$

In particular,

$$\nabla_X = \left.\frac{d}{dt}\right|_{t=0} \overline{G_t}^* . \qquad (3.14)$$

This interprets the fat field $\nabla_X$ to be an infinitesimal fat diffeomorphism of $\overline{M}$ lifted by $\nabla$ from the infinitesimal diffeomorphism $X$ of $M$.

### 3.5.8

Formula (3.14) can be extended to arbitrary one-parameter families of local diffeomorphisms. For simplicity, consider here one-parameter family of (globally defined) diffeomorphisms only. So, $\{\overline{G_t}\}_{t\in\mathbb{I}}$ stands now for a one-parameter family of diffeomorphisms of a manifold $M$ and the corresponding map

$$G : \mathbb{I}_M \to M$$

is defined on $\mathbb{I}_M = M \times \mathbb{I}$.

We are looking for an analogue of the formula

$$\frac{\partial}{\partial t} \circ G^* = \mathbb{G}^* \circ X , \qquad (3.15)$$

with $X \in \mathrm{D}(M)_\mathbb{I}$ corresponding to the time-dependent vector field $\{X_t\}_{t\in\mathbb{I}}$ associated with $\{G_t\}_{t\in\mathbb{I}}$ (see n. 0.6.12).

Let $\nabla$ be a linear connection on $\overline{M}$ and $\{\overline{G_t}\}_{t\in \mathbb{I}}$ the $\nabla$-lift of $\{G_t\}_{t\in \mathbb{I}}$. Denote by
$$\overline{G} : \overline{\mathbb{I}_M} \to \overline{M},$$
the fat map corresponding to $\{\overline{G_t}\}_{t\in\mathbb{I}}$ and let $\square$ be the induced by $\overline{G}$ from $\nabla$ linear connection.

If $\overline{\pi} : \overline{\mathbb{I}_M} \to \overline{M}$ is the canonical fat projection, then, according to the universal property of induced bundles, there exists a unique fat map
$$\overline{\mathbb{G}} : \overline{\mathbb{I}_M} \to \overline{\mathbb{I}_M}$$
over $\mathbb{G}$ such that $\overline{\pi} \circ \overline{\mathbb{G}} = \overline{G}$. Finally, denote by $\nabla^{\mathbb{I}}$ the linear connection associated with $\nabla$ via $\overline{\pi}$, i.e., in notation of Definition 3.4.6
$$\nabla^{\mathbb{I}} \stackrel{\text{def}}{=} \nabla_{\overline{\pi}}.$$

Since $\nabla^{\mathbb{I}}$ is a linear connection along $\overline{\pi}$ and
$$X = \left(\mathbb{G}^{-1}\right)^* \circ \frac{\partial}{\partial t} \circ G^*$$
is a vector field along $\pi$,
$$\nabla^{\mathbb{I}}_X \stackrel{\text{def}}{=} \left(\nabla^{\mathbb{I}}\right)_X$$
is a fat vector field along $\overline{\pi}$.

**Proposition.** The following formula takes place
$$\frac{\partial}{\partial t} \circ \overline{G}^* = \overline{\mathbb{G}}^* \circ \nabla^{\mathbb{I}}_X$$

**Proof.**
$$\frac{\partial}{\partial t} \circ \overline{G}^* \stackrel{\text{Exer. 3.5.4}}{=} \square_{\frac{\partial}{\partial t}} \circ \overline{G}^* \stackrel{\text{Prop. 3.4.9}}{=} \left(\nabla_{\overline{G}}\right)_{\frac{\partial}{\partial t} \circ G^*} \stackrel{(3.15)}{=} \left(\nabla_{\overline{\pi}\circ\overline{\mathbb{G}}}\right)_{\mathbb{G}^* \circ X}$$
$$\stackrel{\text{Exer. 3.4.6}}{=} \overline{\mathbb{G}}^* \circ \nabla^{\mathbb{I}}_X \cdot \square$$

Alternatively, the above formula can be read as
$$\frac{\mathrm{d}}{\mathrm{d}t} \overline{G_t}^* = \overline{G_t}^* \circ \nabla_{X_t}.$$

Indeed,
$$\overline{i_{t_0}}^* \circ \frac{\partial}{\partial t} \circ \overline{G}^* = \overline{i_{t_0}}^* \circ \overline{\mathbb{G}}^* \circ \nabla^{\mathbb{I}}_X = \overline{G_{t_0}}^* \circ \overline{i_{t_0}}^* \circ \nabla^{\mathbb{I}}_X \stackrel{\text{Exer. 3.4.6}}{=} \overline{G_{t_0}}^* \circ \nabla_{X_{t_0}},$$
or, equivalently,
$$\left.\frac{\mathrm{d}}{\mathrm{d}t}\right|_{t=t_0} \overline{G_t}^* = \overline{G_{t_0}}^* \circ \nabla_{X_{t_0}}.$$

### 3.5.9 Local Thickened Forms

The preceding results concern operators acting on $\Gamma(\overline{M}) = \Lambda^0(\overline{M})$. In fact, they extend naturally to all thickened differential forms. To do that it is useful to introduce *local* thickened forms.

Let $\overline{U}$ be an open fat submanifold of $\overline{M}$ and $\overline{i} : \overline{U} \hookrightarrow \overline{M}$ the corresponding fat embedding. In n. 0.5.33 we proved that an $s$-form $\omega$ on $U$ is naturally identified with an $s$-form $\varkappa$ along $i$ by means of the formula

$$\omega\left(X_1|_U, \ldots, X_s|_U\right) = \varkappa(X_1, \ldots, X_s), \quad X_1, \ldots, X_s \in D(M).$$

In other words, we have a natural $C^\infty(U)$-module isomorphism $\Lambda^s(U) \cong \Lambda^s(M)_i$. Since $\Lambda^s(\overline{U}) = \Lambda^s(U) \otimes_{C^\infty(U)} \Gamma(U)$ and $\Lambda^s(\overline{M})_{\overline{i}} = \Lambda^s(M)_i \otimes_{C^\infty(U)} \Gamma(U)$ (see Propositions 0.5.10 and 3.4.1), we get a natural isomorphism $\Lambda^s(\overline{U}) \cong \Lambda^s(\overline{M})_{\overline{i}}$, simply by tensoring the isomorphism $\Lambda^s(U) \cong \Lambda^s(M)_i$ by $\mathrm{id}_{\Gamma(U)}$. So, the thickened analogous of the preceding formula is

$$\overline{\omega}\left(X_1|_U, \ldots, X_s|_U\right) = \overline{\varkappa}(X_1, \ldots, X_s), \quad X_1, \ldots, X_s \in D(M)$$

with $\overline{\omega} \in \Lambda^s(\overline{U})$ and $\overline{\varkappa} \in \Lambda^s(\overline{M})_{\overline{i}}$ corresponding to each other.

**Definition.** A *local thickened form on* $\overline{M}$ is a form along $\overline{i} : \overline{U} \hookrightarrow \overline{M}$, and it is sometimes identified with the corresponding thickened form on $\overline{U}$.

### 3.5.10 Time-dependent Thickened Forms

Let $\overline{\pi} : \overline{T_M} \to \overline{M}$ the fat projection of a $T$-cylinder over $\overline{M}$ and $\overline{\Omega}$ a degree $s$ thickened form along $\overline{\pi}$. By putting

$$\overline{\omega}_t(X_1, \ldots, X_s) = \overline{i}_t^*\left(\overline{\Omega}(X_1, \ldots, X_s)\right), \quad X_1, \ldots, X_s \in D(M),$$

$$\overline{i}_t : \overline{M} \to \overline{T_M}$$

being the fat embedding at $t \in T$, we get the family

$$\{\overline{\omega}_t\}_{t \in T}$$

of degree $s$ thickened forms on $\overline{M}$.

More generally, let $\overline{U}$ be an open fat submanifold of $\overline{T_M}$, $\overline{K}$ an $s$-form along the restriction $\overline{U} \to \overline{M}$ of $\overline{\pi}$, set $\overline{M_t} = \overline{i}_t^{-1}\left(\overline{U}\right)$ and denote again by $\overline{i}_t$ the restriction $\overline{M_t} \to \overline{U}$. Then

$$\overline{\varkappa}_t(X_1, \ldots, X_s) = \overline{i}_t^*\left(\overline{K}(X_1, \ldots, X_s)\right), \quad X_1, \ldots, X_s \in D(M)$$

gives a family

$$\{\overline{\varkappa}_t\}_{t \in T}$$

of local thickened forms on $\overline{M}$.

**Definition.** The families $\{\overline{\omega_t}\}_{t \in T}$ and $\{\overline{\varkappa_t}\}_{t \in T}$ will be called *smooth families*, defined by $\overline{\Omega}$ and $\overline{K}$, respectively.

If $T$ is a nonempty interval, not reduced to a single point, and $\overline{U}$ a relative fat interval, then a family of (possibly local) thickened forms is sometimes called a *time-dependent thickened form on* $\overline{M}$.

**Exercise.** Let $\overline{\omega}$ be a thickened form of degree $s$ on $\overline{U}$. Show that
$$\left\{ \Lambda^s \, \overline{i_t}^*(\overline{\omega}) \right\}_{t \in T}$$
is a smooth family (up to the natural identification introduced in n. 3.5.9).

**Hint.** *Cf.* Exercise 0.6.17.

### 3.5.11 Derivation of Time-dependent Thickened Forms

Let $\{\overline{\omega_t}\}_{t \in \mathbb{I}}$ be a time-dependent thickened form of degree $s$ on a fat manifold $\overline{M}$, defined by a thickened form $\overline{\Omega}$ along the fat projection $\overline{\pi} : \overline{\mathbb{I}_M} \to \overline{M}$, and $\overline{\partial/\partial t}$ be the standard fat field on the relative fat interval $\overline{\mathbb{I}_M}$. Since $\overline{\partial/\partial t}$ is a $C^\infty(M)$–module endomorphism of $\Gamma(\overline{\mathbb{I}_M})$,
$$(X_1, \ldots, X_s) \mapsto \overline{\frac{\partial}{\partial t}} \left( \overline{\Omega}(X_1, \ldots, X_s) \right), \quad X_1, \ldots, X_s \in \mathrm{D}(M)$$
is a thickened $s$-form $\overline{\Omega'}$ along $\overline{\pi}$ too.

**Definition.** The time-dependent thickened form $\left\{\overline{\omega'_t}\right\}_{t \in \mathbb{I}}$ defined by $\overline{\Omega'}$ is called the *derivative of* $\{\overline{\omega_t}\}_{t \in \mathbb{I}}$.

The local thickened form $\overline{\omega'_{t_0}}$, $t_0 \in \mathbb{I}$, will be sometimes denoted by
$$\left. \frac{\mathrm{d}\,\overline{\omega_t}}{\mathrm{d}\,t} \right|_{t=t_0} \quad \text{or} \quad \left. \frac{\mathrm{d}}{\mathrm{d}\,t} \right|_{t=t_0} \overline{\omega_t}.$$

### 3.5.12 Lie Derivative Along Fat Fields

Let $\overline{X}$ be a fat field on a fat manifold without boundary $\overline{M}$ and $\overline{\Phi} : \overline{\mathbb{I}_M} \to \overline{M}$ the generated by $\overline{X}$ fat flow. The one-parameter family $\{\overline{\Phi_t}\}_{t \in \mathbb{R}}$ corresponding to $\overline{\Phi}$ is called the *one-parameter group generated by* $\overline{X}$.

If $\overline{\omega}$ is a thickened form of degree $s$, then
$$\left\{ \Lambda^s \, \overline{\Phi_t}^*(\overline{\omega}) \right\}_{t \in \mathbb{R}}$$

is a smooth time-dependent (local) thickened form. Moreover, since $\overline{\Phi}_0$ is the identity map of $\overline{M}$, the $s$-form
$$\left.\frac{\mathrm{d}}{\mathrm{d}t}\right|_{t=0} \Lambda^s \overline{\Phi}_t^{\,*}(\overline{\omega})$$
is global, i.e., it is actually a thickened form on $\overline{M}$.

**Definition.** The $s$-form
$$\mathcal{L}_{\overline{X}}(\overline{\omega}) \stackrel{\text{def}}{=} \left.\frac{\mathrm{d}}{\mathrm{d}t}\right|_{t=0} \Lambda^s \overline{\Phi}_t^{\,*}(\overline{\omega})$$
is called the *Lie derivative of $\overline{\omega}$ along $\overline{X}$*.

The notation $\mathcal{L}_{\overline{X}}$ will refer to the operator $\Lambda^\bullet(\overline{M}) \to \Lambda^\bullet(\overline{M})$ whose homogeneous components are given by
$$\overline{\omega} \mapsto \mathcal{L}_{\overline{X}}(\overline{\omega}) \,.$$

It is natural to denote the operator
$$\overline{\omega} \mapsto \left.\frac{\mathrm{d}}{\mathrm{d}t}\right|_{t=0} \Lambda^s \overline{\Phi}_t^{\,*}(\overline{\omega}) \,,$$
by
$$\left.\frac{\mathrm{d}}{\mathrm{d}t}\right|_{t=0} \Lambda^\bullet \overline{\Phi}_t^{\,*} \,,$$
so that
$$\mathcal{L}_{\overline{X}} = \left.\frac{\mathrm{d}}{\mathrm{d}t}\right|_{t=0} \Lambda^\bullet \overline{\Phi}_t^{\,*} \,.$$

According to (3.12) for $s = 0$ the Lie derivative $\mathcal{L}_{\overline{X}}$ reduces to $\overline{X}$, that is
$$\mathcal{L}_{\overline{X}}(p) = \overline{X}(p), \quad p \in \Gamma(\overline{M}) \,.$$

Recall that for an ordinary local $s$-form $\omega$ on a relative interval $\mathbb{I}_M$ the derivative of the time-dependent form
$$\{\Lambda^s i_t^*(\omega)\}_{t \in \mathbb{I}} \,,$$
with $i_t : M_t \to \mathbb{I}_M$ being the embedding at $t$, is the time-dependent form
$$\left\{\Lambda^s i_t^* \left(\mathcal{L}_{\frac{\partial}{\partial t}}(\omega)\right)\right\}_{t \in \mathbb{I}} \,,$$
where $\mathcal{L}_{\partial/\partial t} = [i_{\partial/\partial t}, \mathrm{d}]^{(\mathrm{gr})}$ is the Lie derivative along $\partial/\partial t$ (see Proposition 0.6.18 and n. 0.6.20).

Consider now a relative fat interval $\overline{\mathbb{I}_M}$ over $\overline{M}$, $\omega \in \Lambda^s(\mathbb{I}_M)$, $p \in \Gamma(\overline{\mathbb{I}_M})$ and the thickened $s$-form $\omega \wedge p \in \Lambda^s(\overline{\mathbb{I}_M})$.

**Exercise.** Show that the derivative of the time-dependent thickened form
$$\left\{ \Lambda^s \overline{i_t}^* (\omega \wedge p) \right\}_{t \in \mathbb{I}},$$
$\overline{i_t} : \overline{M_t} \to \overline{\mathbb{I}_M}$ being the fat embedding at $t$, is
$$\left\{ \Lambda^s \overline{i_t}^* \left( \mathcal{L}_{\frac{\partial}{\partial t}}(\omega) \wedge p + \omega \wedge \overline{\frac{\partial}{\partial t}} p \right) \right\}_{t \in \mathbb{I}}.$$

**Proposition.** Let $\overline{X}$ be a fat field on a fat manifold $\overline{M}$ without boundary, $\omega \in \Lambda^s(M)$, $p \in \Gamma(\overline{M})$. Then
$$\mathcal{L}_{\overline{X}}(\omega \wedge p) = \mathcal{L}_X(\omega) \wedge p + \omega \wedge \overline{X}(p).$$

**Proof.** It comes easily from the Exercise, taking into account that $\overline{\partial/\partial t}$ and $\overline{X}$ are compatible with respect to the fat flow of $\overline{X}$. $\square$

The above formula completely determines the operator $\mathcal{L}_{\overline{X}}$, because $\Lambda^s(\overline{M}) = \Lambda^s(M) \otimes_{C^\infty(M)} \Gamma(\overline{M})$ and $\omega \otimes p = \omega \wedge p$.

The above result extends naturally to the Leibnitz rule
$$\mathcal{L}_{\overline{X}}(\omega \wedge \overline{\varkappa}) = \mathcal{L}_X(\omega) \wedge \overline{\varkappa} + \omega \wedge \mathcal{L}_{\overline{X}}(\overline{\varkappa}), \quad \omega \in \Lambda^\bullet(M), \overline{\varkappa} \in \Lambda^\bullet(\overline{M}),$$
as it follows easily from n. 0.6.21.

### 3.5.13

Let $\overline{\mathbb{I}_M}$ be a relative fat interval over $\overline{M}$ and $\overline{\omega}$ a thickened $s$-form on it. From Exercise 3.5.12 and Proposition 3.5.12 it follows that the derivative of the time-dependent thickened form
$$\left\{ \Lambda^s \overline{i_t}^* (\overline{\omega}) \right\}_{t \in \mathbb{I}},$$
$\overline{i_t} : \overline{M_t} \to \overline{\mathbb{I}_M}$ being the fat embedding at $t$, is
$$\left\{ \Lambda^s \overline{i_t}^* \left( \mathcal{L}_{\overline{\frac{\partial}{\partial t}}}(\overline{\omega}) \right) \right\}_{t \in \mathbb{I}}.$$

### 3.5.14

**Proposition.** Let $\overline{M}$ be a fat manifold without boundary, $\overline{X} \in \overline{D}(\overline{M})$, $\overline{\omega} \in \Lambda^s(\overline{M})$. Then

(1)
$$\mathcal{L}_{\overline{X}}(\overline{\omega})(X_1, \ldots, X_s) = \overline{X}(\overline{\omega}(X_1, \ldots, X_s))$$
$$- \sum_i \overline{\omega}(X_1, \ldots, [X, X_i], \ldots, X_s), \quad X_1, \ldots, X_s \in D(M);$$

(2) $[\mathcal{L}_{\overline{X}}, \bar{\mathrm{i}}_Y] = \bar{\mathrm{i}}_{[X,Y]}$, $Y \in \mathrm{D}(M)$.

**Proof.** According to the representation $\Lambda^\bullet(\overline{M}) = \Lambda^\bullet(M) \otimes_{C^\infty(M)} \Gamma(\overline{M})$, it suffices to check both formulas on simple thickened forms $\overline{\omega} = \overline{\omega} \otimes p = \omega \wedge p$ only. This is, however, immediate in view of the result of Exercise 0.6.22 and Proposition 3.5.12. $\square$

### 3.5.15 Covariant Lie Derivative

Let $\nabla$ be a linear connection on a fat manifold without boundary $\overline{M}$ and $X$ a vector field on $M$. If $\{\Phi_t\}_{t\in\mathbb{R}}$ is the one-parameter group generated by $X$, then the one-parameter group generated by $\nabla_X$ is the lift $\{\overline{\Phi_t}\}_{t\in\mathbb{R}}$ of it by means of $\nabla$ specified by $\overline{\Phi_0} = \mathrm{id}_{\overline{M}}$. So, it is natural to introduce the following terminology.

**Definition.** The operator

$$\mathcal{L}_{\nabla_X} = \left.\frac{\mathrm{d}}{\mathrm{d}t}\right|_{t=0} \Lambda^\bullet \overline{\Phi_t}^*$$

is called *covariant Lie derivative along* $X$ and denoted by $\mathcal{L}_X^\nabla$.

The following is an analogue of the *Cartan formula* (see Proposition 0.6.20).

**Proposition.** It holds

$$\mathcal{L}_X^\nabla = \left[\bar{\mathrm{i}}_X, \mathrm{d}_\nabla\right]^{(\mathrm{gr})}$$

with $\bar{\mathrm{i}}_X$ being the insertion of $X$ into $\Lambda^\bullet\left(\overline{M}\right)$ operator.

**Proof.** According to the representation $\Lambda^\bullet(\overline{M}) = \Lambda^\bullet(M) \otimes_{C^\infty(M)} \Gamma(\overline{M})$, it suffices to verify that the action of operators in question coincide on simple thickened forms $\overline{\omega} = \overline{\omega} \otimes p = \omega \wedge p$. But this is straightforward in view of Propositions 0.6.20, 3.2.11, 3.5.12 and the Leibnitz rule for insertion operators (it is easy to extend Proposition 0.5.19 to thickened forms). $\square$

The Cartan formula interpreted as a definition allows to extend the notion of covariant Lie derivative to linear connections defined on an arbitrary module over a commutative algebra. In particular, it is defined for fat manifolds with boundary.

### 3.5.16  Natural Actions on the Space of Linear Connections

Recall that linear connections on a fat manifold $\overline{M}$ constitute an affine space modelled over the module $\Lambda^1\left(\mathrm{End}\left(\Gamma(\overline{M})\right)\right)$ (see n. 2.1.9). The group of fat diffeomorphisms of $\overline{M}$ acts naturally on this space. Namely, the correspondent $T_{\overline{f}}(\nabla)$ of $\nabla$ by the action of $\overline{f} : \overline{M} \to \overline{M}$ is defined to be the induced by $\overline{f}$ from $\nabla$ linear connection. Compatibility condition (2) in n. 3.4.11 gives an explicit formula for this action in terms of the zeroth component of the covariant differential:

$$\mathrm{d}_{T_{\overline{f}}(\nabla)} = \Lambda^1 \overline{f}^* \circ \mathrm{d}_{\nabla} \circ \overline{f}^{*-1}, \qquad (3.16)$$

or, equivalently,

$$\left(T_{\overline{f}}(\nabla)\right)_X = \overline{f}^* \circ \nabla_{f_*(X)} \circ \overline{f}^{*-1}, \quad X \in \mathrm{D}(M) .$$

In particular, the above formulas define an action of the gauge group $\mathrm{GL}(\overline{M})$ of fat identity maps on the affine space of all connections on $\overline{M}$.

As in many similar situations there is an infinitesimal version of the above action. Namely, the action of a fat vector field $\overline{X}$ on a connection $\nabla$ is defined as

$$\mathcal{L}_{\overline{X}}(\nabla) = \left.\frac{\mathrm{d}}{\mathrm{d}t}\right|_{t=0} (\nabla_t) ,$$

where $\nabla_t$ stands for the induced by $\overline{\Phi}_t : \overline{M_t} \to \overline{M}$ from $\nabla$ linear connection with $\{\overline{\Phi}_t\}_{t \in \mathbb{R}}$ being the fat flow generated by $\overline{X}$. A clarity must be, however, introduced into this heuristic definition in order to make it formally rigorous. More exactly, the derivative in this right hand side is to be duly defined. The only thing to be done on this concern is to define the increment $\nabla_{t+\Delta t} - \nabla_t$. But since connections form an affine space this difference is well defined and belongs to $\Lambda^1\left(\mathrm{End}\left(\Gamma(\overline{M})\right)\right)$ (see n. 2.1.9). A natural interpretation of the right-hand side of the heuristic definition in terms of covariant differential, namely,

$$\left.\frac{\mathrm{d}}{\mathrm{d}t}\right|_{t=0} (\mathrm{d}_{\nabla_t}) : \Gamma(\overline{M}) \to \Lambda^1(\overline{M}), \quad p \mapsto \left.\frac{\mathrm{d}}{\mathrm{d}t}\right|_{t=0} [\mathrm{d}_{\nabla_t}(p|_{M_t})] ,$$

allows to avoid this problem and we shall adopt it. It is not difficult to check smoothness of $\mathrm{d}_{\nabla_t}(p|_{M_t})$ with respect to $t$, so that the above formula is well defined.

In this connection note that for all $a \in C^\infty(M)$ and $p \in \Gamma(\overline{M})$,

$$\left.\frac{\mathrm{d}}{\mathrm{d}t}\right|_{t=0} [\mathrm{d}_{\nabla_t}((ap)|_{M_t})] = \left.\frac{\mathrm{d}}{\mathrm{d}t}\right|_{t=0} \left(\mathrm{d}a|_{M_t} \wedge p|_{M_t} + a|_{M_t} \mathrm{d}_{\nabla_t}(p|_{M_t})\right)$$

$$= a \left.\frac{\mathrm{d}}{\mathrm{d}t}\right|_{t=0} [\mathrm{d}_{\nabla_t}(p|_{M_t})]$$

because $da \wedge p$ is constant with respect to $t$. Hence $(d/dt)|_{t=0}(d_{\nabla_t})$ is a $C^\infty(M)$-module homomorphism $\Gamma(\overline{M}) \to \Lambda^1(\overline{M})$, which is naturally identified with an element $\mathcal{L}_{\overline{X}}(\nabla) \in \Lambda^1\left(\mathrm{End}\left(\Gamma(\overline{M})\right)\right)$, as expected.

An explicit formula for $\mathcal{L}_{\overline{X}}(\nabla)$ can be heuristically deduced from (3.16) in the following way:

$$\frac{d}{dt}\bigg|_{t=0} \left(\Lambda^1 \Phi_t^* \circ d_\nabla \circ \Phi_t^{*-1}\right)$$
$$= \left(\frac{d}{dt}\bigg|_{t=0} \Lambda^1 \Phi_t^*\right) \circ d_\nabla \circ \Phi_0^{*-1} + \Lambda^1 \Phi_0^* \circ d_\nabla \circ \left(\frac{d}{dt}\bigg|_{t=0} \Phi_t^{*-1}\right)$$
$$= \mathcal{L}_{\overline{X}} \circ d_\nabla - d_\nabla \circ \overline{X},$$

which is the zeroth degree component of the commutator $[\mathcal{L}_{\overline{X}}, d_\nabla]^{(\mathrm{gr})}$. The following two exercises give a due rigor to this procedure.

### 3.5.17

Let $\nabla$ be a linear connection on $\overline{M}$, $\overline{\mathbb{I}_M}$ a relative fat interval and $\overline{\omega}$ a thickened $s$-form on it. Put $\overline{\omega_t} = \Lambda^s \overline{i_t}^*(\overline{\omega})$ with $\overline{i_t}: \overline{M_t} \to \overline{\mathbb{I}_M}$ being the fat embedding at $t$. If $\square$ is induced from $\nabla$ by the fat projection $\overline{\mathbb{I}_M} \to \overline{M}$ connection, then $\Lambda^{s+1} \overline{i_t}^*(d_\square \overline{\omega}) = d_{\nabla|_{\overline{M_t}}}(\overline{\omega_t})$. Hence $\{d_{\nabla|_{\overline{M_t}}}(\overline{\omega_t})\}_{t\in \mathbb{I}}$ is a (smooth) time-dependent thickened form.

**Exercise.** Show that the derivative of $\{d_{\nabla|_{\overline{M_t}}}(\overline{\omega_t})\}_{t\in \mathbb{I}}$ is the family

$$\left\{d_{\nabla|_{\overline{M_t}}}\left(\overline{\omega_t'}\right)\right\}_{t\in \mathbb{I}},$$

with $\left\{\overline{\omega_t'}\right\}_{t\in \mathbb{I}}$ being the derivative of $\{\overline{\omega_t}\}_{t\in \mathbb{I}}$.

**Hint.** According to the representation $\Lambda^s(\overline{\mathbb{I}_M}) = \Lambda^s(\mathbb{I}_M) \otimes_{C^\infty(M)} \Gamma(\overline{M})$, it suffices to check the assert for forms $\overline{\omega} = \omega \wedge \overline{\pi}^*(p)$, $\omega \in \Lambda^s(\mathbb{I}_M)$ and $p \in \Gamma(\overline{M})$, only. Note that in this case Exercise 3.5.12 gives

$$\overline{\omega_t'} = \Lambda^s \overline{i_t}^* \left(\mathcal{L}_{\partial/\partial t}(\omega) \wedge \overline{\pi}^*(p)\right).$$

### 3.5.18

Let $\{\overline{\Phi_t}\}_{t\in \mathbb{R}}$ be the fat flow generated by a fat field $\overline{X}$ (suppose that $\overline{M}$ is without boundary). Denote by $\overline{\Psi_t}$ the restriction $\overline{M_t} \xrightarrow{\sim} \overline{\Phi_t(M_t)} = \overline{M_{-t}}$ of $\overline{\Phi_t}$ on its range. Suppose that $\{\overline{\omega_t}\}_{t\in \mathbb{R}}$ is a time-dependent thickened

$s$-form such that $\overline{\omega_t}$ is a thickened form on $\overline{M_{-t}}$ for all $t$ and let $\left\{\overline{\omega'_t}\right\}_{t\in\mathbb{R}}$ be its derivative.

**Exercise.** Show that
$$\left.\frac{\mathrm{d}}{\mathrm{d}t}\right|_{t=0}\left[\Lambda^s\,\overline{\Psi_t}^*(\overline{\omega_t})\right] = \mathcal{L}_{\overline{X}}(\overline{\omega_0}) + \overline{\omega'_0}\,.$$

**Hint.** Take into account n. 0.6.15.

### 3.5.19

If $\nabla_t$ is as in n. 3.5.16 and $\overline{\Psi_t}$ as in n. 3.5.18, then
$$\mathrm{d}_{\nabla_t}(p|_{M_t}) = \Lambda^1\,\overline{\Psi_t}^*\left(\mathrm{d}_{\nabla|_{M_t}}\left(\overline{\Phi_{-t}}^*(p)\right)\right),\quad p\in\Gamma(\overline{M})\,.$$

Since $\left\{\overline{\Phi_{-t}}\right\}_{t\in\mathbb{R}}$ is the flow generated by $-\overline{X}$, it follows directly from Exercises 3.5.17 and 3.5.18 that
$$\left.\frac{\mathrm{d}}{\mathrm{d}t}\right|_{t=0}\Lambda^1\,\overline{\Psi_t}^*\left(\mathrm{d}_{\nabla|_{M_t}}\left(\overline{\Phi_{-t}}^*(p)\right)\right) = \mathcal{L}_{\overline{X}}\left(\mathrm{d}_\nabla(p)\right) - \mathrm{d}_\nabla\left(\overline{X}(p)\right),\quad p\in\Gamma(\overline{M})\,.$$

By evaluating the so-obtained thickened form on $Y\in D(M)$ we get:
$$\overline{\mathrm{i}}_Y\left(\mathcal{L}_{\overline{X}}(\mathrm{d}_\nabla(p)) - \mathrm{d}_\nabla\left(\overline{X}(p)\right)\right) \stackrel{\mathrm{Prop.3.5.14}}{=} \overline{X}(\nabla_Y(p)) - \nabla_{[X,Y]}(p) - \nabla_Y(\overline{X}(p))$$
$$= \left([\overline{X},\nabla_Y] - \nabla_{[X,Y]}\right)(p)\,.$$

This shows that $\nabla$ under the action of $\overline{X}$ goes to the $\mathrm{End}\left(\Gamma(\overline{M})\right)$-valued 1-form
$$\mathcal{L}_{\overline{X}}(\nabla):\quad Y \mapsto [\overline{X},\nabla_Y] - \nabla_{[X,Y]}\,. \tag{3.17}$$

In particular, if $\overline{X} = \nabla_X$, then
$$\mathcal{L}_{\nabla_X}(\nabla)(Y) = R^\nabla(X,Y)\,,$$
that is
$$\mathcal{L}_{\nabla_X}(\nabla) = \overline{\mathrm{i}}_X^{\mathrm{End}}(R^\nabla)\,,$$
where $\overline{\mathrm{i}}^{\mathrm{End}}$ stands for the insertion into $\Lambda^\bullet\left(\mathrm{End}\left(\Gamma(\overline{M})\right)\right)$ operator.

It is not difficult to generalize this fact to higher degrees. If $\mathrm{d}_\nabla$ stands now for the extended covariant differential, then $[\mathcal{L}_{\nabla_X},\mathrm{d}_\nabla] = [\mathcal{L}_X^\nabla,\mathrm{d}_\nabla]$ coincides with the wedge multiplication by the form $\overline{\mathrm{i}}_X^{\mathrm{End}}(R^\nabla)$ operator (see n. 3.2.9).

## 3.5.20

The particular case of (3.17) that concerns the action of an infinitesimal symmetry $\varphi \in \mathrm{gl}\,(\overline{M}) \stackrel{\text{def}}{=} \mathrm{End}(\Gamma\,(\overline{M}))$ (*cf.* n. 1.7.4) is worth special mentioning. Since $\varphi$ is a fat field over the zero field on $M$, formula (3.17) reads as

$$\mathcal{L}_\varphi(\nabla) : Y \mapsto [\varphi, \nabla_Y] - \nabla_{[0,Y]},$$

that is,

$$\mathcal{L}_\varphi(\nabla) = -\,\mathrm{d}_{\nabla^{\mathrm{End}}}(\varphi).$$

## 3.5.21

The formula (3.13) allows to describe the Lie algebra $\mathrm{hol}\,(\nabla, m)$ of $\mathrm{Hol}\,(\nabla, m)$. Let $X$ and $Y$ be commuting vector fields on $M$, $\{F_t\}_{t \in \mathbb{R}}$ and $\{G_s\}_{s \in \mathbb{R}}$ the generated by them one-parameter groups, respectively, and $\{\overline{F_t}\}_{t \in \mathbb{R}}$ and $\{\overline{G_s}\}_{s \in \mathbb{R}}$ their lifts by a linear connection $\nabla$, corresponding to the initial data $\overline{F_0} = \overline{G_0} = \mathrm{id}_{\overline{M}}$, respectively. Then $F_t$ and $G_s$ locally commute as well and four curves

$$\gamma_1 : \tau \mapsto F_\tau(m), \qquad 0 \leq \tau \leq t,$$

$$\gamma_2 : \tau \mapsto G_\tau(F_t(m)), \qquad 0 \leq \tau \leq s,$$

$$\gamma_3 : \tau \mapsto F_{-\tau}(G_s(F_t(m))), \, 0 \leq \tau \leq t,$$

and

$$\gamma_4 : \tau \mapsto G_{-\tau}(F_{-t}(G_s(F_t(m)))), \, 0 \leq \tau \leq s$$

form a piecewise smooth loop at $m$. So, the (local) fat diffeomorphism $\overline{G_{-s}} \circ \overline{F_{-t}} \circ \overline{G_s} \circ \overline{F_t}$ sends $\overline{m}$ into itself. Obviously, this automorphism of $\overline{m}$ belongs to $\mathrm{Hol}\,(\nabla, m)$ for all $m \in M$. On the other hand, the restriction of the endomorphism

$$\left.\frac{\partial^2}{\partial t\,\partial s} \left(\overline{G_{-s}} \circ \overline{F_{-t}} \circ \overline{G_s} \circ \overline{F_t}\right)^*\right|_{t=s=0} : \Gamma\,(\overline{M}) \to \Gamma\,(\overline{M})$$

of a fat point $\overline{m}$ is, as it is not difficult to see, an element of $\mathrm{hol}\,(\nabla, m)$. Moreover, since $[X, Y] = 0$,

$$\left.\frac{\partial^2}{\partial t\,\partial s} \left(\overline{G_{-s}} \circ \overline{F_{-t}} \circ \overline{G_s} \circ \overline{F_t}\right)^*\right|_{t=s=0} = [\nabla_X, \nabla_Y] = R^\nabla(X, Y).$$

Indeed, according to (3.13),

$$\left.\frac{\mathrm{d}}{\mathrm{d}t}\left(\overline{G_{-s}}\circ\overline{F_{-t}}\circ\overline{G_s}\circ\overline{F_t}\right)^*\right|_{t=0} = \nabla_X - \overline{G_s}^*\circ\nabla_X\circ\overline{G_{-s}}^*$$

and, similarly,

$$\left.\frac{\mathrm{d}}{\mathrm{d}s}\left(\nabla_X - \overline{G_s}^*\circ\nabla_X\circ\overline{G_{-s}}^*\right)\right|_{s=0} = [\nabla_X, \nabla_Y]\,.$$

Since any two vectors $\xi, \eta \in T_m M$ can be extended to commuting fields $X, Y$, this shows that $R_m^\nabla(\xi, \eta) \in \mathrm{hol}\,(\nabla, m)$, $R_m^\nabla$ being the value of $R^\nabla$ at $m$.

Obviously, the parallel translation with respect to the connection $\nabla^{\mathrm{End}}$ along a curve connecting a point $m' \in M$ with $m$ identifies $\mathrm{hol}\,(\nabla, m')$ with $\mathrm{hol}\,(\nabla, m)$. By this reason such a translation carries the curvature operator $R_{m'}^\nabla(\xi', \eta')$, $\xi', \eta' \in T_{m'}M$, into an element of $\mathrm{hol}\,(\nabla, m)$.

The following assertion is intuitively clear, and is not used in the sequel. So, we omit a proof.

**Proposition.** The Lie algebra $\mathrm{hol}\,(\nabla, m)$ is generated by all possible curvature operators $R_{m'}^\nabla(\xi', \eta')$, $m' \in M$, $\xi', \eta' \in T_{m'}M$, transferred to $m$ by means of the connection $\nabla^{\mathrm{End}}$.

## 3.6 Gauge/Fat Structures and Linear Connections

In this section we shall consider inner structures on fat manifolds in full generality by developing the preliminary idea discussed in Sect. 1.7. Such a structure is *fat* if it is *the same* in all fat points. Here 'the same' means that such an inner structure induces equivalent structures on single fat points. In particular, all these individual structures are equivalent to a *model* one. The totality of all equivalences among individual structures of fat points and the model structure constitutes a principal bundle. This bundle is, therefore, a a construction materializing the idea of identity of individual inner structures of fat points. For Lie algebra structures details of this constructions were discussed in n. 1.7.8 and, in fact, they are common for all kinds of inner structures.

However, the formalism of principal bundles is rather cumbersome form purely mathematical point of view, to say nothing about its other shortcomings. Another mechanism of establishing equivalence of individual inner structures comes from the theory of connections. Namely, parallel translations with respect to a connection $\nabla$ in $\overline{M}$ identify fat points of $\overline{M}$ if $M$

is connected. Moreover, if $\nabla$ preserves the considered inner structure on $\overline{M}$, then these translations identify individual inner structures of single fat points of $\overline{M}$. This motivates to call *fat* those inner structures on $\overline{M}$ that admit preserving them connections.

From physical point of view the last approach is much preferred because it introduces not only physical quantities (fat inner structures) but also physical means (connections) to observe them. This is the main idea of the modern theory of gauge fields, although formulated non in the very standard manner.

From mathematical point of view the approach to fat structures based on connections has the advantage that it works well in a much wider algebraic context where the use of principal bundles is highly problematic. Below the necessary details of this approach are given and we shall use the term *gauge* (inner structure, quantity, *etc.*) instead of fat ones in order to be in conformity with the established terminology (see also n. 1.7.8).

### 3.6.1

We start with an example completing the discussion in n. 2.5.3 which illustrates equivalence of approaches based on principal bundles and connections, respectively.

**Proposition.** Let $\overline{M}$ be a fat manifold. A $C^\infty(M)$–Lie algebra structure in $P = \Gamma\left(\overline{M}\right)$ admitting a Lie connection is a gauge Lie algebra in the sense of n. 1.7.8.

**Proof.** Let $\overline{X}$ be a derivation of the considered $C^\infty(M)$–Lie bracket $\langle \cdot , \cdot \rangle$ in $P$ and $\{\overline{G_t}\}_{t \in \mathbb{R}}$ be the generated by it one-parameter group. We prove that

$$\left\langle \overline{G_t}^*(s_1), \overline{G_t}^*(s_2) \right\rangle = \overline{G_t}^*(\langle s_1, s_2 \rangle), \qquad s_1, s_2 \in P. \tag{3.18}$$

By simplifying the notation we set $s_i(t) = \overline{G_t}^*(s_i)$, $i = 1, 2$, $K_1(t) = \overline{G_t}^*(\langle s_1, s_2 \rangle)$ and $K_2(t) = (\langle s_1(t), s_2(t) \rangle)$. Then, according to

$$\frac{\mathrm{d}}{\mathrm{d}\, t} K_1(t) = \overline{X}(K_1(t))$$

and

$$\frac{\mathrm{d}}{\mathrm{d}\, t} K_2(t) = \left\langle \frac{\mathrm{d}\, s_1(t)}{\mathrm{d}\, t}, s_2(t) \right\rangle + \left\langle s_1(t), \frac{\mathrm{d}\, s_2(t)}{\mathrm{d}\, t} \right\rangle$$
$$= \left\langle \overline{X}(s_1(t)), s_2(t) \right\rangle + \left\langle s_1(t), \overline{X}(s_2(t)) \right\rangle = \overline{X}(\langle s_1(t), s_2(t) \rangle) = \overline{X}(K_2(t)) \ .$$

This shows that $K_1(t)$ and $K_2(t)$ are solutions of the equation

$$\frac{d}{dt} K(t) = \overline{X}(K(t)), \qquad K(t) \in P$$

such that $K_1(0) = \langle s_1, s_2 \rangle = K_2(0)$. Hence by uniqueness $K_1(t) = K_2(t)$, i.e., (3.18) holds.

Since $\nabla$ is a Lie connection, i.e., preserves the structure $\langle \cdot, \cdot \rangle$, all covariant derivatives $\nabla_X$ are derivations of this bracket. Hence (3.18) is valid for the generated by $\nabla_X$ one-parameter group $\{\overline{G_t}\}_{t \in \mathbb{R}}$ for all $X \in \mathrm{D}(M)$. In particular, this implies that a parallel translation with respect to $\nabla$ of one fat point to another is an isomorphism of the corresponding Lie algebras. $\square$

### 3.6.2

Passing to a general situation we, first, recall some elementary facts concerning the tensor algebra of a finite dimensional vector space $E$ over a field $k$ of zero characteristic. Denote it by

$$\mathrm{T}(E) = \bigoplus_{p,q \geq 0} E_q^p, \qquad E_q^p = \underbrace{E \otimes \cdots \otimes E}_{p \text{ factors}} \otimes \underbrace{E^\vee \otimes \cdots \otimes E^\vee}_{q \text{ factors}}.$$

$\mathrm{GL}(E)$ will be used for the group of automorphisms of $E$. An endomorphism $g$ of $E$ extends naturally to a homomorphism of the algebra $\mathrm{T}(E)$, also denoted by $g$, i.e., $g(\alpha \otimes \beta) = g(\alpha) \otimes g(\beta)$, $\alpha, \beta \in \mathrm{T}(E)$. It leaves invariant symmetric, skew-symmetric and similar parts of $\mathrm{T}(E)$ ([2]). The notation like $g^{\otimes p}$ refers to the restriction of $g$ to $E^{\otimes p}$, etc. Symmetric and wedge products there are denoted by '·' and '∧', respectively.

The group $\mathrm{GL}(E)$ acts on $\mathrm{T}(E)$ and the orbits of this action are called *tensor types* and tensor belonging to the same orbit are said to be *equivalent*. Standard identifications such as $\mathrm{End}(E) = \mathrm{Hom}_k(E, E)$ and $E \otimes E^\vee$ we shall use without special mentioning. For instance, a (Lie, associative, *etc.*) algebra structure on $E$, i.e., an element of $\mathrm{Hom}_k(E \otimes E, E)$, is identified with an element of $E_2^1 = E \otimes E^\vee \otimes E^\vee$. This way the isomorphism class of such an algebra is identified with the tensor type of the corresponding to it element of $E_2^1$.

---

[2] Since the characteristic of $k$ is zero, we can consider the symmetric and the exterior algebras as subspaces (not subalgebras) of $\mathrm{T}(E)$.

### 3.6.3

All the above said extends with due modifications to a module $P$ over a commutative algebra $A$. For instance,

$$\mathrm{T}(P) = \bigoplus_{p,q\geq 0} P_q^p, \qquad P_q^p = \underbrace{P \otimes \cdots \otimes P}_{p \text{ factors}} \otimes \underbrace{P^\vee \otimes \cdots \otimes P^\vee}_{q \text{ factors}}$$

stands for the tensor algebra of $P$. It should be stressed that the identifications like that of $\mathrm{End}(P)$ and $P \otimes P^\vee$ do not, generally take place for arbitrary $P$. But they remain valid for finitely generated projective modules. This is the case of fat manifolds, *i.e.*, $P = \Gamma\left(\overline{M}\right)$.

**Definition.** An *inner structure* of a fat manifold $\overline{M}$ is an element of the module $\mathrm{T}(P)$, $P = \Gamma\left(\overline{M}\right)$.

We stress that in a general algebraic situation the idea of an inner structure in $P$ should be realized in a wider context of multilinear algebra over $P$.

### 3.6.4

A linear connection $\nabla$ in $P$ extends naturally to a linear connection $\nabla^{\mathrm{T}}$ of $\mathrm{T}(P)$. Indeed, we have already associated with $\nabla$ a connection $\nabla^\vee$ on $P^\vee$ (see n. 2.4.4) and defined the tensor product of two linear connections (see n. 2.4.5). So, by suitably tensoring connections $\nabla$ and $\nabla^\vee$ one can construct connections $\nabla_q^p$ on $P_q^p$ and hence a connection $\nabla^{\mathrm{T}}$ on $\mathrm{T}(P)$, whose restriction to $P_q^p$ coincides with $\nabla_q^p$. In its turn, connections $\nabla_q^p$ restrict to 'smaller' connections on *natural submodules* of $P_q^p$. For instance, $\mathrm{d}_{\nabla_q^p}$ leaves invariant the $\Lambda(A)$-submodule $\Lambda^\bullet\left(\mathrm{S}^p(P)\right)$ of $\Lambda^\bullet\left(P_0^p\right)$ and hence $\mathrm{d}_{\nabla^{\mathrm{S}^p}} \stackrel{\mathrm{def}}{=} \mathrm{d}_{\nabla_0^k}\Big|_{\Lambda^\bullet(\mathrm{S}^p(P))}$ gives a linear connection in $\mathrm{S}^p(P)$. Similarly, $\nabla_q^0$ restricts to a connection $\nabla^{\mathrm{Poly}}$ on $\mathrm{S}^q\left(P^\vee\right)$, whose elements we interpret as polynomials on $P$, etc.

**Definition.** A linear connection in a fat manifold $\overline{M}$ *preserves* an inner structure $\Xi$ in $\overline{M}$ if

$$\mathrm{d}_{\nabla^{\mathrm{T}}}(\Xi) = 0 \,.$$

Obviously this is equivalent to the fact that

$$\nabla_X^{\mathrm{T}}(\Xi) = 0, \quad \forall X \in \mathrm{D}(A) \,.$$

### 3.6.5 Gauge/Fat Structures

The following definition is key.

**Definition.** An inner structure in a fat manifold admitting a preserving it linear connection is called *gauge* (informally, *fat*).

In this book we do not discuss the existence problem of connections preserving inner structures of a given kind. The following example-exercise illustrates the question.

**Exercise.** Prove that a generic $\Xi \in \mathrm{End}(P) = P_1^1$ ($\Xi \in P_0^2$) does not admit a preserving it connection even locally.

### 3.6.6

Now we have to show that inner structures at single fat points are 'the same' in the case of a gauge structure. The proof is based on a rather simple fact. Let $\pi_1$ and $\pi_2$ be vector bundles over a manifold $M$, $P_i = \Gamma(\pi_i)$, $i = 1, 2$, and $\Box_1, \Box_2$ be der-operators over the same vector field $X \in \mathrm{D}(M)$ in $P_1$ and $P_2$, respectively. Consider the one-parameter groups $\{\overline{G_t}\}_{t \in \mathbb{R}}$, $\{\overline{G_t}'\}_{t \in \mathbb{R}}$ and $\{\overline{G_t}''\}_{t \in \mathbb{R}}$ generated by $\Box_1 \boxtimes \Box_2$, $\Box_1$ and $\Box_2$, respectively. By abusing slightly the notation, here we use $\overline{G_t}'^* \otimes \overline{G_t}''^*$ for the natural map

$$p_1 \otimes p_2 \mapsto \overline{G_t}'^*(p_1) \otimes \overline{G_t}''^*(p_2)$$

(*cf.* n. 3.3.9). Accordingly, denote the corresponding regular morphism of vector bundles by $\overline{G_t}' \otimes \overline{G_t}''$.

**Exercise.** Show that $\overline{G_t}' \otimes \overline{G_t}'' = \overline{G_t}$.

**Hint.** Mind that

$$\frac{\mathrm{d}}{\mathrm{d}t} \overline{G_t}^* = \overline{G_t}^* \circ (\Box_1 \boxtimes \Box_2)$$

and use the Leibnitz rule for

$$\frac{\mathrm{d}}{\mathrm{d}t}\left(\overline{G_t}'^* \otimes \overline{G_t}''^*\right).$$

By applying the result of the above exercise to the one-parameter group $\{\overline{G_t}\}_{t \in \mathbb{R}}$ generated by $\nabla_X$ and one parameter groups generated by $\left(\nabla_q^p\right)_X$'s we have the following fact.

**Proposition.** Let $\overline{m}_t = \overline{G}_t(\overline{m})$, $m \in M$, and $\alpha_t = \overline{G}_t|_{\overline{m},\overline{m}_t} : \overline{m} \to \overline{m}_t$. The isomorphism $\mathrm{T}(\alpha_t) : \mathrm{T}(\overline{m}) \to \mathrm{T}(\overline{m}_t)$ of tensor algebras generated by the isomorphism $\alpha_t$ coincides with the isomorphism of the one-parameter group generated by $(\nabla^\mathrm{T})_X$, understood as the family of one-parameter groups generated by $(\nabla^p_q)_X$'s.

Thus the above exercise allows to identify the one-parameter group generated by $(\nabla^p_q)_X$ with $\left\{(\overline{G}_t)^p_q\right\}_{t \in \mathbb{R}}$ if $\{\overline{G}_t\}_{t \in \mathbb{R}}$ is the one-parameter group generated by $\nabla_X$.

Let now $\Xi \in P^p_q$, $P = \Gamma(\overline{M})$, be a gauge structure in $\overline{M}$ and $\nabla$ a preserving it linear connection. Then for all $X \in \mathrm{D}(M)$ the one-parameter group generated by $(\nabla^p_q)_X$ leaves $\Xi$ invariant. This is equivalent to say that

$$(\overline{G}_t)^p_q(\Xi_m) = \Xi_{m_t} \tag{3.19}$$

where $\Xi_m \in \overline{m}^p_q$ stands for the value of $\Xi$ at $m \in M$.

**Theorem.** Let $\Xi$ be a gauge structure on $\overline{M}$ and $\nabla$ a preserving it linear connection. Then

$$\left(\mathbf{T}^\gamma_{t_0,t_1}\right)^p_q\left(\Xi_{\gamma(t_0)}\right) = \Xi_{\gamma(t_1)}$$

for all curves $\gamma : \mathbb{I} \to M$ and $t_0, t_1 \in \mathbb{I}$.

**Proof.** Apply (3.19) to the induced by $\overline{\gamma}$ from $\nabla$ connection with $\overline{\gamma}$ being the $\nabla$-lift of $\gamma$. $\square$

**Corollary.** If $M$ is connected and $\Xi$ is a gauge structure on $\overline{M}$, then tensors $\Xi_m \in \overline{m}^p_q$ are equivalent to each other.

### 3.6.7

The above corollary justifies the following definition.

**Definition.** A tensor $\theta \in F^p_q$ ($F$ is the general fiber of $\overline{M}$) is called *general* (or *model*) for a gauge structure $\Xi$ on $\overline{M}$, if $\theta$ is equivalent to all $\Xi_m$, $m \in M$. The tensor type of $\theta$ is called the *gauge type of* $\Xi$.

For instance, if $b$ is a gauge bilinear form on $\overline{M}$, then a bilinear form $\beta$ on $F$ that is equivalent to $b_m$ for all $m \in M$ is called *general* (or *model*) for $b$.

### 3.6.8

Let $\nabla$ be a linear connection in an $A$-module $P$ and $\omega \in \Lambda^\bullet(A)$. If $\omega$ is understood to be the multiplication by $\omega$ operator $\theta \mapsto \omega \wedge \theta$, $\theta \in \Lambda^\bullet(P)$, then $[\mathrm{d}_\nabla, \omega]^{(\mathrm{gr})} = \mathrm{d}\omega : \theta \mapsto \mathrm{d}\omega \wedge \theta$, *i.e.*, the multiplication by $\mathrm{d}\omega$ operator, and hence $\left[[\mathrm{d}_\nabla, \theta]^{(\mathrm{gr})}, \rho\right]^{(\mathrm{gr})} = 0$, $\rho \in \Lambda^\bullet(A)$. This equality tells that $\mathrm{d}_\nabla$ is a graded first order differential operator (see n. 3.6.9 below).

The differential $\mathrm{d} : \Lambda^\bullet(A) \to \Lambda^\bullet(A)$ is a 'graded vector field' over $\Lambda^\bullet(A)$, *i.e.*, a derivation of the graded algebra $\Lambda^\bullet(A)$. Then the Leibnitz rule for $\mathrm{d}_\nabla$ shows that it is a graded der-operator over the graded vector field $\mathrm{d}$ (see n. 3.6.9). Moreover $[\mathrm{d}_\nabla, \mathrm{d}_\nabla]^{(\mathrm{gr})} = 2\mathrm{d}_\nabla^2$ is a graded der-operator over the graded vector field $[\mathrm{d}, \mathrm{d}]^{(\mathrm{gr})} = 2\mathrm{d}^2 = 0$. Hence $\mathrm{d}_\nabla^2$ is a der-operator over the zero vector field, *i.e.*, a graded endomorphism of the $\Lambda^\bullet(A)$–module $\Lambda^\bullet(P)$. This explains why $\mathrm{d}_\nabla^2$ is a zero-order differential operator over $A$ and not of the second order as, generally, should be.

The graded Jacobi identity (see n. 3.6.9) for differential operators $\Delta_1 = \Delta_2 = \Delta_3 = \mathrm{d}_\nabla$ becomes

$$0 = 3\left[[\mathrm{d}_\nabla, \mathrm{d}_\nabla]^{(\mathrm{gr})}, \mathrm{d}_\nabla\right]^{(\mathrm{gr})} = 6\left[R^\nabla, \mathrm{d}_\nabla\right]^{(\mathrm{gr})} = 6\,\mathrm{d}_{\nabla^{\mathrm{End}}}\left(R^\nabla\right).$$

This proves the Bianchi identity (*cf.* n. 3.2.14) by showing it to be a very particular case of the Jacobi identity.

The above observations reveal naturalness of differential calculus over the graded commutative algebra $\Lambda(A)$ when dealing with connections. We shall sketch some basic points of this approach.

### 3.6.9 Graded Differential Operators

Let $\mathcal{A}$ be a graded $k$-algebra, $k$ being a field, and $\mathcal{P}_1$, $\mathcal{P}_2$ be graded $\mathcal{A}$-modules. A graded $k$-linear map $\Delta : \mathcal{P}_1 \to \mathcal{P}_2$ is called a *graded differential operator of order $\leq s$* if

$$\left[\ldots\left[[\Delta, a_0]^{(\mathrm{gr})}, a_1\right]^{(\mathrm{gr})}, \ldots, a_s\right]^{(\mathrm{gr})} = 0, \qquad \forall a_0, a_1, \ldots, a_s \in \mathcal{A}$$

where homogeneous $a_i$'s are understood to be the multiplication by $a_i$ operators. This is the graded version of n. 0.1.2. Below it is mainly applied to $\mathcal{A} = \Lambda^\bullet(A)$ and $\mathcal{P} = \Lambda^\bullet(P)$.

The *graded Jacobi identity* of graded $k$-linear maps looks as

$$(-1)^{\deg X \deg Z}\left[[X,Y]^{(\mathrm{gr})},Z\right]^{(\mathrm{gr})}$$
$$+(-1)^{\deg Z \deg Y}\left[[Y,Z]^{(\mathrm{gr})},X\right]^{(\mathrm{gr})}$$
$$+(-1)^{\deg Y \deg X}\left[[Z,X]^{(\mathrm{gr})},Y\right]^{(\mathrm{gr})}=0\,.$$

A graded $k$-linear map

$$\mathcal{P}\otimes\mathcal{P}\xrightarrow{\langle\cdot,\cdot\rangle}\mathcal{P}$$

is a *graded Lie algebra structure* in $\mathcal{P}$ if

- $\deg\langle p_1,p_2\rangle = \deg p_1 + \deg p_2$;
- the bracket $\langle\cdot,\cdot\rangle$ satisfies graded Jacobi identity.

A graded $k$-linear operator $\Delta:\mathcal{P}\to\mathcal{P}$ is a *graded der-operator* over a graded derivation $X:\mathcal{A}\to\mathcal{A}$ if

$$\Delta(ap) = X(a)p + (-1)^{r\deg X}a\Delta(p),\qquad a\in\mathcal{A}_r, p\in\mathcal{P}\,.$$

This, obviously, implies $\deg X = \deg \Delta$ and shows $\Delta$ to be a first order graded differential operator over $\mathcal{A}$.

It is easy to see that the totality of all graded der-operators gives a graded $\mathcal{A}$-module with respect to the left multiplication by elements of $\mathcal{A}$ and a graded $k$-Lie algebra with respect to the graded commutator operation. For instance, if $\nabla$ and $\square$ are linear connections in $P$, then $\mathrm{d}_\nabla$ and $\mathrm{d}_\square$ are graded der-operators over $\mathrm{d}$ and $[\mathrm{d}_\nabla,\mathrm{d}_\square]^{(\mathrm{gr})} = \mathrm{d}_\nabla\circ\mathrm{d}_\square + \mathrm{d}_\square\circ\mathrm{d}_\nabla$ is a der-operator over $[\mathrm{d},\mathrm{d}]^{(\mathrm{gr})} = 2\,\mathrm{d}^2 = 0$ and hence a degree two endomorphism of $\Lambda^\bullet(P)$.

The operator $\bar{\mathrm{i}}_X$, $X\in\mathrm{D}(A)$, (see n. 0.5.18) is a degree $-1$ graded der-operator in $\Lambda^\bullet(P)$ over the graded derivation $\mathrm{i}_X$ of $\Lambda^\bullet(A)$. Note that both $\mathrm{i}_X$ and $\bar{\mathrm{i}}_X$ are $A$-linear but not $\Lambda^\bullet(A)$-linear. Recall in this connection that $\mathcal{L}_X^\nabla = \left[\bar{\mathrm{i}}_X,\mathrm{d}_\nabla\right]^{(\mathrm{gr})}$ (Proposition 3.5.14, (2)).

### 3.6.10 $\Lambda$-*Extension of scalars*

Assume $P$ to be projective and finitely generated. Every tensor $\Xi\in\mathrm{T}(P)$ extends naturally to a 'graded' tensor $\Xi^\Lambda$ in the $\Lambda^\bullet(A)$–module $\Lambda^\bullet(P)$. For instance, if $\varphi\in\mathrm{End}(P)$, then $\varphi^\Lambda$ is the graded endomorphism

$$\omega\wedge p\mapsto \omega\wedge\varphi(p),\qquad \omega\in\Lambda^\bullet(A), p\in P\,.$$

Similarly, if $b$ is a bilinear form on $P$, then

$$b^\Lambda(\omega_1 \wedge p_1, \omega_2 \wedge p_2) = \omega_1 \wedge \omega_2\, b(p_1, p_2), \qquad \omega_i \in \Lambda^\bullet(A), p_i \in P.$$

These $\Lambda$-extensions may be viewed as elements of $\Lambda^\bullet(\mathrm{T}(P))$. For instance, the $\Lambda$-extension of $\mathrm{End}(P)$ is $\Lambda^\bullet(\mathrm{End}(P))$, and the extended action of $\Lambda^\bullet(\mathrm{End}(P))$ on $\Lambda^\bullet(P)$ (*cf.* n. 3.2.9) looks as

$$(\omega \wedge \varphi)(\varrho \wedge p) = \omega \wedge \varrho \wedge \varphi(p), \qquad \omega, \varrho \in \Lambda^\bullet(A), \varphi \in \mathrm{End}(P), p \in P.$$

Similarly, the extended action of $\Lambda^\bullet(\mathrm{End}(P))$ on $b^\Lambda$, $b$ being a bilinear form on $P$, reads

$$\Big((\omega \wedge \varphi)(b^\Lambda)\Big)(\theta_1, \theta_2) = -\omega \wedge \Big(b^\Lambda\big(\varphi^\Lambda(\theta_1), \theta_2\big) + b^\Lambda\big(\theta_2, \varphi^\Lambda(\theta_1)\big)\Big)$$

with $\theta_i = \omega_i \wedge p_i$, $\omega_i \in \Lambda^\bullet(A)$, $p_i \in P$, $i = 1, 2$. In particular,

$$-R^\nabla(b^\Lambda)(\theta_1, \theta_2) = b^\Lambda\big(R^\nabla(\theta_1), \theta_2\big) + b^\Lambda\big(\theta_1, R^\nabla(\theta_2)\big).$$

Obviously, the extended commutator in $\Lambda^\bullet(\mathrm{End}(P))$ is

$$[\omega_1 \wedge \varphi_1, \omega_2 \wedge \varphi_2]^{(\mathrm{gr})} = \omega_1 \wedge \omega_2 \wedge [\varphi_1, \varphi_2],$$

$$\omega_1, \omega_2 \in \Lambda^\bullet(A), \varphi_1, \varphi_2 \in \mathrm{End}(P).$$

Chapter 4

# Cohomological Aspects of Linear Connections

In this chapter we discuss some cohomological aspects of the theory of linear connections. The basic is the cohomology of complexes associated with flat linear connections. They are interpreted naturally as the *de Rham cohomology with twisted coefficients* and play an important role in many situations. This cohomology is illustrated by some simple examples at the beginning of the chapter and then some basic techniques of computations are introduced. Then Maxwell's equations are discussed from this point of view. Also, some useful elements of Homological Algebra, like the long exact cohomology sequence of a 'flat pair' and the 'fat' homotopy formula, adapted to the context, are described. In the final section of this chapter we develop the theory of characteristic classes of gauge, or, 'fat', structures.

## 4.1 An Introductory Example

Various natural questions appearing in the context of connections have a cohomological nature. Some of them are related with the classification problem. To provide an instructive example, this point is developed in this section starting from simplest fat manifolds, *i.e.*, trivial fat manifolds of type 1. From now on in this chapter, $A$ stands for a commutative $k$-algebra, $k$ being a field.

### 4.1.1 Gauge Equivalence Between Linear Connections

In the gauge theory two fields are physically indistinguishable if one of them is obtained from another by a gauge transformation, *i.e.*, by a fat identity map. In conformity with that two connections are to be declared indistinguishable if one is obtained from another by a fat identity map.

This is is natural from a mathematical point of view as well.

**Definition.** Let $\nabla$ and $\square$ be linear connections in an $A$-module $P$. They are called *gauge equivalent* if $\square$ and $\nabla$ are 'conjugated' by an automorphism $\varphi$ of $P$, i.e.,

$$\square_X = \varphi \circ \nabla_X \circ \varphi^{-1}, \quad X \in \mathrm{D}(A) .$$

In particular, two connections in a fat manifold are gauge equivalent if one is induced from another by a fat identity map.

It is worth stressing that two connections in a module $P$ (resp., in a fat manifold $\overline{M}$) are gauge equivalent if and only if they are compatible with respect to an automorphism of $P$ (resp., a fat identity map of $\overline{M}$)).

Below this definition will be applied to free $A$-modules of rank 1. In this situation automorphisms of $P$ are just multiplication by invertible elements of $A$, in particular, by nowhere vanishing functions on $M$ if $A = \mathrm{C}^\infty(M)$.

### 4.1.2

Let $\nabla$ be a linear connection in $P = A$.

**Definition.** The 1-form

$$\mathrm{d}_\nabla (1) \in \Lambda^1 (P) = \Lambda^1 (A)$$

is called *associated with* $\nabla$.

**Proposition.** If $\varrho$ is the 1-form associated with $\nabla$, then

$$(\mathrm{d}_\nabla - \mathrm{d})(\omega) = \varrho \wedge \omega, \quad \omega \in \Lambda^\bullet (A),$$

d being the exterior differential.

*Proof.* We can assume that $\omega$ is homogeneous. Since $(\Lambda^\bullet (A), \mathrm{d}_\nabla)$ is a cd-module over the cd-algebra $(\Lambda^\bullet (A), \mathrm{d})$ (see Example 3.3.5), we have

$$\begin{aligned}(\mathrm{d}_\nabla - \mathrm{d})(\omega) &= \mathrm{d}_\nabla (\omega \wedge 1) - \mathrm{d}(\omega) \\ &= \mathrm{d}(\omega) \wedge 1 + (-1)^{\deg \omega} \omega \wedge \mathrm{d}_\nabla (1) - \mathrm{d}(\omega) \\ &= (-1)^{\deg \omega} \omega \wedge \varrho = (-1)^{\deg \omega \deg \varrho} \omega \wedge \varrho = \varrho \wedge \omega .\end{aligned}$$
$\square$

### 4.1.3

The formula in the above proposition may be written as

$$d_\nabla - d = \varrho,$$

if $\varrho$ is understood to be the multiplication by $\varrho$ operator. Similarly can be interpreted the formula

$$\nabla - D = \varrho, \qquad (4.1)$$

where $D$ is the trivial connection. Namely, $\varrho$ is a map

$$D(A) \to A.$$

By composing it with the natural homomorphism

$$\iota : A \to \mathrm{End}_A(A) \subseteq \mathrm{Der}(A)$$

that sends $a$ into the multiplication by $a$ operator we get a map

$$\varrho' : D(A) \to \mathrm{Der}(A).$$

Moreover, if $X \in D(A)$ and $a \in A$, then

$$(\nabla_X - D_X)(a) = ((d_\nabla - d)(a))(X) = (\varrho \wedge a)(X) = a\varrho(X) = \varrho'(X)(a),$$

i.e.,

$$\nabla - D = \varrho'$$

as elements of $\mathrm{Hom}(D(A), \mathrm{Der}(A))$ (cf. n. 2.1.9). So, by identifying $\varrho$ with $\varrho'$ we get (4.1).

### 4.1.4

**Proposition.** For a given $\varrho \in \Lambda^1(A)$ there exists a unique linear connection $\nabla$ whose associated 1-form is $\varrho$.

**Proof.** In view of nn. 2.1.9 and 4.1.3, it suffices to set

$$\nabla \stackrel{\text{def}}{=} D + \varrho,$$

with $D$ being the trivial connection and $\varrho$ being considered as an element of $\mathrm{Hom}(D(A), \mathrm{Der}(A))$. □

This proposition together with the previous results establish a one-to-one correspondence between linear connections in $P = A$ and 1-forms. In particular, it allows to supply the totality of all linear connections in $A$ with an $A$-module structure.

### 4.1.5

Now it is natural to wonder what kind of linear connections correspond to closed and exact forms.

**Proposition.** The curvature of a linear connection $\nabla$ in $A$ with associated 1-form $\varrho$ is (naturally identified with) $\mathrm{d}\,\varrho$.

**Proof.** According to n. 4.1.3

$$\mathrm{d}_\nabla = \mathrm{d} + \varrho\,.$$

Since $\varrho \wedge \varrho = 0$, $\varrho^2 = \varrho \circ \varrho = 0$ if $\varrho$ is understood as the multiplication by $\varrho$ operator. Moreover, the Leibnitz formula

$$\mathrm{d}\,(\varrho \wedge \omega) = \mathrm{d}\,\varrho \wedge \omega - \varrho \wedge \mathrm{d}\,\omega\,,$$

may be read as an equality of operators in $\Lambda^\bullet(A)$:

$$\mathrm{d} \circ \varrho = \mathrm{d}\,\varrho - \varrho \circ \mathrm{d}\,.$$

Hence

$$\mathrm{d}_\nabla^2 = \mathrm{d}^2 + \varrho \circ \mathrm{d} + \mathrm{d} \circ \varrho + \varrho^2 = \varrho \circ \mathrm{d} + \mathrm{d}\,\varrho - \varrho \circ \mathrm{d} = \mathrm{d}\,\varrho\,,$$

and the result follows from Proposition 3.2.10. □

**Corollary.** A linear connection in $A$ is flat if and only if the associated with it 1-form is closed.

**Exercise.** Let $\nabla$ be a linear connection in the $A$-module $A$. Let $\varkappa \in \Lambda^s(A)$ and set $\varkappa' = \iota \circ \varkappa$, with $\iota : A \to \mathrm{End}_A(A)$ being as in n. 4.1.3, so that $\varkappa = \varkappa'$ when regarded as multiplication operators in $\Lambda^\bullet(A)$ (as $\mathrm{d}\,\rho$ and $R^\nabla$ before). Check that $\mathrm{d}\,\varkappa = \mathrm{d}_{\nabla^{\mathrm{End}}} \varkappa'$ (as multiplication operators).

### 4.1.6

It is natural to call *gauge trivial* a connection in a standard fat manifold that is gauge equivalent to the trivial one.

**Proposition.** Let $\overline{M}$ be a fat manifold with $\Gamma\left(\overline{M}\right) = \mathrm{C}^\infty(M)$. Then a linear connection $\nabla$ in $\overline{M}$ is gauge trivial if and only if the associated with $\nabla$ 1-form $\varrho$ is exact.

**Proof.** Assume that $\nabla$ is induced from the trivial connection by a fat identity map $\overline{f} : \overline{M} \to \overline{M}$. Obviously, $\overline{f}^*$ is the multiplication by $a_{\overline{f}}$ operator with $a_{\overline{f}} = \overline{f}^*(1) \in C^\infty(M)$. Note that $a_{\overline{f}}$ is invertible, i.e., nowhere vanishing function on $M$.

The extension $\Lambda^\bullet \overline{f}^* : \Lambda^\bullet(\overline{M}) \to \Lambda^\bullet(\overline{M})$ of $\overline{f}^*$ is the multiplication by $a_{\overline{f}}$ as well, but in $\Lambda^\bullet(\overline{M})$ (see Definition 3.3.8).

The trivial connection $D$ is $\overline{f}$-compatible with $\nabla$ if and only if

$$\mathrm{d} \circ \Lambda^\bullet \overline{f}^* = \Lambda^\bullet \overline{f}^* \circ \mathrm{d}_\nabla,$$

with d being the exterior differential in $\Lambda^\bullet(\overline{M}) = \Lambda^\bullet(M)$. Since

$$\Lambda^\bullet \overline{f}^* \circ \mathrm{d}_\nabla \stackrel{\text{Prop. 4.1.2}}{=} a_{\overline{f}}\mathrm{d} + a_{\overline{f}}\varrho$$

and

$$\mathrm{d} \circ \Lambda^\bullet \overline{f}^* = a_{\overline{f}}\mathrm{d} + \mathrm{d}\left(a_{\overline{f}}\right)$$

(as operators in $\Lambda^\bullet(\overline{M})$), $\nabla$ and $D$ are $\overline{f}$-compatible if and only if

$$\varrho = \frac{\mathrm{d}\left(a_{\overline{f}}\right)}{a_{\overline{f}}}$$

i.e., $\varrho = \mathrm{d}(b_{\overline{f}})$ with $b_{\overline{f}} \stackrel{\text{def}}{=} \ln\left|a_{\overline{f}}\right|$. $\square$

The above result holds for every standard trivial fat manifold $\overline{M}$ of type 1, since in this case $\Gamma(\overline{M})$ is identified with $C^\infty(M)$ by means of a canonical isomorphism preserving the trivial connection.

### 4.1.7

We have established the following one-to-one correspondences for a standard trivial fat manifold of type 1:

$$\text{closed 1-forms} \longleftrightarrow \text{flat linear connections}$$

$$\text{exact 1-forms} \longleftrightarrow \text{gauge trivial connections}.$$

What about an arbitrary trivial fat manifold $\overline{M}$ of type 1, that is, $\Gamma(\overline{M}) \cong C^\infty(M)$, but no such isomorphism is canonically fixed?

**Exercise.** Let $\Delta$ and $\nabla$ be linear connection on a trivial fat manifold $\overline{M}$ of type 1. Note that $\mathrm{d}_\nabla - \mathrm{d}_\Delta$ is the multiplication operator by a differential form $\varrho \in \Lambda^1(M)$. Prove that

(1) $\Delta$ and $\nabla$ have the same curvature if and only if $\varrho$ is closed;
(2) $\Delta$ and $\nabla$ are gauge equivalent if and only if $\varrho$ is exact.

If a linear connection on a trivial fat manifold $\overline{M}$ corresponds to a gauge-trivial connection through an isomorphism $\Gamma(\overline{M}) \xrightarrow{\sim} C^\infty(M)$, then the same is true for any other such isomorphism. Hence the notion of a gauge-trivial linear connection is well defined also for arbitrary trivial fat manifolds.

We conclude that the totality of gauge equivalence classes of flat connections on a trivial fat manifold of type 1 is naturally identified with

$$H^1(M) = \frac{\mathrm{Ker}\left(\mathrm{d}: \Lambda^1(M) \to \Lambda^2(M)\right)}{\mathrm{d}\,\Lambda^0(M)}$$

(it is the first de Rham cohomology space of the base $M$; we shall recall the general definition later, in n. 4.2.3).

## 4.2 Cohomology of Flat Linear Connections

### 4.2.1

**Definition.** If $\overline{i}: \overline{N} \hookrightarrow \overline{M}$ is a closed fat submanifold, then the form

$$\overline{\omega}|_N \stackrel{\mathrm{def}}{=} \Lambda^\bullet \overline{i}^*(\overline{\omega}) \in \Lambda^\bullet\left(\overline{N}\right), \quad \overline{\omega} \in \Lambda^\bullet\left(\overline{M}\right),$$

is called the *restriction of* $\overline{\omega} \in \Lambda^\bullet\left(\overline{M}\right)$ *to* $\overline{N}$.

If, according to n. 3.3.10, a thickened form $\overline{\omega} \in \Lambda^s\left(\overline{M}\right)$ is understood as the field $\overline{m} \mapsto \overline{\omega}_m$ on $M$, then the field $\overline{n} \mapsto (\overline{\omega}|_N)_n$ on $N$ is defined by

$$(\overline{\omega}|_N)_n = \overline{\omega}_n|_{(T_nN)^s},$$

where $T_nN$ is identified naturally with the corresponding to it subspace of $T_nM$.

### 4.2.2

**Definition.** A thickened form $\overline{\omega} \in \Lambda^\bullet\left(\overline{M}\right)$ is called *relative with respect to* a closed fat submanifold $\overline{N}$ if

$$\overline{\omega}|_N = 0.$$

The notation $\Lambda^\bullet(M, N)$ and $\Lambda^\bullet\left(\overline{M}, \overline{N}\right)$ will be used for sets of relative with respect to $N$ forms and relative with respect to $\overline{N}$ thickened forms, respectively. Obviously, $\Lambda^\bullet(M, N)$ and $\Lambda^\bullet\left(\overline{M}, \overline{N}\right)$ are submodules of graded

$\Lambda^\bullet(M)$–modules $\Lambda^\bullet(M)$ and $\Lambda^\bullet\left(\overline{M}\right)$, respectively, and $\Lambda^\bullet\left(\overline{M},\overline{N}\right)$ is the kernel of $\Lambda^\bullet(M)$–module homomorphism

$$\Lambda^\bullet\, \overline{i}^* : \Lambda^\bullet\left(\overline{M}\right) \to \Lambda^\bullet\left(\overline{N}\right).$$

The map $\Lambda^\bullet\, \overline{i}^*$ is surjective as it is easy to see from nn. 0.5.15 and 0.2.20.

**Proposition.** Let $\nabla$ be a linear connection in $\overline{M}$ and $\nabla|_{\overline{N}}$ the induced linear connection in a closed fat submanifold $\overline{i} : \overline{N} \to \overline{M}$. Then the submodule $\Lambda^\bullet\left(\overline{M},\overline{N}\right) \subseteq \Lambda^\bullet\left(\overline{M}\right)$ is stable with respect to $d_\nabla$ and

$$0 \to \Lambda^\bullet\left(\overline{M},\overline{N}\right) \longrightarrow \Lambda^\bullet\left(\overline{M}\right) \xrightarrow{\Lambda^\bullet\, \overline{i}^*} \Lambda^\bullet\left(\overline{N}\right) \to 0,$$

is an exact sequence of cd-modules.

***Proof.*** It directly follows from the definition of the induced linear connection that $\Lambda^\bullet\, \overline{i}^*$ is a cd-module homomorphism. □

Written in details the above exact sequence becomes a commutative diagram with exact rows:

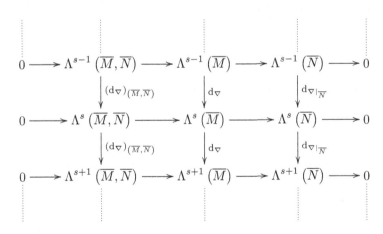

where $(d_\nabla)_{(\overline{M},\overline{N})}$ stands for $d_\nabla$ restricted to $\Lambda^\bullet\left(\overline{M},\overline{N}\right)$.

**4.2.3**

According to Corollary 3.2.10 and Proposition 2.3.4, if $\nabla$ is a flat connection, then cd-modules from the above exact sequence are complexes and below we fix the notation for their cohomologies.

**Definition.** Let $A$ be a $k$-algebra, $k$ being a field, and $P$ be an $A$-module. The graded vector space (over $k$) of cohomology of the complex $(\Lambda^\bullet(P), d_\nabla)$ is denoted by $H_\nabla^\bullet(P)$ and called the *cohomology of* $\nabla$. Its degree $s$ component is denoted by $H_\nabla^s(P)$, i.e.,

$$H_\nabla^s(P) = \frac{\operatorname{Ker}(d_\nabla)_s}{\operatorname{Im}(d_\nabla)_{s-1}},$$

and called the *s-th cohomology (space) of* $\nabla$. For $P = \Gamma(\overline{M})$ these cohomologies are denoted by $H_\nabla^\bullet(\overline{M})$ and $H_\nabla^s(\overline{M})$, respectively.

The cohomology of the complex $\left(\Lambda^\bullet(\overline{M}, \overline{N}), (d_\nabla)_{(\overline{M}, \overline{N})}\right)$ is called the *relative cohomology of $\nabla$ with respect to $\overline{N}$* and denoted by $H_\nabla^\bullet(\overline{M}, \overline{N})$. Its $s$-th component is denoted by $H_\nabla^s(\overline{M}, \overline{N})$ and called the *s-th relative cohomology of $\nabla$ with respect to $\overline{N}$*.

It is worth stressing that if $A = C^\infty(M)$, $P = A$ and $D$ is the trivial connection, then $H_D^\bullet(A)$ is nothing but the standard *de Rham cohomology of $M$* which will be denoted, as usually, by $H^\bullet(M)$. More generally, for every $k$-algebra $A$, the *de Rham cohomology of $A$* will be the space $H^\bullet(A) \stackrel{\text{def}}{=} H_D^\bullet(A)$, with $D$ being the trivial connection on $A$.

### 4.2.4

It is easy to understand what are zeroth cohomology of a flat connection. Indeed, by definition $H^0(M) = \operatorname{Ker}(d_\nabla)_0$. So, in view of n. 3.2.7, we have

**Proposition.** $H_\nabla^0(P)$ coincides with the space of $\nabla$-constant sections of $P$.

Effective computations of higher cohomologies of flat connections require a developed techniques which are basically the same as for the de Rham cohomology. One of them exploits long exact cohomology sequences associated with 'flat pairs' and is introduced below.

### 4.2.5

Let $\overline{N}$ be a closed fat submanifold of a fat manifold $\overline{M}$ and $\nabla$ a flat linear connection on $\overline{M}$. So all cd-modules in the sequence

$$0 \to \Lambda^\bullet(\overline{M}, \overline{N}) \to \Lambda^\bullet(\overline{M}) \to \Lambda^\bullet(\overline{N}) \to 0$$

introduced in Proposition 4.2.2 are complexes. The long exact cohomology sequence canonically associated with this short exact sequence of complexes (see for instance [Hilton and Stammbach (1971), Chap. IV, Theorem 2.1]) looks in the above notation as follows:

$$0 \to H^0_{\overline{\nabla}}(\overline{M}, \overline{N}) \to H^0_{\overline{\nabla}}(\overline{M}) \to H^0_{\overline{\nabla}|_{\overline{N}}}(\overline{N})$$

$$\to H^1_{\overline{\nabla}}(\overline{M}, \overline{N}) \to H^1_{\overline{\nabla}}(\overline{M}) \to H^1_{\overline{\nabla}|_{\overline{N}}}(\overline{N})$$

$$\to H^2_{\overline{\nabla}}(\overline{M}, \overline{N}) \to \quad \cdots \quad .$$

We shall illustrate how long exact sequences do work by deducing a fat analogue of the Newton-Leibnitz formula.

### 4.2.6

We start with a cohomological interpretation of the ordinary Newton-Leibnitz formula.

Let $\mathbb{I} = [a, b] \subseteq \mathbb{R}$, $\partial \mathbb{I} = \{a, b\}$, $a < b$. First, describe the segment

$$0 \to H^0(\mathbb{I}, \partial \mathbb{I}) \to H^0(\mathbb{I}) \to H^0(\partial \mathbb{I}) \to H^1(\mathbb{I}, \partial \mathbb{I}) \to H^1(\mathbb{I})$$

of the long de Rham cohomology sequence of the pair $\{\mathbb{I}, \partial \mathbb{I}\}$.

Let $t$ be the canonical coordinate function on $\mathbb{I}$, that is, the inclusion $\mathbb{I} \hookrightarrow \mathbb{R}$. Denote by d the exterior differential on $\mathbb{I}$ and by $d_{(\mathbb{I}, \partial \mathbb{I})}$ its restriction to $\Lambda^\bullet(\mathbb{I}, \partial \mathbb{I})$. Then

$$\Lambda^1(C^\infty(\mathbb{I})) = \{f \, dt : f \in C^\infty(\mathbb{I})\}$$

and we have

$$H^0(\mathbb{I}, \partial \mathbb{I}) = \frac{\operatorname{Ker}(d_{(\mathbb{I}, \partial \mathbb{I})})_0}{\{0\}} \cong \operatorname{Ker}(d_{(\mathbb{I}, \partial \mathbb{I})})_0$$

$$= \{f \in C^\infty(\mathbb{I}) : f|_{\partial \mathbb{I}} = 0 \text{ and } df = 0\}$$

$$= \{f \in C^\infty(\mathbb{I}) : f|_{\partial \mathbb{I}} = 0 \text{ and } f \text{ is constant}\} = \{0\}.$$

The zero degree of the exterior differential

$$d_0 : \Lambda^0(\mathbb{I}) \to \Lambda^1(\mathbb{I}), \quad f \mapsto f' \, dt$$

is surjective, since every smooth function on $\mathbb{I}$ admits a smooth primitive. This shows that $H^1(\mathbb{I}) = 0$.

Further we see that

$$H^0(\mathbb{I}) = \frac{\operatorname{Ker} d_0}{\{0\}} \cong \operatorname{Ker} d_0 = \{f \in C^\infty(\mathbb{I}) : f \text{ is constant}\}$$

and a canonical isomorphism

$$H^0(\mathbb{I}) \xrightarrow{\sim} \mathbb{R}$$

that sends each constant function to its value takes place.

Since $\partial\mathbb{I}$ is zero-dimensional,
$$\Lambda^\bullet(\partial\mathbb{I}) = \Lambda^0(\partial\mathbb{I}) = C^\infty(\partial\mathbb{I}),$$
and hence the exterior differential vanishes. So,
$$H^0(\partial\mathbb{I}) = \frac{C^\infty(\partial\mathbb{I})}{\{0\}} \cong C^\infty(\partial\mathbb{I}) = \{f : \{a, b\} \to \mathbb{R}\}$$
and the map $H^0(\partial\mathbb{I}) \xrightarrow{\sim} \mathbb{R} \oplus \mathbb{R}$ that sends each function
$$f : \{a, b\} \to \mathbb{R}$$
to the pair
$$(f(a), f(b)),$$
is a (canonical) isomorphism.

Finally, the embedding $i : \partial\mathbb{I} \hookrightarrow \mathbb{I}$ induces the homomorphism
$$H^0(i) : H^0(\mathbb{I}) \to H^0(\partial\mathbb{I}), \quad [f] \longmapsto [i^*(f)],$$
where square brackets stands for the corresponding cohomology classes. By combining it with the above isomorphisms
$$\mathbb{R} \xrightarrow{\sim} H^0(\mathbb{I}) \xrightarrow{H^0(i)} H^0(\partial\mathbb{I}) \xrightarrow{\sim} \mathbb{R} \oplus \mathbb{R}$$
one gets a map $\mathbb{R} \to \mathbb{R}\oplus\mathbb{R}$ which is nothing but the diagonal homomorphism
$$\Delta : \mathbb{R} \to \mathbb{R} \oplus \mathbb{R}, \quad \lambda \mapsto (\lambda, \lambda).$$

Thus the segment
$$H^0(\mathbb{I}, \partial\mathbb{I}) \to H^0(\mathbb{I}) \to H^0(\partial\mathbb{I}) \to H^1(\mathbb{I}, \partial\mathbb{I}) \to H^1(\mathbb{I})$$
of the long de Rham cohomology sequence is identified, via the above canonical isomorphisms, with the short exact sequence
$$0 \to \mathbb{R} \xrightarrow{\Delta} \mathbb{R} \oplus \mathbb{R} \to H^1(\mathbb{I}, \partial\mathbb{I}) \to 0.$$

### 4.2.7

Obviously, every 1-form on $\mathbb{I}$ is also a relative one with respect to $\partial\mathbb{I}$, *i.e.*,
$$\Lambda^1(\mathbb{I}, \partial\mathbb{I}) = \Lambda^1(\mathbb{I})$$
and
$$\operatorname{Ker} d_1 = \operatorname{Ker} d_{(\mathbb{I}, \partial\mathbb{I})} = \Lambda^1(\mathbb{I}).$$

Therefore,
$$H^1(\mathbb{I}, \partial\mathbb{I}) = \frac{\Lambda^1(\mathbb{I})}{B^1(\mathbb{I}, \partial\mathbb{I})}, \qquad (4.2)$$

where $B^1(\mathbb{I}, \partial\mathbb{I})$ is the image of $d_{(\mathbb{I},\partial\mathbb{I})_0} : \Lambda^0(\mathbb{I}, \partial\mathbb{I}) \to \Lambda^1(\mathbb{I}, \partial\mathbb{I})$.

On the other hand, it follows from the exact sequence

$$0 \to \mathbb{R} \xrightarrow{\Delta} \mathbb{R} \oplus \mathbb{R} \to H^1(\mathbb{I}, \partial\mathbb{I}) \to 0$$

that

$$H^1(\mathbb{I}, \partial\mathbb{I}) \cong \frac{\mathbb{R} \oplus \mathbb{R}}{\operatorname{Im}\Delta}. \qquad (4.3)$$

Obviously, $\operatorname{Im}\Delta \cong \mathbb{R}$ and

$$\frac{\mathbb{R} \oplus \mathbb{R}}{\operatorname{Im}\Delta} \cong \mathbb{R}. \qquad (4.4)$$

Among isomorphisms (4.4) there are two 'privileged' ones, namely, $\sigma_1 = (\pi \circ \iota_1)^{-1}$ and $\sigma_2 = (\pi \circ \iota_2)^{-1}$ where

$$\pi : \mathbb{R}^2 \to \frac{\mathbb{R}^2}{\operatorname{Im}\Delta}$$

is the canonical projection and homomorphisms

$$\iota_1, \iota_2 : \mathbb{R} \to \mathbb{R}^2,$$

are defined by

$$\lambda \mapsto (0, \lambda) \quad \text{and} \quad \lambda \mapsto (\lambda, 0),$$

respectively. It is easy to see that

$$\sigma_1([(x,y)]) = y - x \quad \text{and} \quad \sigma_2([(x,y)]) = x - y,$$

where square brackets stand for cosets modulo $\operatorname{Im}\Delta$.

Thus canonical isomorphism (4.3) composed with $\sigma_i, i = 1, 2$, leads to canonical isomorphisms

$$\int_i : H^1(\mathbb{I}, \partial\mathbb{I}) \to \mathbb{R}, \quad i = 1, 2.$$

### 4.2.8 Cohomological Interpretation of Newton-Leibnitz Formula

Canonical isomorphisms $\int_i$, $i = 1, 2$, mark relative cohomology classes of 1-form on $\mathbb{I}$, *i.e.*, elements of $\mathrm{H}^1(\mathbb{I}, \partial \mathbb{I})$, with real numbers. This allows to associate a real number with a 1-form $\omega$ on $\mathbb{I}$:

$$\omega \Rightarrow [\omega] \in \mathrm{H}^1(\mathbb{I}, \partial \mathbb{I}) \Rightarrow \int_i ([\omega]) \in \mathbb{R}.$$

Obviously,

$$\int_1 ([\omega]) = - \int_2 ([\omega]).$$

Now put

$$\int_1 ([\omega]) \stackrel{\text{def}}{=} {\rlap{/}{\int}}_a^b \omega, \qquad \int_2 ([\omega]) \stackrel{\text{def}}{=} {\rlap{/}{\int}}_b^a \omega.$$

**Definition.** ${\rlap{/}{\int}}_a^b \omega$ (respectively, ${\rlap{/}{\int}}_b^a \omega$) is called the *cohomological definite integral of* $\omega$ over the oriented from $a$ to $b$ (respectively, from $b$ to $a$) interval $\mathbb{I}$.

### 4.2.9

Recall that the isomorphism (4.3) comes from the cohomology exact sequence (see n. 4.2.5) by means of the *co-boundary homomorphism*

$$\mathrm{H}^0(\partial \mathbb{I}) \stackrel{\sim}{\to} \mathrm{H}^1(\mathbb{I}, \partial \mathbb{I}).$$

By definition this homomorphism sends a pair $(p, q) \in \mathrm{H}^0(\partial \mathbb{I})$ to $[\mathrm{d} f] \in \mathrm{H}^1(\mathbb{I}, \partial \mathbb{I})$ where $f \in \mathrm{C}^\infty(\mathbb{I})$ is a function such that

$$f(a) = p, \quad f(b) = q.$$

On the other hand,

$$\sigma_1([p, q]) = q - p = f(b) - f(a).$$

Therefore,

$${\rlap{/}{\int}}_a^b \mathrm{d} f = f(b) - f(a).$$

This way we get the *cohomological Newton-Leibnitz formula* which is nothing but a description of the co-boundary operator in the long cohomology exact sequence of the pair $(\mathbb{I}, \partial \mathbb{I})$ in terms of canonical numerical marks of cohomology classes from $\mathrm{H}^1(\mathbb{I}, \partial \mathbb{I})$. It is worth stressing that this formula corresponds to the orientation of $\mathbb{I}$ from $a$ to $b$.

### 4.2.10

By comparing the cohomological Newton-Leibnitz formula with the standard we see that the *cohomological integral* coincides with the ordinary one, i.e.,

$$\unicode{x2A0D}_a^b df = \int_a^b df.$$

This apparently simple, if not banal, fact has, nevertheless, a deep philosophical meaning. Indeed, it indicates clearly that conceptually *integrals are cohomology classes marked by (real) numbers* and, therefore, puts in question the commonly accepted status of measure theory as the theory of integral. It is not, however, our intention to discuss this important problem here and we send the reader to [Moreno et al. (in preparation)] for further arguments in favor of cohomology origins of integrals. Below the Newton-Leibnitz formula is generalized to linear connections on the basis of the cohomological approach.

### 4.2.11 *Newton-Leibnitz formula for Linear Connections*

The mentioned generalization is obtained by substituting the cohomology of a flat linear connection for that of de Rham in our previous considerations.

Let $\nabla$ be a linear connection on a fat manifold $\bar{\mathbb{I}}$ over $\mathbb{I} = [a,b]$, $a < b$. Since the base dimension is 1, $\nabla$ is automatically flat and the corresponding long exact cohomology sequence of the pair $(\bar{\mathbb{I}}, \overline{\partial \mathbb{I}})$ is well-defined. As in n. 4.2.6 describe the following segment of it:

$$H_\nabla^0(\bar{\mathbb{I}}, \overline{\partial \mathbb{I}}) \to H_\nabla^0(\bar{\mathbb{I}}) \to H_\nabla^0(\overline{\partial \mathbb{I}}) \to H_\nabla^1(\bar{\mathbb{I}}, \overline{\partial \mathbb{I}}) \to H_\nabla^1(\bar{\mathbb{I}}) \ .$$

The kernel $\mathcal{F}$ of $(d_\nabla)_0$ consists of $\nabla$-constant sections of $\bar{\mathbb{I}}$. Therefore, $H_\nabla^0(\bar{\mathbb{I}}, \overline{\partial \mathbb{I}}) = 0$ and there is a canonical isomorphism $H_\nabla^0(\bar{\mathbb{I}}) \cong \mathcal{F}$. It is also easy to see that $H_\nabla^0(\overline{\partial \mathbb{I}})$ is the direct sum $\bar{a} \oplus \bar{b}$ and the homomorphism

$$H_\nabla^0(\bar{\mathbb{I}}) \to H_\nabla^0(\overline{\partial \mathbb{I}})$$

corresponds to

$$h: \mathcal{F} \to \bar{a} \oplus \bar{b}, \quad p \mapsto (p(a), p(b)) \ .$$

According to Proposition 2.2.2, $\nabla$ is compatible with a trivial connection $D$ with respect to a suitable fat identity map $\bar{\mathbb{I}}' \to \bar{\mathbb{I}}$ and hence $H_\nabla^1(\bar{\mathbb{I}})$ is isomorphic to $H_D^1(\bar{\mathbb{I}}')$ which is, obviously, trivial.

Thus the segment

$$H_\nabla^0(\bar{\mathbb{I}}, \overline{\partial \mathbb{I}}) \to H_\nabla^0(\bar{\mathbb{I}}) \to H_\nabla^0(\overline{\partial \mathbb{I}}) \to H_\nabla^1(\bar{\mathbb{I}}, \overline{\partial \mathbb{I}}) \to H_\nabla^1(\bar{\mathbb{I}})$$

is naturally identified with the short exact sequence
$$0 \to \mathcal{F} \xrightarrow{h} \overline{a} \oplus \overline{b} \to \mathrm{H}^1_\nabla\left(\overline{\mathbb{I}}, \overline{\partial \mathbb{I}}\right) \to 0,$$
which gives rise to an isomorphism
$$\frac{\overline{a} \oplus \overline{b}}{h(\mathcal{F})} \xrightarrow{\sim} \mathrm{H}^1_\nabla\left(\overline{\mathbb{I}}, \overline{\partial \mathbb{I}}\right).$$

Denote by $\mathbf{T}^\nabla_{t_0,t_1} : \overline{t_0} \to \overline{t_1}$, $t_0, t_1 \in \mathbb{I}$, the parallel translation isomorphism. Then
$$h(\mathcal{F}) = \{(\mathbf{v}_a, \mathbf{v}_b) : \mathbf{v}_b = \mathbf{T}^\nabla_{a,b}(\mathbf{v}_a)\}$$
since $\mathcal{F}$ is composed of $\nabla$-constant sections.

Now by combining this description of $h$ with natural embeddings
$$\iota_a : \overline{a} \hookrightarrow \overline{a} \oplus \overline{b} \quad \text{and} \quad \iota_b : \overline{b} \hookrightarrow \overline{a} \oplus \overline{b}$$
one gets canonical isomorphisms
$$\frac{\overline{a} \oplus \overline{b}}{h(\mathcal{F})} \xrightarrow{\sim} \overline{a} \quad \text{and} \quad \frac{\overline{a} \oplus \overline{b}}{h(\mathcal{F})} \xrightarrow{\sim} \overline{b},$$
given by
$$[(\mathbf{v}, \mathbf{w})] \mapsto \mathbf{v} - \mathbf{T}^\nabla_{b,a}(\mathbf{w}) \quad \text{and} \quad [(\mathbf{v}, \mathbf{w})] \mapsto \mathbf{w} - \mathbf{T}^\nabla_{a,b}(\mathbf{v}),$$
respectively. This way cohomology classes from $\mathrm{H}^1_\nabla\left(\overline{\mathbb{I}}, \overline{\partial \mathbb{I}}\right)$ can be canonically marked either with vectors from $\overline{a}$, or from $\overline{b}$.

But $\Lambda^1\left(\overline{\mathbb{I}}, \overline{\partial \mathbb{I}}\right) = \Lambda^1\left(\overline{\mathbb{I}}\right)$ and hence
$$\mathrm{H}^1_\nabla\left(\overline{\mathbb{I}}, \overline{\partial \mathbb{I}}\right) = \frac{\Lambda^1\left(\overline{\mathbb{I}}\right)}{B^1\left(\overline{\mathbb{I}}, \overline{\partial \mathbb{I}}\right)}.$$

So, it is natural to define *cohomological definite $\nabla$-integrals* of a thickened 1-form $\overline{\omega}$
$$\boxed{\nabla\!\!\int}_b^a \overline{\omega} \in \overline{a} \quad \text{and} \quad \boxed{\nabla\!\!\int}_a^b \overline{\omega} \in \overline{b}$$
as markers of the cohomology class $[\overline{\omega}]$. Now, exactly as in n. 4.2.9, the *cohomological Newton-Leibnitz formulas*
$$\boxed{\nabla\!\!\int}_b^a d_\nabla(p) = p(a) - \mathbf{T}^\nabla_{b,a}(p(b))$$
and
$$\boxed{\nabla\!\!\int}_a^b d_\nabla(p) = p(b) - \mathbf{T}^\nabla_{a,b}(p(a))$$
follow immediately from the definition of the co-boundary operator in the above long exact sequence for $\nabla$-cohomology.

## 4.3 Maxwell's Equations

The electromagnetic field $(\mathbf{E}, \mathbf{B})$, where $\mathbf{E} = (E_x, E_y, E_z)$ is the electric field and $\mathbf{B} = (B_x, B_y, B_z)$ is the magnetic field, is described by the famous Maxwell's equations:

$$\operatorname{div} \mathbf{B} = 0 , \quad \operatorname{rot} \mathbf{E} = -\frac{\partial \mathbf{B}}{\partial t} , \tag{4.5}$$

$$\operatorname{div} \mathbf{E} = \rho , \quad \operatorname{rot} \mathbf{B} - \frac{\partial \mathbf{E}}{\partial t} = \mathbf{j} , \tag{4.6}$$

where $\rho$ and $\mathbf{j} = (j_x, j_y, j_z)$ are densities of charge and current, respectively.

Much later after Maxwell it was observed that quantities $\mathbf{E}$ and $\mathbf{B}$ are naturally interpreted as coefficients of the differential 2-form

$$\begin{aligned} F = &\, E_x \, \mathrm{d}\, x \wedge \mathrm{d}\, t + E_y \, \mathrm{d}\, y \wedge \mathrm{d}\, t + E_z \, \mathrm{d}\, z \wedge \mathrm{d}\, t \\ &+ B_z \, \mathrm{d}\, x \wedge \mathrm{d}\, y - B_y \, \mathrm{d}\, x \wedge \mathrm{d}\, z + B_x \, \mathrm{d}\, y \wedge \mathrm{d}\, z . \end{aligned} \tag{4.7}$$

Indeed, the first pair (4.5) tells that the form $F$ is closed, *i.e.*, $\mathrm{d} F = 0$. However, on the basis of this observation we cannot be sure that $F$ is a true differential form. Namely, in view of our previous experience (Sect. 4.2), (4.7) could be the coordinate expression of the curvature tensor of a connection and the differential d in $\mathrm{d} F = 0$ could be the covariant differential of it. Some more delicate physical considerations that we cannot report here show that this is the case by giving the origin of gauge field theory. In this section we describe the mathematical part of the question.

### 4.3.1 *Minkowski Space*

The background of the classical electromagnetic theory is the Minkowski spacetime $M^4$. It is an affine space over $\mathbb{R}$ in which the affine structure may be described as four dimensional abelian sub-Lie algebra $\mathfrak{a}$ of $\mathrm{D}\left(M^4\right)$, whose fields are all complete, *i.e.*, the corresponding flows are defined on the whole of $\mathbb{R} \times M^4$. Denote by $G_t^{\mathbf{v}}$ diffeomorphisms of the one-parameter group corresponding to a vector field $\mathbf{v} \in \mathfrak{a}$. $G_1^{\mathbf{v}}$ is interpreted as the affine translation of $M^4$ by $\mathbf{v}$. Fix a point $m \in M^4$. Then the map

$$\mathfrak{a} \ni \mathbf{v} \mapsto G_1^{\mathbf{v}}(m) \in M^4$$

is, obviously, a bijection and $G_1^{\mathbf{v}}(m)$ is interpreted as the result of application of the vector $\mathbf{v}$ to $m$.

A cartesian frame in $M^4$ is composed of a point $m_0$ and a basis $(\mathbf{e}_0, \mathbf{e}_1, \mathbf{e}_2, \mathbf{e}_3)$ in $\mathfrak{a}$. The coordinates of a point $m \in M^4$ with respect to this frame are coordinates of the vector **v** in the basis $(\mathbf{e}_0, \mathbf{e}_1, \mathbf{e}_2, \mathbf{e}_3)$ in $\mathfrak{a}$ assuming that $m = G_1^{\mathbf{v}}(m_0)$.

Let $\beta$ be a nondegenerate symmetric bilinear form in the algebra $\mathfrak{a}$ understood as a vector space. This form determines a pseudo-metric $\eta$ in $M^4$

$$\eta_m(\mathbf{v}_m, \mathbf{w}_m) = \beta(\mathbf{v}, \mathbf{w}), \qquad \forall m \in M^4.$$

Vector fields $\mathbf{e}_0, \ldots, \mathbf{e}_3$ form a basis of $C^\infty(M^4)$-module $D(M^4)$ as well. So, if $X, Y \in D(M^4)$ then $X = \sum_i \mu_i \mathbf{e}_i$, $Y = \sum_i \nu_i \mathbf{e}_i$ and

$$\eta(X, Y) = \sum_{i,j} \mu_i \nu_j \beta(\mathbf{e}_i, \mathbf{e}_j).$$

Assume now that $\beta$ is of signature $(1,3)$. The affine space $M^4$ supplied with the pseudo-metric $\eta$ corresponding to such a $\beta$ is called the *Minkowski spacetime*. Let $(\mathbf{e}_0, \mathbf{e}_1, \mathbf{e}_2, \mathbf{e}_3)$ be orthonormal, i.e., $\beta(\mathbf{e}_i, \mathbf{e}_j) = \epsilon_{ij}$, with $\epsilon_{00} = 1$, $\epsilon_{ii} = -1$, $i > 0$, and $\epsilon_{ij} = 0$ if $i \neq j$. Then the Minkowski metric $\eta$ in the corresponding Cartesian frame reads

$$\eta = dt^2 - dx^2 - dy^2 - dz^2,$$

with $t, x, y, z$ being the corresponding coordinates.

### 4.3.2

Now, to proceed on we need some facts concerning fat manifolds of type 2 supplied with a gauge metric. The general (model) fiber in this case is $\mathbb{R}^2$ and the standard scalar product on it is the model metric $\beta$. Note that there is no difference between inner and gauge metrics on fat manifolds because all symmetric positive bilinear forms of the same dimension are equivalent.

So, assume $\overline{M}$ be such a fat manifold with connected $M$ and $g$ a gauge metric on it. $\overline{M}$ is *fat orientable* if the second exterior power $\overline{M}^{\wedge 2}$ of the vector bundle $\overline{M} \to M$ is trivial (1-dimensional) bundle. From geometrical point of view this means that each fat point $\overline{m}$ of $\overline{M}$ can be oriented in a manner that the orentation varies continuously when passing from a point to another.

The model Lie algebra for the gauge Lie algebra $o(\overline{M}, g)$ is so(2). So, $o(\overline{M}, g)$ is a rank 1 projective $C^\infty(M)$-module (the fibers of the corresponding vector bundle are one-dimensional).

Assume that $\overline{M}$ is fat oriented. Denote by $J_m : \overline{m} \to \overline{m}$ the $g_m$-orthogonal operator which is the rotation by $\pi/2$ in the clock-wise direction with respect to the chosen orientation in $\overline{m}$. The family of rotations $m \mapsto J_m$ determine an operator $J \in \mathrm{o}\left(\overline{M}, g\right)$ such that $J^2 = -\mathrm{id}$. This show that $\overline{M}$ is fat orientable if and only if $\mathrm{o}\left(\overline{M}, g\right)$ is a trivial (1-dimensional) $\mathrm{C}^\infty(M)$-module. Below we simplify the notation by writing $\mathfrak{o}$ for $\mathrm{o}\left(\overline{M}, g\right)$.

**Proposition.** $\overline{M}$ is trivial if and only if it has a nowhere vanishing section and is fat orientable.

**Proof.** Let $s \in \Gamma\left(\overline{M}\right)$ be such a section. Since $\overline{M}$ is fat orientable the operator $J$ is globally defined. Obviously, $s$ and $J(s)$ are $\mathrm{C}^\infty(M)$-independent and hence form a base of $\mathrm{C}^\infty(M)$-module $\Gamma\left(\overline{M}\right)$. The converse is obvious. □

In the general case the operator $J$ is well-defined locally by choosing a local fat orentation of $\overline{M}$. Since the $\mathrm{C}^\infty(M)$-module $\mathfrak{o}$ has constant rank one, $J$ is a local base of it and hence a local base of $\Lambda^\bullet(M)$-module $\Lambda^\bullet(\mathfrak{o})$. If $\overline{M}$ is fat oriented, then $J$ is a base of both $\mathfrak{o}$ and $\Lambda^\bullet(\mathfrak{o})$.

According to Exercise 2.5.2, (2) if a linear connection $\nabla$ preserves $g$ then $\nabla^{\mathrm{End}}$ preserves $\mathfrak{o}$. This means that (locally) $\mathrm{d}_{\nabla^{\mathrm{End}}}(J) = \omega \wedge J$, $\omega \in \Lambda^1(M)$, since $J$ is a local base of $\Lambda^\bullet(\mathfrak{o})$.

**Exercise.** Prove that $\omega = 0$, *i.e.*, $\mathrm{d}_{\nabla^{\mathrm{End}}}(J) = 0$.

As a consequence we see that $J$ defines a (local) gauge complex structure in $\overline{M}$.

### 4.3.3

Another important consequence is the following.

**Proposition.** If $\overline{M}$ is fat orientable and $\nabla$ preserves $g$, then

$$\Lambda^\bullet(M) \ni \omega \mapsto \omega \wedge J \in \Lambda^\bullet(\mathfrak{o})$$

is an isomorphism of cd-modules and of complexes $(\Lambda^\bullet(M), \mathrm{d})$ and $(\Lambda^\bullet(\mathfrak{o}), \mathrm{d}_{\nabla^\mathfrak{o}})$, with $\nabla^\mathfrak{o} = \nabla^{\mathrm{End}}\big|_\mathfrak{o}$.

If $\overline{M}$ is fat orientable and $\nabla$ preserves $g$, then $R^\nabla \in \Lambda^2(\mathfrak{o})$ and, so, (locally) $R^\nabla = \Omega \wedge J$. Moreover, $\Omega$ is a closed 2-form as it immediately follows from the Bianchi identity.

### 4.3.4

Assume $\overline{M}$ be fat orientable and $(\overline{M}, g, J)$ as before. Fix a preserving $g$ linear connection $\Delta$. Exercise 2.5.2, (1) implies that $\Lambda^\bullet(M) \to \Lambda^\bullet(\mathrm{o})$

$$\varrho \mapsto \nabla \stackrel{\text{def}}{=} \Delta + \varrho \wedge J$$

is a one-to-one correspondence between 1-forms on $M$ and linear connections that preserves $g$.

The following proposition is key in understanding which kind of mathematical quantities are electromagnetic fields. The notation $\varrho \wedge J$ in it stands for the multiplication operator in $\Lambda^\bullet(\overline{M})$.

**Proposition.** Let $(\overline{M}, g, J)$ b as above. If $\Delta$ and $\nabla$ are preserving $g$ linear connections, and, therefore, $\nabla = \Delta + \varrho \wedge J$ for a uniquely determined 1-form $\varrho$, then

(1) $d_\nabla = d_\Delta + \varrho \wedge J$;
(2) $R^\nabla = R^\Delta + d\varrho \wedge J$;
(3) $\Delta$ is induced from $\nabla$ by $\overline{f} \in \mathrm{SO}(\overline{M}, g)$ if and only if $\varrho = d\theta$ with $\overline{f}^* = \exp(\theta J) \stackrel{\text{def}}{=} (\cos \theta) \operatorname{id} + (\sin \theta) J$.

**Proof.** (1). In zeroth degree the equality is quite obvious. Namely, for all $p \in \Gamma(\overline{M})$ and $X \in \mathrm{D}(M)$, we have

$$d_\nabla(p)(X) = \nabla_X(p) = \Delta_X(p) + (\varrho \wedge J)(X)(p) = d_\Delta(X)(p) + \varrho(X)(J(p))$$
$$= d_\Delta(X)(p) + (\varrho \wedge J)(X)(p).$$

For higher degrees, it suffices to check the equality on $\omega \wedge p$, with $\omega \in \Lambda^s(M)$ and $p \in \Gamma(\overline{M})$:

$$d_\nabla(\omega \wedge p) = d\omega \wedge p + (-1)^s \omega \wedge d_\nabla(p) = d\omega \wedge p + (-1)^s \omega \wedge d_\Delta(p)$$
$$+ (-1)^s \omega \wedge (\varrho \wedge J) \wedge p \stackrel{n.\ 3.2.9}{=} d_\Delta(\omega \wedge p) + (\varrho \wedge J) \wedge (\omega \wedge p).$$

(2). First note that the Leibniz rule defining a cd-module $(\mathcal{P}, \overline{d})$ over $(\mathcal{A}, d)$ (see Definition 3.3.2), rewritten in terms of operators on $\mathcal{P}$, reads

$$\overline{d} \circ w_r = dw_r + (-1)^{lr} w_r \circ \overline{d},$$

with $l$ being the degree of $\overline{d}$ and $w_r \in \mathcal{A}_r$. Similarly, the Leibniz rule of Proposition 3.2.12 reads

$$d_\nabla \circ T = d_{\nabla^{\mathrm{End}}} T + (-1)^r T \circ d_\nabla.$$

In particular, since $\Delta$ preserves $J$, we also have

$$d_\Delta \circ J = J \circ d_\Delta.$$

Therefore,
$$R^\nabla = d_\nabla \circ d_\nabla = d_\Delta \circ d_\Delta + \varrho \circ J \circ d_\Delta + d_\Delta \circ \varrho \circ J + \varrho \circ J \circ \varrho \circ J$$
$$= R^\Delta + \varrho \circ d_\Delta \circ J + d\varrho \circ J - \varrho \circ d_\Delta \circ J + \varrho \circ \varrho \circ J \circ J$$
$$= R^\Delta + d\varrho \circ J = R^\Delta + d\varrho \wedge J .$$

(3). We have
$$\Lambda^\bullet \overline{f}^* \circ d_\nabla = \exp(\theta J) \circ d_\Delta + \exp(\theta J) \circ \varrho \circ J$$
and
$$d_\Delta \circ \Lambda^\bullet \overline{f}^* = d_{\Delta\text{End}} (\exp(\theta J)) + \exp(\theta J) \circ d_\Delta .$$

Moreover,
$$d_{\Delta\text{End}} (\exp(\theta J)) = d_{\Delta\text{End}} (\cos\theta \, \text{id} + \sin\theta J)$$
$$\stackrel{\text{Proposition 3.2.12}}{=} -\sin\theta \, d\theta \wedge \text{id} + \cos\theta \, d\theta \wedge J = d\theta \wedge (\sin\theta J \circ J + \cos\theta J)$$
$$= d\theta \wedge (\exp(\theta J) \circ J) ,$$

which in terms of operators reads
$$d_{\Delta\text{End}} (\exp(\theta J)) = d\theta \circ \exp(\theta J) \circ J$$

Therefore,
$$\Lambda^\bullet \overline{f}^* \circ d_\nabla = d_\Delta \circ \Lambda^\bullet \overline{f}^*$$

if and only if
$$\exp(\theta J) \circ \varrho \circ J = d\theta \circ \exp(\theta J) \circ J .$$

This condition is equivalent to $\varrho = d\theta$ because $\exp(\theta J) \circ \varrho = \varrho \circ \exp(\theta J)$ (according to n. 3.2.9) and $\exp(\theta J) \circ J$ is invertible.  □

## 4.3.5

Let $(\overline{M}, g, J)$ be as before. Two $g$-preserving linear connections in $\overline{M}$ are called *g-gauge equivalent* if one is induced from another by some $\overline{f} \in \text{SO}(\overline{M}, g)$. This terminology is applied to any fat manifold and any inner pseudo-metric in it.

Assume that there exists $s \in \Gamma(\overline{M})$ such that $g(s, s) = 1$. According to Proposition 4.3.2, this is the case if and only if $\overline{M}$ is trivial. Define a linear connection □ by conditions
$$d_\square(s) = d_\square (J(s)) = 0 .$$

Obviously, this connection preserves $g$. A $g$-preserving linear connection in $\overline{M}$ is called *g-gauge trivial* if it is $g$-gauge equivalent to □.

**Exercise.** Prove that $g$-gauge triviality is well-defined, *i.e.*, it does not depend on the choice of $s$.

Since any $g$-preserving linear connection is of the form $\nabla_{\Box,\varrho} \stackrel{\text{def}}{=} \Box + \varrho \wedge J$. By Proposition 4.3.4, the assignment

$$\varrho \mapsto \nabla_{\Box,\varrho} \tag{4.8}$$

establishes one-to-one correspondences

closed 1-forms $\longleftrightarrow$ $g$-preserving flat connections

exact 1-forms $\longleftrightarrow$ $g$-gauge trivial connections.

This induces a one-to-one correspondence between first de Rham cohomology classes of $M$ and $g$-gauge equivalence classes of flat linear connections in $\overline{M}$. As it is easy to see, this correspondence is canonical, *i.e.*, does not depend on the choice of $\Box$.

### 4.3.6

Denote by $\operatorname{Conn} g$ the totality of all $g$-preserving connections and observe that the correspondence (4.8) allows to interpret $\operatorname{Conn} g$ as an affine space modeled over $\Lambda^1(M)$. The map

$$\operatorname{Conn} g \ni \nabla \mapsto R^\nabla \in \Lambda^2(\mathfrak{o}) \tag{4.9}$$

is not injective. Indeed, Proposition 4.3.4 tells that $R^\nabla = R^\Delta$ if and only if $d_\nabla = d_\Delta + \varrho \wedge J$ with $d \varrho = 0$. If, additionally, $H^1(M) = 0$, *i.e.*, closed 1-forms are exact, then (4.9) induces a one-to-one correspondence between $g$-gauge equivalence classes and $d_{\nabla^\circ}$-exact forms in $\Lambda^2(\mathfrak{o})$. The Bianchi identity completely characterizes these 2-forms if $H^2_{\nabla^\circ}(\mathfrak{o}) = 0$. By Proposition 4.3.3 this is equivalent to vanishing of $H^2(M)$.

### 4.3.7

Now we are ready to encode what electromagnetic fields are from mathematical point of view. To this end, in our previous considerations take for $M$ the Minkowski spacetime $M^4$. Observe that in this case $\overline{M}$ is automatically trivial and all $g$-preserving flat connections are $g$-gauge trivial. Moreover, since $H^2(M^4) = 0$, the curvature tensor $R^\nabla$ completely characterizes $g$-gauge equivalence class of $\nabla$. In its turn, curvature tensors are completely characterized in $\Lambda^2(\mathfrak{o})$ as $d_{\nabla^\circ}$-closed forms. The isomorphism of complexes $(\Lambda^\bullet(M), \mathrm{d})$ and $(\Lambda^\bullet(\mathfrak{o}), \mathrm{d}_{\nabla^\circ})$ (Proposition 4.3.3) shows that homogeneous Maxwell's equations (4.5), p. 247, which are equivalent to $\mathrm{d}\, F = 0$ (see (4.7)), translate to the Bianchi identity for $F \wedge J$.

Thus homogeneous Maxwell's equations reveal the fundamental correspondence

$$\text{Electromagnetic tensors} \longleftrightarrow g\text{-gauge equivalence classes} .$$

In its turn, the remaining Maxwell's equations (4.6) describe the dynamic of $g$-equivalence classes of $g$-preserving connections. This is, however, a physical question we cannot touch here.

To conclude, we remark that similar considerations lead to famous Yang-Mills equations.

## 4.4 Homotopy Formula for Linear Connections

Given a fat map $\overline{f} : \overline{N} \to \overline{M}$, let $\nabla$ and $\square$ be $\overline{f}$-compatible flat linear connections on $\overline{M}$ and $\overline{N}$, respectively. Then the homomorphism

$$\mathrm{H}^\bullet_\nabla \left(\overline{f}\right) : \mathrm{H}^\bullet_\nabla \left(\overline{M}\right) \to \mathrm{H}^\bullet_\nabla \left(\overline{N}\right), \quad [\varpi] \mapsto \left[\Lambda^\bullet \overline{f}^* (\varpi)\right]$$

is well-defined. Like the ordinary situation, a smooth 'compatibility-preserving' deformation of $\overline{f}$ does not affect $\mathrm{H}^\bullet_\nabla \left(\overline{f}\right)$. This important fact is properly formalized in terms of the concept of $\nabla$-*homotopy* between fat maps introduced below. It is worth stressing that this notion makes sense and is useful for arbitrary, not necessarily flat, linear connections.

### 4.4.1 *Homotopic Fat Maps*

**Definition.** Fat maps

$$\overline{f}, \overline{g} : \overline{N} \to \overline{M}$$

are called *homotopic* if there exists a one-parameter family $\{\overline{G_t}\}_{t \in [0,1]}$ ([1]) of fat maps such that

$$\overline{f} = \overline{G}_0 \quad \text{and} \quad \overline{g} = \overline{G}_1 ,$$

and the family $\{\overline{G_t}\}_{t \in [0,1]}$ is called a *(smooth) fat homotopy between* $\overline{f}$ *and* $\overline{g}$.

---

[1] For simplicity, the reader may assume that fat manifolds in this section are without boundary (*cf.* the footnote in n. 1.6.13).

Sometimes the term 'fat homotopy' will also refer to the corresponding fat map

$$\overline{G} : \overline{\mathbb{I}_N} \to \overline{M}, \quad \mathbb{I} = [0,1] .$$

### 4.4.2  $\nabla$-homotopy

Assume $\overline{f}$, $\overline{g}$ and $\overline{G}$ to be as above.

**Definition.** Let $\nabla$ be a linear connection on $\overline{M}$ and $\square$ the induced by $\overline{f} \circ \overline{\pi}$ linear connection, $\overline{\pi} : \overline{\mathbb{I}_N} \to \overline{N}$ being the fat projection. If $\square$ and $\nabla$ are $\overline{G}$-compatible, then the fat homotopy $\{\overline{G_t}\}_{t \in [0,1]}$ is called a $\nabla$-*homotopy* and $\overline{f}$ and $\overline{g}$ are called $\nabla$-*homotopic*.

Recall that $\overline{G_t} = \overline{G} \circ \overline{i_t}$ with $\overline{i_t} : \overline{N} \to \overline{\mathbb{I}_N}$ being the fat embedding at $t \in [0,1]$. If $\{\overline{G_t}\}_{t \in [0,1]}$ is a $\nabla$-homotopy, then the induced by $\overline{f}$ linear connection $\nabla^{\overline{N}}$ is $\overline{G_t}$-compatible with $\nabla$ for all $t$ as it follows easily from the definition of $\square$ and the fact that $\overline{\pi} \circ \overline{i_t}$ is the identity map.

Moreover, $\{\overline{G_t}\}_{t \in [0,1]}$ is a lift of $\{G_t\}_{t \in [0,1]}$ by $\nabla$, because the induced from $\square$ by a fat embedding $\overline{j_n}$ linear connection (see n. 1.6.13) is trivial.

Also, it follows easily from n. 3.3.19 that the two above conditions characterize a $\nabla$-homotopy $\{\overline{G_t}\}_{t \in [0,1]}$.

If $\nabla$ is flat and $\{\overline{G_t}\}_{t \in [0,1]}$ is a lift of $\{G_t\}_{t \in [0,1]}$ by $\nabla$, it is not difficult to deduce from n. 3.5.19 that $\nabla^{\overline{N}}$ is $\overline{G_t}$-compatible with $\nabla$ for all $t$. We do not report details on this, since it will not be used. An interesting consequence is that if $\nabla$ is flat, then a fat homotopy $\{\overline{G_t}\}_{t \in [0,1]}$ is a $\nabla$-homotopy between $\overline{f} = \overline{G_0}$ and $\overline{g} = \overline{G_1}$ if and only if it is a lift by $\nabla$ of a family $\{G_t\}_{t \in [0,1]}$.

### 4.4.3  Integral of Time-dependent Forms

It is not difficult to define time integral of a family $\{\overline{\omega_t}\}_{t \in \mathbb{I}}$, $\mathbb{I} = [a,b]$, of thickened $s$-forms (*cf.* [Berger and Gostiaux (1988), 0.3.15.3]) on $\overline{M}$. Fix a fat point $\overline{m}$. Then

$$t \mapsto (\overline{\omega_t})_m$$

defines a function

$$\alpha_{\overline{m}} : \mathbb{I} \to \text{Alt}^s (T_m M, \overline{m}) ,$$

where $\text{Alt}^s(T_m M, \overline{m})$ is the space of alternating $s$-linear functions on $T_m M$ with values in $\overline{m}$. It is a finite-dimensional vector space over $\mathbb{R}$ and, so, the integral

$$\overline{\theta_{\overline{m}}} = \int_a^b \alpha_{\overline{m}}(t)\,\mathrm{d}t$$

is well-defined if, for instance, $\alpha_{\overline{m}}$ is continuous. Note also that if $\xi_1, \ldots, \xi_s \in T_m M$ are tangent vectors, then

$$\overline{\theta_{\overline{m}}}(\xi_1, \ldots, \xi_s) = \int_a^b \alpha_{\overline{m}}(t)(\xi_1, \ldots, \xi_s)\,\mathrm{d}t.$$

The smooth dependence of the integral on parameters theorem (see, *e.g.*, [Berger and Gostiaux (1988), 0.4.8]) implies that if $\{\overline{\omega}\}_{t \in \mathbb{I}}$ is smooth, then the field $\{\overline{\theta_{\overline{m}}}\}_{m \in M}$ is smooth. Thus we have a thickened $s$-form $\overline{\theta}$ on $\overline{M}$.

**Definition.** The thickened $s$-form $\overline{\theta}$ is called the *integral of* $\{\overline{\omega_t}\}_{t \in \mathbb{I}}$, $\mathbb{I} = [a, b]$, and is denoted by

$$\int_a^b \overline{\omega_t}\,\mathrm{d}t.$$

It follows directly from the definition that

$$\left(\int_a^b \overline{\omega_t}\,\mathrm{d}t\right)(X_1, \ldots, X_s) = \int_a^b \overline{\omega_t}(X_1, \ldots, X_s)\,\mathrm{d}t.$$

**Exercise.** Let $\{\overline{\omega_t}\}_{t \in [a,b]}$ be a time-dependent thickened $s$-form on $\overline{M}$ and $\{\mathrm{d}\overline{\omega_t}/\mathrm{d}t\}_{t \in \mathbb{I}}$ its derivative (see Definition 3.5.11). Show that

$$\int_a^b \frac{\mathrm{d}\overline{\omega_t}}{\mathrm{d}t}\,\mathrm{d}t = \overline{\omega}_b - \overline{\omega}_a$$

and

$$\int_a^b \mathrm{d}_\nabla(\overline{\omega_t})\,\mathrm{d}t = \mathrm{d}_\nabla\left(\int_a^b \overline{\omega_t}\,\mathrm{d}t\right)$$

for a linear connection $\nabla$ on $\overline{M}$.

### 4.4.4 *Homotopy Operator*

Let $\{\overline{G_t}\}_{t \in \mathbb{I}}$, $\mathbb{I} = [0, 1]$, be a fat homotopy between fat maps $\overline{f}, \overline{g} : \overline{N} \to \overline{M}$. As usual, denote by $\overline{G} : \overline{\mathbb{I}_N} \to \overline{M}$ the corresponding fat map, by $\partial/\partial t$ the standard 'time' vector field on $N \times \mathbb{I}$. For each $t \in \mathbb{I}$ consider the operator

$$r_t = \Lambda^\bullet \, \overline{i_t}^* \circ \bar{\mathrm{i}}_{\frac{\partial}{\partial t}} \circ \Lambda^\bullet \, \overline{G}^* : \Lambda^\bullet(\overline{M}) \to \Lambda^\bullet(\overline{N}),$$

$\overline{i}_t : \overline{N} \to \overline{\mathbb{I}_N}$ being the fat embedding at $t \in \mathbb{I}$. According to Exercise 3.5.10, for a given $\overline{\omega} \in \Lambda^s(\overline{M})$

$$\{r_t(\overline{\omega})\}_{t \in \mathbb{I}}$$

is a time-dependent thickened form on $\overline{N}$ of degree $s$.

**Definition.** The $C^\infty(M)$-module homomorphism of degree $-1$
$$h : \Lambda^\bullet(\overline{M}) \to \Lambda^\bullet(\overline{N})$$
whose degree $s$ component is given by

$$\overline{\omega} \mapsto \int_0^1 r_t(\overline{\omega}) \, dt, \quad \overline{\omega} \in \Lambda^s(\overline{M})$$

is called the *homotopy operator associated with* $\{\overline{G}_t\}_{t \in \mathbb{I}}$.

### 4.4.5 Homotopy Formula

**Proposition.** Let $\nabla$ be a linear connection on a fat manifold $\overline{M}$. If $\{\overline{G}_t\}_{t \in [0,1]}$ is a $\nabla$-homotopy between fat maps $\overline{f}, \overline{g} : \overline{N} \to \overline{M}$, with associated homotopy operator $h$, then

$$\Lambda^\bullet \overline{g}^* - \Lambda^\bullet \overline{f}^* = h \circ d_\nabla + d_{\nabla^{\overline{N}}} \circ h, \qquad (4.10)$$

with $\nabla^{\overline{N}}$ being as in n. 4.4.2.

**Proof.** Let $\overline{G} : \overline{\mathbb{I}_N} \to \overline{M}$ be the fat map corresponding to $\{\overline{G}_t\}_{t \in [0,1]}$, $\overline{i}_t : \overline{N} \to \overline{\mathbb{I}_N}$ the fat embedding at $t \in \mathbb{I} = [0,1]$ and $\partial/\partial t$ the standard 'time' field on $\mathbb{I}_N = N \times \mathbb{I}$. If $\square$ is the induced by $\overline{G}$ from $\nabla$ linear connection, then

$$\Lambda^\bullet \overline{g}^*(\overline{\omega}) - \Lambda^\bullet \overline{f}^*(\overline{\omega})$$

$$\stackrel{\text{Exer. 4.4.3}}{=} \int_0^1 \frac{d}{dt} \Lambda^\bullet \overline{G}_t^*(\overline{\omega}) \, dt = \int_0^1 \frac{d}{dt} \Lambda^\bullet \overline{i}_t^* \left( \Lambda^\bullet \overline{G}^*(\overline{\omega}) \right) dt$$

$$\stackrel{\text{Exer. 3.5.4, n. 3.5.13, Prop. 3.5.15, n. 4.4.2}}{=}$$

$$\int_0^1 \Lambda^\bullet \overline{i}_t^* \left( \overline{i}_{\frac{\partial}{\partial t}} \left( d_\square \left( \Lambda^\bullet \overline{G}^*(\overline{\omega}) \right) \right) + d_\square \left( \overline{i}_{\frac{\partial}{\partial t}} \left( \Lambda^\bullet \overline{G}^*(\overline{\omega}) \right) \right) \right) dt$$

$$\stackrel{\text{n. 4.4.2}}{=} \int_0^1 \Lambda^\bullet \overline{i}_t^* \left( \overline{i}_{\frac{\partial}{\partial t}} \left( \Lambda^\bullet \overline{G}^*(d_\nabla(\overline{\omega})) \right) \right) dt$$

$$+ \int_0^1 d_{\nabla^{\overline{N}}} \left( \Lambda^\bullet \overline{i}_t^* \left( \overline{i}_{\frac{\partial}{\partial t}} \left( \Lambda^\bullet \overline{G}^*(\overline{\omega}) \right) \right) \right) dt$$

$$\stackrel{\text{n. 4.4.4}}{=} \int_0^1 r_t(d_\nabla(\overline{\omega})) \, dt + \int_0^1 d_{\nabla^{\overline{N}}}(r_t(\overline{\omega})) \, dt \stackrel{\text{Exer. 4.4.3}}{=} h(d_\nabla(\overline{\omega})) + d_{\nabla^{\overline{N}}}(h(\overline{\omega})),$$

as required. $\square$

**Corollary.** Let $\nabla$ be a flat linear connection on a fat manifold $\overline{M}$. If $\overline{f}, \overline{g} : \overline{N} \to \overline{M}$ are $\nabla$-homotopic, then the induced homomorphisms $H^\bullet_\nabla(\overline{f}), H^\bullet_\nabla(\overline{g}) : H^\bullet_\nabla(\overline{M}) \to H^\bullet_\nabla(\overline{N})$ coincide.

**Proof.** In this case (4.10) tells that the homotopy operator $h$ associated with a $\nabla$-homotopy connecting $\overline{f}$ and $\overline{g}$ is a cochain homotopy connecting cochain maps $\Lambda^\bullet \overline{f}^*$ and $\Lambda^\bullet \overline{g}^*$ of the complex $(\Lambda^\bullet(\overline{M}), d_\nabla)$ to $(\Lambda^\bullet(\overline{N}), d_{\nabla^N})$ (cf., e.g., [Hilton and Stammbach (1971), Chap. IV, Proposition 3.1]). □

## 4.5 Characteristic Classes

Cd-modules of the form $(\Lambda^\bullet(P), d_\nabla)$ are not generally cochain complexes. So, no cohomology can be associated with them straightforwardly. Nevertheless, there are various cohomologies related with a connection. They may be subdivided into two classes formed by 'fine' and 'rough' cohomologies, respectively. 'Fine' cohomologies characterize a connection itself and change when passing from one connection to another. On the contrary, 'rough' cohomologies do not depend on a concrete connection preserving an internal structure in a fat manifold. Hence such cohomologies characterize fat manifolds supplied with an internal structure and are called *characteristic classes*. Below we show how both kind of cohomologies related with connections arise and the corresponding basic constructions.

Construction of characteristic classes requires a reciprocal action of various structures of the 'connection calculus' that were introduced and discussed previously. So, this section is instructive also in this sense.

### 4.5.1

The starting point is the following observation. Let $P$ and $Q$ be $A$-modules supplied with linear connections $\nabla$ and $\square$, respectively. Assume additionally that $\square$ is flat. If $\varphi : \Lambda^\bullet(P) \to \Lambda^\bullet(Q)$ is a homomorphism of cd-modules over $\Lambda^\bullet(A)$ (see n. 3.3.3), then for all $\overline{\omega} \in \operatorname{Ker} d_\nabla$ $\varphi(\overline{\omega})$ is cocycle of the complex $(\Lambda^\bullet(Q), d_\square)$. Denote by $[\varphi(\overline{\omega})]_\square$ its cohomology class in $H^\bullet_\square(Q)$, i.e., the coset $\varphi(\overline{\omega}) + \operatorname{Im} d_\square$. But $\varphi(\operatorname{Im} d_\nabla) \subseteq \operatorname{Im} d_\square$, and hence

$$\varphi(\overline{\omega} + \operatorname{Im} d_\nabla) \subseteq \varphi(\overline{\omega}) + \operatorname{Im} d_\square \subseteq \varphi(\overline{\omega}) + \operatorname{Ker} d_\square .$$

So, although $\operatorname{Ker} d_\nabla + \operatorname{Im} d_\nabla \not\subseteq \operatorname{Ker} d_\nabla$ (generally), $\varphi$ induces a map

$$\frac{\operatorname{Ker} d_\nabla + \operatorname{Im} d_\nabla}{\operatorname{Im} d_\nabla} \to \frac{\operatorname{Ker} d_\square}{\operatorname{Im} d_\square} = H^\bullet_\square(Q) .$$

Now we observe that $(\operatorname{Ker} d_\nabla + \operatorname{Im} d_\nabla)/\operatorname{Im} d_\nabla$ is the cohomology of the complex $(\operatorname{Coker} d_\nabla^2, \delta_\nabla)$, where $\delta_\nabla = d_\nabla \bmod d_\nabla^2$. Denote it by $K_\nabla^\bullet$. Thus any homomorphism of the cd-module $(\Lambda^\bullet(P), d_\nabla)$ in a flat cd-module $(\Lambda^\bullet(Q), d_\square)$ induces a homomorphism $K_\nabla^\bullet \to \operatorname{H}_\square^\bullet(Q)$. In this the cohomology $K_\nabla^\bullet$ is universal with respect to cd-homomorphisms of $(\Lambda^\bullet(P), d_\nabla)$ to flat cd-modules over $\Lambda^\bullet(A)$. It is worth mentioning that $\operatorname{Coker} d_\nabla^2$ is a cd-module over $\Lambda^\bullet(A)$ due to the fact that $d_\nabla^2 = R^\nabla$ is a $\Lambda^\bullet(A)$-homomorphism. Obviously, $K_\nabla^\bullet = \operatorname{H}_\nabla^\bullet(P)$ if $\nabla$ is flat.

A natural isomorphism

$$\frac{\operatorname{Ker} d_\nabla + \operatorname{Im} d_\nabla}{\operatorname{Im} d_\nabla} \cong \frac{\operatorname{Ker} d_\nabla}{\operatorname{Ker} d_\nabla \cap \operatorname{Im} d_\nabla}$$

shows that $K_\nabla^\bullet$ may be interpreted as the cohomology of the complex $(\operatorname{Ker} d_\nabla^2, d_\nabla)$.

### 4.5.2

More generally, for a given nonnegative $s$ consider complexes

$$\left(\frac{\operatorname{Im} d_\nabla^s}{\operatorname{Im} d_\nabla^{s+2}}, \delta_{\nabla, s}\right) \quad \text{and} \quad \left(\operatorname{Im} d_\nabla^s \cap \operatorname{Ker} d_\nabla^2, d_\nabla\right)$$

where $\delta_{\nabla, s}$ is naturally induced by $d_\nabla$.

**Exercise.** Show that cohomologies of these complexes are isomorphic.

Denote them by $K_{\nabla, s}^\bullet$ and note that a homomorphism of cd-modules $\varphi : (\Lambda^\bullet(P), d_\nabla) \to (\Lambda^\bullet(Q), d_\square)$ induces a map $K_{\nabla, s}^\bullet \to K_{\square, s}^\bullet$. Cohomologies $K_{\nabla, s}^\bullet$, $s = 0, 1, \ldots$, are 'fine', *i.e.*, generally change when passing from one connection to another. They were not studied systematically and their geometrical meaning is still not very clear.

### 4.5.3

Cohomologies of the above kind can be, in fact, associated with any sequence of linear maps and, in particular, with a linear operator. If $\delta : V \to V$ is a linear operator, $V$ being a finite dimensional vector space over a field $k$, then the dimension of the cohomology space of the complex $(\operatorname{Im} \delta^{s-1} \cap \operatorname{Ker} \delta^2, \delta)$ is equal to the number of nilpotent Jordan $s \times s$ cells in the Jordan decomposition of $\delta$.

### 4.5.4

In the case when $Q = A$ and $d_\nabla = d$, the above construction gives a map $K_\nabla^\bullet \to H^\bullet(A)$ (de Rham cohomology of $A$) induced by a cd-module homomorphism $\varphi : \Lambda^\bullet(P) \to \Lambda^\bullet(A)$. In particular, if $\nabla$ is a linear connection in a fat manifold $\overline{M}$, then this construction gives a map $K_\nabla^\bullet \to H^\bullet(M)$ (de Rham cohomology of $M$). So, this procedure allows us to associate with $\nabla$ some cohomology classes of $M$. They become global invariants of $\nabla$ if $\varphi$ can be chosen in a canonical manner, as well as some cohomology classes in $K_\nabla^\bullet$. At this point it is very important noticing that we can get a cohomological characterization of $\nabla$ by applying the above procedure to connections associated with $\nabla$, such as $\nabla^{\text{End}}$, $\nabla^{\text{Bil}}$, etc., as well. For instance, the Bianchi identity (Proposition 3.2.14) tells us that $R^\nabla$ is $d_{\nabla^{\text{End}}}$–closed and hence defines a cohomology class $\zeta_\nabla \in K_{\nabla^{\text{End}}}^2$ canonically associated with $\nabla$. Then we need a linear map $\varphi : \text{End}\, P \to A$. A natural candidate for a such one is $\varphi = \text{tr}$, i.e., the map that associates with an endomorphism its trace. Below this idea is implemented.

### 4.5.5

Let $V$ be a $k$-vector space of dimension $r$ and $L : V \to V$ a linear operator. The operator $L$ extends uniquely to the whole exterior algebra $\bigwedge^\bullet V$ of $V$ as a derivation $\partial_L$ of zeroth degree, i.e.,

$$\partial_L(w_1 \wedge w_2) = \partial_L(w_1) \wedge w_2 + w_1 \wedge \partial_L(w_2)$$

for $w_1, w_2 \in \bigwedge^\bullet V$ and $\partial_L(v) = L(v)$ if $v \in V$. The $r$-th exterior power of $V$ is 1-dimensional. So, the restriction of $\partial_L$ to $\bigwedge^r V$ is multiplication by a number called the *trace* of $L$ and denoted by $\text{tr}\, L$.

**Exercise.** Check that this (conceptual) definition of the trace coincides with the standard one, and that $[\partial_L, \partial_{L'}] = \partial_{[L,L']}$, $L, L' \in \text{End}\, V$.

### 4.5.6

From now on, in order to avoid improper algebraic digressions, we make the following assumptions: the $k$-algebra $A$ has zero characteristic and the $A$-module $P$ is projective, finitely generated and has constant rank $r$.

The definition of trace given in n. 4.5.5 directly extends to $\Lambda^\bullet(A)$–modules of the form $\Lambda^\bullet(P)$.

Namely, let $L : \Lambda^\bullet(P) \to \Lambda^\bullet(P)$ be a graded $\Lambda^\bullet(A)$–module endomorphism (see n. 0.1.1). Consider the exterior algebra $P^\wedge \stackrel{\text{def}}{=} \bigwedge^\bullet P$. Its top component $P_r^\wedge = \bigwedge^r P$, is a projective $A$-module of constant rank 1. Hence any $\Lambda^\bullet(A)$–module endomorphism of $\Lambda^\bullet(P_r^\wedge) = \Lambda^\bullet(A) \otimes P_r^\wedge$ (see n. 0.5.10) is multiplication by a form $\omega \in \Lambda^\bullet(A)$ (cf. n. 0.1.5, (10)).

The wedge product '$\wedge$' in $\Lambda^\bullet(A)$ and the wedge product '$\wedge_P$' in $P^\wedge$ are combined in a common product $\bar{\wedge}$ in $\Lambda^\bullet(P^\wedge) = \Lambda^\bullet(A) \otimes_A P^\wedge = \bigoplus_{s,t} \Lambda^s\left(\bigwedge^t P\right)$. Explictly:

$$(\omega \wedge \Pi) \,\bar{\wedge}\, (\omega' \wedge \Pi') = (\omega \wedge \omega') \wedge (\Pi \wedge_P \Pi'), \quad \omega,\omega' \in \Lambda^\bullet(A), \Pi, \Pi' \in P^\wedge .$$

With this product $\Lambda^\bullet(P^\wedge)$ becomes a *bigraded commutative algebra*:

$$\overline{\varkappa} \,\bar{\wedge}\, \overline{\varkappa}' = (-1)^{ss'+tt'} \overline{\varkappa}' \,\bar{\wedge}\, \overline{\varkappa} ,$$

where $\overline{\varkappa} \in \Lambda^s\left(\bigwedge^t P\right), \overline{\varkappa}' \in \Lambda^{s'}\left(\bigwedge^{t'} P\right)$.

The endomorphism $L$ extends to $\Lambda^\bullet(P^\wedge)$ as a graded $\Lambda^\bullet(A)$–derivation $\partial_L$:

$$\partial_L \left(\overline{\varkappa} \,\bar{\wedge}\, \overline{\varkappa}'\right) = \partial_L \left(\overline{\varkappa}\right) \,\bar{\wedge}\, \overline{\varkappa}' + (-1)^{s \deg L} \overline{\varkappa} \,\bar{\wedge}\, \partial_L \left(\overline{\varkappa}'\right) ,$$

$$\partial_L|_{\Lambda^\bullet(A)} = 0, \quad \partial_L|_{\Lambda^\bullet(P)} = L$$

([2]); or, also,

$$\partial_L \left(\omega \wedge \Pi \wedge_P \Pi'\right)$$
$$= (-1)^{s \deg L} \omega \wedge \partial_L (\Pi) \wedge_P \Pi' + (-1)^{s \deg L} \omega \wedge \Pi \wedge_P \partial_L (\Pi')$$

(here $\overline{\varkappa} \in \Lambda^s(P^\wedge), \overline{\varkappa}' \in \Lambda^\bullet(P^\wedge), \omega \in \Lambda^s(A), \Pi, \Pi' \in P^\wedge$). Then $\partial_L|_{\Lambda^\bullet(P_r^\wedge)}$ is a $\Lambda^\bullet(A)$–endomorphism of $\Lambda^\bullet(P_r^\wedge)$ and, as it was noticed above, is a multiplication by a form, denoted by $\mathrm{tr}(L)$, *i.e.*,

$$\partial_L(\Theta) = \mathrm{tr}(L) \wedge \Theta, \quad \Theta \in \Lambda^\bullet(P_r^\wedge), \mathrm{tr}(L) \in \Lambda^\bullet(A) .$$

**Exercise.** Check that

$$[\partial_L, \partial_{L'}]^{(\mathrm{gr})} = \partial_{[L, L']^{(\mathrm{gr})}}, \quad L, L' \in \mathrm{End}^{(\mathrm{gr})}_{\Lambda^\bullet(A)}(\Lambda^\bullet(P)) ,$$

where $[\cdot, \cdot]^{(\mathrm{gr})}$ stands for the graded commutator with respect to the $\Lambda^\bullet(A)$–grading.

---

[2] Here $\Lambda^\bullet(A)$ and $\Lambda^\bullet(P)$ are regarded as a subset of $\Lambda^\bullet(P^\wedge)$ due to the identifications $A = P_0^\wedge$, $P = P_1^\wedge$.

### 4.5.7

Let now $\overline{Y}$ be a *graded der-operator* acting in the $\Lambda^\bullet(A)$–module $\Lambda^\bullet(P)$ over a graded derivation $Y : \Lambda^\bullet(A) \to \Lambda^\bullet(A)$. This means that degrees of $\overline{Y}$ and $Y$ coincide and

$$\overline{Y}(\omega \wedge \overline{\varkappa}) = Y(\omega) \wedge \overline{\varkappa} + (-1)^{r \deg Y} \omega \wedge \overline{Y}(\overline{\varkappa}), \quad \omega \in \Lambda^r(A), \overline{\varkappa} \in \Lambda^\bullet(P).$$

As in n. 4.5.6, $\overline{Y}$ extends to $\Lambda^\bullet(P^\wedge)$ as a graded derivation $\partial_{\overline{Y}}$, i.e.,

$$\partial_{\overline{Y}}(\overline{\varkappa} \wedge \overline{\varkappa}') = \partial_{\overline{Y}}(\overline{\varkappa}) \wedge \overline{\varkappa}' + (-1)^{s \deg \overline{Y}} \overline{\varkappa} \wedge \partial_{\overline{Y}}(\overline{\varkappa}'),$$

$$\partial_{\overline{Y}}|_{\Lambda^\bullet(A)} = Y, \qquad \partial_{\overline{Y}}|_{\Lambda^\bullet(P)} = \overline{Y};$$

or, also,

$$\partial_{\overline{Y}}(\omega \wedge \Pi) = Y(\omega) \wedge \Pi + (-1)^{s \deg \overline{Y}} \omega \wedge \partial_{\overline{Y}}(\Pi),$$

$$\partial_{\overline{Y}}(\Pi \wedge_P \Pi') = \partial_{\overline{Y}}(\Pi) \wedge_P \Pi' + \Pi \wedge_P \partial_{\overline{Y}}(\Pi')$$

$$\partial_{\overline{Y}}|_{\Lambda^\bullet(P)} = \overline{Y}$$

(here $\overline{\varkappa} \in \Lambda^s(P^\wedge)$, $\overline{\varkappa}' \in \Lambda^\bullet(P^\wedge)$, $\omega \in \Lambda^s(A)$, $\Pi, \Pi' \in P^\wedge$).

**Exercise.** With $\overline{Y}$ being as above and $L$ as in n. 4.5.6, check that $[\overline{Y}, L]^{\mathrm{gr}}$ is an endomorphism of the $\Lambda^\bullet(A)$–module $\Lambda^\bullet(P^\wedge)$ and

$$[\partial_{\overline{Y}}, \partial_L]^{\mathrm{gr}} = \partial_{[\overline{Y},L]^{\mathrm{gr}}}. \tag{4.11}$$

By restricting (4.11) to $\Lambda^\bullet(P_r^\wedge)$ we obtains the following important formula

$$\mathrm{tr}\left([\overline{Y},L]^{\mathrm{gr}}\right) = Y(\mathrm{tr}(L)). \tag{4.12}$$

Indeed, if $\Theta \in \Lambda^\bullet(P_r^\wedge)$, then

$$[\partial_{\overline{Y}}, \partial_L]^{\mathrm{gr}}(\Theta) = \partial_{\overline{Y}}(\partial_L(\Theta)) - (-1)^{\deg \overline{Y} \deg L} \partial_L(\partial_{\overline{Y}}(\Theta))$$

$$= \partial_{\overline{Y}}(\mathrm{tr}(L) \wedge \Theta) - (-1)^{\deg \overline{Y} \deg L} \mathrm{tr}(L) \wedge \partial_{\overline{Y}}(\Theta)$$

$$= Y(\mathrm{tr}(L)) \wedge \Theta + (-1)^{\deg \overline{Y} \deg L} \mathrm{tr}(L) \wedge \partial_{\overline{Y}}(\Theta) - (-1)^{\deg \overline{Y} \deg L} \mathrm{tr}(L) \wedge \partial_{\overline{Y}}(\Theta)$$

$$= Y(\mathrm{tr}(L)) \wedge \Theta;$$

on the other side,

$$\partial_{[\overline{Y},L]^{\mathrm{gr}}}(\Theta) = \mathrm{tr}\left([\overline{Y},L]^{\mathrm{gr}}\right) \wedge \Theta.$$

Now, by applying (4.12) to $\overline{Y} = d_\nabla$, $Y = d$ and a zeroth degree $\Lambda^\bullet(A)$–endomorphism of $\Lambda^\bullet(P)$ $\varphi$ for $L$, we get $[d_\nabla, \varphi]^{\mathrm{gr}} = d \operatorname{tr}(\varphi)$. But $[d_\nabla, \varphi]^{\mathrm{gr}} = d_{\nabla \mathrm{End}}(\varphi)$. So, $\operatorname{tr}(d_{\nabla \mathrm{End}}(\varphi)) = d \operatorname{tr}(\varphi)$ for all $\varphi$. This proves that

$$\operatorname{tr} \circ d_{\nabla \mathrm{End}} = d \circ \operatorname{tr}, \tag{4.13}$$

*i.e.*, that the map $\operatorname{tr} : \Lambda^\bullet(\mathrm{End}(P)) \to \Lambda^\bullet(A)$ is a homomorphism of cd-modules. As such, $\operatorname{tr}$ induces a map in cohomology

$$\mathrm{H}_{\operatorname{tr}} : \mathrm{K}^\bullet_{\nabla \mathrm{End}} \mapsto \mathrm{H}^\bullet(A)$$

and we obtain the cohomology class $\operatorname{tr}(\nabla) \stackrel{\mathrm{def}}{=} \mathrm{H}_{\operatorname{tr}}(\zeta_\nabla) \in \Lambda^2(A)$ (see n. 4.5.4). If $P = \Gamma(\overline{M})$, then $\operatorname{tr}(\nabla) \in \mathrm{H}^2(M)$.

**4.5.8**

The last question to be asked now is: whether the class $\operatorname{tr}(\nabla)$ is non-trivial. The answer is apparently disappointing. Indeed, a more detailed analysis (see Proposition 4.5.29) shows that it is trivial. Nevertheless, the ideas we previously used in constructing $\operatorname{tr}(\nabla)$ remain valid and give a positive result if one will extract from them more than it was done up to now. One of possible suggestions on how it could be done comes from the observation that a linear invariant function on $\mathrm{End}(P)$, namely, the trace, leads almost automatically to the necessary cd-module homomorphism $\Lambda^\bullet(\mathrm{End}(P)) \to \Lambda^\bullet(A)$. But endomorphisms posses many other invariants which are polynomial and hence cannot be used directly for construction of the needed cd-module homomorphism. However, this is not a difficulty because a polynomial can be treated as a linear function on a suitable symmetric power of the base vector space. Even more, one can immediately select a $d_\square$–closed element in $\mathrm{K}^\bullet_\square$, where $\square$ is the induced by $\nabla$ connection in this symmetric power $S^k(\mathrm{End}(P))$. This element is nothing but the $l$-th power of $R^\nabla$. As invariant polynomials one may take, for instance, coefficients of the characteristic polynomial. The details are as follows.

### 4.5.9 Linear Functions and Connections

First, we describe some constructions that allow to duly formalize what was said before.

Let $\overline{X} \in \overline{\mathrm{D}}(\overline{M})$ and $\{\overline{G_t}\}_{t \in \mathbb{R}}$ be the one-parameter group generated by $\overline{X}$. A *linear function on* $\overline{M}$ is a $\mathrm{C}^\infty(M)$–linear map $L : \Gamma(\overline{M}) \to \mathrm{C}^\infty(M)$, *i.e.*, $L \in \overline{M}^\vee$. Geometrically $L$ may be interpreted as a family $L_m : \overline{m} \to \mathbb{R}$, $m \in M$, of of linear functions on fat points. The map $L$ is called *invariant with respect to* $\overline{X}$ if (locally)

$$L \circ \overline{G_t}^* = G_t^* \circ L, \quad t \in \mathbb{R}. \tag{4.14}$$

The equivalent infinitesimal version of (4.14) is obtained by applying to it the operator $(\mathrm{d}/\mathrm{d}t)|_{t=0}$:

$$L \circ \overline{X} = X \circ L. \tag{4.15}$$

**Definition.** Let $\nabla$ be a linear connection in $P$. An $A$-linear function $L : P \to A$ is called *invariant with respect to* $\nabla$ if

$$L \circ \nabla_X = X \circ L, \quad \forall X \in \mathrm{D}(A),$$

*i.e.*, if

$$\mathrm{d}_{\nabla^\vee}(L) = 0.$$

The function $L$ induces a homomorphism of $\Lambda^\bullet(A)$–modules

$$\Lambda^\bullet(L) : \Lambda^\bullet(P) \to \Lambda^\bullet(A), \quad \omega \otimes p \mapsto L(p)\omega$$

with $\omega \in \Lambda^\bullet(A), p \in P$.

**Proposition.** The function $L$ is invariant with respect to $\nabla$ if and only if

$$\Lambda^\bullet(L) \circ \mathrm{d}_\nabla = \mathrm{d} \circ \Lambda^\bullet(L).$$

*Proof.* Since $\mathrm{i}_X \circ \Lambda^\bullet(L) = \Lambda^\bullet(L) \circ \overline{\mathrm{i}}_X$ the result follows from $\nabla_X = \overline{\mathrm{i}}_X \circ \mathrm{d}_\nabla|_P$ and $X = \mathrm{i}_X \circ \mathrm{d}|_A$. □

In particular, this proposition holds for fat manifolds.

### 4.5.10 *Polynomials and Connections*

An $l$-th degree polynomial on an $A$-module $P$ is a function $f : P \to A$ such that $f(q) = \Pi(p,\ldots,p)$, where $\Pi : P \times \cdots \times P \to A$ ($l$ factors) is a symmetric $A$-multilinear function on $P$. Formula

$$\Pi(p_1,\ldots,p_l) = \frac{1}{l!} \sum_{1 \leq i_1 < \ldots < i_s \leq l} (-1)^{l-s} f(p_{i_1} + \cdots + p_{i_s})$$

shows that not only $\Pi$ determines $f$ but, *vice versa*, $f$ completely determines $\Pi$. The polynomial determined by a symmetric $A$-multilinear $\Pi$ will be denoted by $f_\Pi$ and, conversely, such a function corresponding to a polynomial

$f$ will be denoted by $\Pi_f$. The function $\Pi$ can be identified with an element of $\mathrm{Poly}(P) \stackrel{\mathrm{def}}{=} \mathrm{Hom}\left(\mathrm{S}^l(P), A\right)$, i.e., a linear function on $\mathrm{S}^l(P)$. Since we are assuming that $P$ is projective and finitely generated, $\mathrm{Hom}\left(\mathrm{S}^l(P), A\right)$ is canonically identified with $\mathrm{S}^l(P^\vee)$. A linear connection $\nabla$ in $P$ induces a linear connection $\nabla^{\mathrm{Poly}}$ in $\mathrm{Poly}(P)$:

$$\mathrm{d}_{\nabla^{\mathrm{Poly}}}(\Pi)(p_1, \ldots, p_l) = \mathrm{d}\,\Pi(p_1, \ldots, p_l) - \sum_{s=1}^{l} \Pi^\Lambda(p_1, \ldots, \mathrm{d}_\nabla p_s, \ldots, p_l)$$

(recall that, according to n. 3.6.10, $\Pi^\Lambda : \Lambda^\bullet\left(\mathrm{S}^l(P)\right) \to \Lambda^\bullet(A)$ is the extension of $\Pi : \mathrm{S}^l(P) \to A$, and hence can be thought as a graded symmetric $\Lambda^\bullet(A)$-multilinear function on $\Lambda^\bullet(P)$).

**Definition.** A polynomial $f$ on $P$ is called *invariant* with respect to a linear connection $\nabla$ in $P$ if $\mathrm{d}_{\nabla^{\mathrm{Poly}}}(\Pi_f) = 0$.

It immediately follows from the definition of $\nabla^{\mathrm{Poly}}$ that

$$\nabla^{\mathrm{Poly}}(\Pi) = \Pi \circ \mathrm{d}_{\nabla^{\mathrm{Sym}^l}} - \mathrm{d} \circ \Pi .$$

So, invariance of $f$ with respect to $\nabla$ is equivalent to equality $\Pi_f \circ \mathrm{d}_{\nabla^{\mathrm{Sym}^l}} = \mathrm{d} \circ \Pi_f$ and we have the following fact.

**Proposition.** If a polynomial $f$ on $P$ is invariant with respect to $\nabla$, then $\Pi_f^\Lambda : \Lambda^\bullet\left(\mathrm{S}^l(P)\right) \to \Lambda^\bullet(A)$ is a homomorphism of cd-modules.

### 4.5.11 Invariant Polynomials from Geometrical Point of View

Let $\overline{M}$ be a fat manifold and $f$ be a polynomial on $P = \Gamma\left(\overline{M}\right)$. Recall that $f$ can be interpreted as a smooth function on $\overline{M}$ which is polynomial on any fat point $\overline{m}$ (cf. the footnote in n. 0.3.24). Denote this interpretation by $f_{\mathrm{geom}}$, i.e.,

$$f_{\mathrm{geom}}(y) = f(s)(m), \quad \text{if } y = s(m) \in \overline{m} \text{ and } s \in \Gamma\left(\overline{M}\right) .$$

Let $\nabla$ be a linear connection in $\overline{M}$. Denote by $\mathbb{X}_\nabla$ the vector field on $\overline{M}$ corresponding to the fat field $\nabla_X$ (see n. 1.4.5) and by $f_{\overline{m}}$ the restriction of $f_{\mathrm{geom}}$ to the fat point $\overline{m}$ understood as a vector space.

**Exercise.** Prove that $f$ is invariant with respect to $\nabla$ if one of the following two assertions holds:

(1) $f_{\text{geom}} \circ \mathbb{X}_\nabla = X \circ f_{\text{geom}}, \forall X \in D(M)$;
(2) if the fat flow generated by $\nabla_X$ sends $\overline{m}$ to $\overline{m_t}$ 'in $t$ seconds', then the corresponding to $\overline{m} \to \overline{m_t}$ pull-back sends $f_{\overline{m_t}}$ to $f_{\overline{m}}$.

### 4.5.12 Characteristic Polynomial of an Endomorphism

An endomorphism $\varphi$ of $P$ naturally extends to an endomorphism $\varphi^\wedge = \bigwedge^\bullet \varphi$ of the algebra $P^\wedge = \bigwedge^\bullet P$:

$$\varphi^\wedge (p_1 \wedge \cdots \wedge p_l) = \varphi(p_1) \wedge \cdots \wedge \varphi(p_l), \quad p_1, \ldots, p_l \in P.$$

Set $\varphi_s^\wedge = \bigwedge^s \varphi$ and recall that every endomorphism of $P_r^\wedge$ is multiplication by an element of $A$. Denote such an element for the endomorphism $\varphi_r^\wedge$ by $\det \varphi$. In other words, $\varphi_r^\wedge(\hat{p}) = (\det \varphi)\hat{p}$ for all $\hat{p} \in P_r^\wedge$.

**Exercise.** Let $h \in |A|$ and $\varphi \in \text{End}(P)$. According to n. 0.1.5, (9) (and in notation of n. 0.3.1), $\varphi + \text{Ker } h \in (\text{End}(P))_h$ naturally corresponds to an endomorphism $\varphi_h : P_h \to P_h$. Prove that $\det \varphi_h = h(\det \varphi)$.

In particular, if $P = \Gamma(\overline{M})$, then $\varphi \in \text{End}(P)$, $P = \Gamma(\overline{M})$, is interpreted geometrically as a family of linear operators $\varphi_m : \overline{m} \to \overline{m}$, where $\overline{m}$ is understood as a vector space. Then $(\det \varphi)_m = \det \varphi_m$.

The characteristic polynomial $\text{ch}_\varphi(t)$ of $\varphi \in \text{End}(P)$ is defined as

$$\text{ch}_\varphi(t) = \det(\varphi - t \, \text{id}_P) .$$

Set

$$\text{ch}_\varphi(t) = \sum_{l=0}^{r}(-1)^{r-l}C_{r,l}(\Pi)t^{r-l} .$$

Then $C_{r,l} : \text{End}(P) \to A$ is a polynomial of degree $l$ on $P$.

### 4.5.13

**Exercise.** Prove that $C_{r,l} = \text{tr } \varphi_l^\wedge$.

### 4.5.14 Multilinear Functions Corresponding to Determinants

Now we describe explicitly a symmetric $A$-multilinear function on $\text{End } P$ that naturally corresponds to $\det = \det_r$, with $r$ being the constant rank of the projective $A$-module $P$ (see n. 4.5.6). To this end, associate with endomorphisms $\varphi_1, \ldots, \varphi_s \in \mathrm{End}(P)$ the endomorphism $\Delta(\varphi_1, \ldots, \varphi_s) \in \mathrm{End}\,(P_s^\wedge)$ defined as

$$\Delta(\varphi_1, \ldots, \varphi_s)(p_1, \ldots, p_s) \stackrel{\mathrm{def}}{=} \varphi_1(p_1) \wedge \cdots \wedge \varphi_s(p_s).$$

Obviously, $\Delta$ is an $s$-linear function on $\mathrm{End}(P)$ with values in $\mathrm{End}\,(P_s^\wedge)$ (not skew-symmetric!). Since the $A$-module $P_r^\wedge$ is projective, finitely generated and has constant rank 1, any its endomorphism is multiplication by an element of $A$. By denoting $\delta(\varphi_1, \ldots, \varphi_r)$ such an element for $\Delta(\varphi_1, \ldots, \varphi_r)$, we have

$$\Delta(\varphi_1, \ldots, \varphi_r)(\hat{p}) = \delta(\varphi_1, \ldots, \varphi_r)\hat{p}, \qquad \hat{p} \in P_r^\wedge. \qquad (4.16)$$

$A$-multilinearity of $\Delta(\varphi_1, \ldots, \varphi_r)$ immediately implies $A$-multilinearity of $\delta(\varphi_1, \ldots, \varphi_r)$. Observe also that $\varphi_r^\wedge = \delta(\varphi, \ldots, \varphi)$ ($r$ times). So, $\det \varphi = \delta(\varphi, \ldots, \varphi)$. By this reason the symmetrization of $\delta$ gives the symmetric $A$-multilinear function on $\mathrm{End}(P)$, denoted by $\mathrm{Det}_r$, that corresponds to the polynomial function $\det$:

$$\mathrm{Det}_r(\varphi_1, \ldots, \varphi_r) = \frac{1}{r!} \sum_{\sigma \in S_r} \delta\left(\varphi_{\sigma(1)}, \ldots, \varphi_{\sigma(r)}\right), \qquad (4.17)$$

with $S_r$ being the group of all permutations of $\{1, \ldots, r\}$.

In the sequel natural extensions of $A$-multilinear functions $\Delta$, $\delta$ and $\mathrm{Det}_r$ on $P_r^\wedge$ to $\Lambda^\bullet(A)$–multilinear functions on $\mathrm{End}_{\Lambda^\bullet(A)}(\Lambda^\bullet(P_r^\wedge))$ will be denoted by the same symbols. Formulas like (4.16) or (4.17) remain valid for these extensions. For instance,

$$\Delta(\varphi_1, \ldots, \varphi_r)(\omega \otimes \hat{p}) = \delta(\varphi_1, \ldots, \varphi_r)\omega \otimes \hat{p}, \qquad \omega \in \Lambda^\bullet(A), \hat{p} \in P_r^\wedge.$$

### 4.5.15 *Invariance of* $\mathrm{Det}_r$

Now we are ready to prove invariance of $\mathrm{Det}_r$ with respect to $\mathrm{d}_{(\nabla^{\mathrm{End}})^{\mathrm{S}^r}}$, $\nabla$ being a linear connection in $P$. To this end, we need the formula

$$\mathrm{d}\left(\delta(\varphi_1, \ldots, \varphi_r)\right) = \sum_{i=1}^{r} \delta\left(\varphi_1, \ldots, \mathrm{d}_{\nabla^{\mathrm{End}}}(\varphi_i), \ldots, \varphi_r\right), \qquad (4.18)$$

a proof of which is as follows. First, note that

$$\varphi \circ \mathrm{d}_\nabla = \mathrm{d}_\nabla \circ \varphi + [\varphi, \mathrm{d}_\nabla] = \mathrm{d}_\nabla \circ \varphi - \mathrm{d}_{\nabla^{\mathrm{End}}}(\varphi), \qquad \varphi \in \mathrm{End}(P).$$

Then we have

$$\delta(\varphi_1,\ldots,\varphi_r)\,\mathrm{d}_{\nabla^\wedge{}^r} = \Delta(\varphi_1,\ldots,\varphi_r)\circ \mathrm{d}_{\nabla^\wedge{}^r}$$
$$= \sum_{i=1}^{r}\Delta(\varphi_1,\ldots,\varphi_i\circ \mathrm{d}_\nabla,\ldots,\varphi_r)$$
$$= \sum_{i=1}^{r}\Delta(\varphi_1,\ldots,\mathrm{d}_\nabla\circ\varphi_i,\ldots,\varphi_r) - \sum_{i=1}^{r}\Delta(\varphi_1,\ldots,\mathrm{d}_{\nabla^{\mathrm{End}}}(\varphi_i),\ldots,\varphi_r)$$
$$= \mathrm{d}_{\nabla^\wedge{}^r}\circ\Delta(\varphi_1,\ldots,\varphi_r) - \sum_{i=1}^{r}\Delta(\varphi_1,\ldots,\mathrm{d}_{\nabla^{\mathrm{End}}}(\varphi_i),\ldots,\varphi_r)$$
$$= \mathrm{d}(\delta(\varphi_1,\ldots,\varphi_r))\,\mathrm{id}_{P_r^\wedge} + \delta(\varphi_1,\ldots,\varphi_r)\,\mathrm{d}_{\nabla^\wedge{}^r}$$
$$- \sum_{i=1}^{r}\delta(\varphi_1,\ldots,\mathrm{d}_{\nabla^{\mathrm{End}}}(\varphi_i),\ldots,\varphi_r)\,\mathrm{id}_{P_r^\wedge}\ .$$

Second,

$$\mathrm{d}_{\nabla^\wedge{}^r}(\Delta(\varphi_1,\ldots,\varphi_r)(\hat p)) = \mathrm{d}_{\nabla^\wedge{}^r}(\delta(\varphi_1,\ldots,\varphi_r)\hat p)$$
$$= \mathrm{d}(\delta(\varphi_1,\ldots,\varphi_r))\hat p + \delta(\varphi_1,\ldots,\varphi_r)\,\mathrm{d}_{\nabla^\wedge{}^r}(\hat p),\qquad \hat p\in P_r^\wedge\ .$$

In other words,

$$\mathrm{d}_{\nabla^\wedge{}^r}\circ\Delta(\varphi_1,\ldots,\varphi_r)$$
$$= \mathrm{d}(\delta(\varphi_1,\ldots,\varphi_r))\,\mathrm{id}_{\mathrm{S}^r(\Lambda^\bullet(\mathrm{End}(P)))} + \delta(\varphi_1,\ldots,\varphi_r)\,\mathrm{d}_{\nabla^\wedge{}^r}\ .$$

This gives the desired result.

**Proposition.** *The $A$-multilinear function $\mathrm{Det}(\varphi_1,\ldots,\varphi_r)$ on $\mathrm{S}^r(\mathrm{End}(P))$ is invariant with respect to the connection $(\nabla^{\mathrm{End}})^{\mathrm{S}}$, $\nabla$ being a linear connection in $P$.*

*Proof.* By combining formulas (4.17) and (4.18) one easily obtains that

$$\mathrm{d}(\mathrm{Det}_r(\varphi_1,\ldots,\varphi_r)) = \sum_{i=1}^{r}\mathrm{Det}_r(\varphi_1,\ldots,\mathrm{d}_{\nabla^{\mathrm{End}}}(\varphi_i),\ldots,\varphi_r)\ .$$

On the other hand,

$$\sum_{i=1}^{r}\mathrm{Det}_r(\varphi_1,\ldots,\mathrm{d}_{\nabla^{\mathrm{End}}}(\varphi_i),\ldots,\varphi_r) = \left(\mathrm{Det}_r\circ \mathrm{d}_{(\nabla^{\mathrm{End}})^{\mathrm{S}^r}}\right)(\varphi_1,\ldots,\varphi_r)\ .$$

Hence,

$$\mathrm{d}(\mathrm{Det}_r(\varphi_1,\ldots,\varphi_r)) = \left(\mathrm{Det}_r\circ \mathrm{d}_{(\nabla^{\mathrm{End}})^{\mathrm{S}^r}}\right)(\varphi_1,\ldots,\varphi_r)$$

and, therefore,

$$\mathrm{d}\circ \mathrm{Det}_r = \mathrm{Det}_r\circ \mathrm{d}_{(\nabla^{\mathrm{End}})^{\mathrm{S}^r}}\ . \qquad (4.19)$$

□

### 4.5.16 Invariance of $\mathrm{Det}_{r,l}$

Denote by $\mathrm{Det}_{r,l}$ the symmetric $A$-multilinear function on $\mathrm{End}(P)$ that corresponds to the polynomial $C_{r,l}$. As it is easy to see,

$$\mathrm{Det}_{r,l} = \binom{l}{r}^{-1} \mathrm{Det}_r \circ \Upsilon_{r,l}, \quad 1 \leq l \leq r,$$

where

$$\Upsilon_{r,l} : \mathrm{S}^l\left(\mathrm{End}(P)\right) \to \mathrm{S}^r\left(\mathrm{End}(P)\right), \quad \varphi_1 \cdots \varphi_r \mapsto \varphi_1 \cdots \varphi_r \cdot \mathrm{id}_P^{r-l}.$$

Since

$$\mathrm{d}_{(\nabla^{\mathrm{End}})^{\mathrm{S}^l}} = \mathrm{d}_{(\nabla^{\mathrm{End}})^{\mathrm{S}^r}} \circ \Upsilon_{l,r}$$

we prove the following assertion by composing (4.19) from the right by $\binom{l}{r}^{-1} \Upsilon_{l,r}$.

**Proposition.** The symmetric $A$-multilinear function $\mathrm{Det}_{r,l}$ on the module $\mathrm{S}^l\left(\mathrm{End}(P)\right)$ is invariant with respect to the connection

$$\left(\nabla^{\mathrm{End}}\right)^{\mathrm{S}^l}$$

for every linear connection $\nabla$ in $P$, i.e.,

$$\mathrm{d} \circ \mathrm{Det}_{r,l} = \mathrm{Det}_{r,l} \circ \mathrm{d}_{(\nabla^{\mathrm{End}})^{\mathrm{S}^r}}. \tag{4.20}$$

### 4.5.17 Geometric Approach to Invariance

An advantage of the above algebraic proof of the invariance of $\mathrm{Det}_{r,l}$ is that it is valid in wider algebraic context and gives explicit formulas that can be used in actual computations. For fat manifolds the invariance can be proved geometrically in a simple direct way as it is explained below. This simplicity is due to the equivalence of 'finite' and 'infinitesimal' versions of the notion of invariance (see n. 4.5.9).

**Definition.** Let $\nabla$ be a linear connection on a fat manifold $\overline{M}$. A function $f \in \mathrm{C}^\infty\left(\overline{M}\right)$ is called $\nabla$-*constant* if

$$f_{\gamma(t_0)} = f_{\gamma(t_1)} \circ \mathbf{T}^\gamma_{t_0,t_1}, \quad \forall \gamma, t_0, t_1, \tag{4.21}$$

where $f_m$ stands for the restriction of $f$ to the fiber $\overline{m}$.

**Proposition.** A $\nabla$-constant linear function $L$ on $\overline{M}$ is invariant with respect to $\nabla$.

**Proof.** Let $\overline{X} = \nabla_X$ and $\gamma$ be a trajectory of $X$. Then relation (4.14) is obviously, equivalent to (4.21). On the other hand, (4.14) implies (4.15) (see n. 4.5.9). □

To proceed on we need two elementary facts formulated in the subsequent two exercises.

### 4.5.18

Let $\tau : V \to W$ be an isomorphism of finite-dimensional vector spaces and $\alpha$ be a $l$-th degree polynomial on $W$. Denote by $\Pi_\alpha$ the linear function on $l$-th symmetric tensor power $\mathrm{S}^l(W)$ of $W$ corresponding to $\alpha$ (see n. 4.5.10).

**Exercise.** Prove that

$$\Pi_\alpha \circ \mathrm{S}^l \tau = \Pi_{\alpha \circ \tau}. \tag{4.22}$$

### 4.5.19

Another geometrically obvious fact we leave to the reader is the following.

**Exercise.** Let $\mathbf{T} \stackrel{\text{def}}{=} \mathbf{T}^\gamma_{t_0, t_1}$ be a parallel translation operator via a linear connection connection $\nabla$ on a fat manifold $\overline{M}$. Then $\mathrm{S}^l(\mathbf{T})$ is the parallel translation operator (from $t_0$ to $t_1$ along $\gamma$) via the connection $\nabla^{\mathrm{S}^l}$.

**Hint.** Reduce the problem to the fat interval supplied by the induced by $\gamma$ connection.

### 4.5.20

A direct consequence of (4.22) and the last exercise is the following.

**Proposition.** If $f$ is an $l$-th degree $\nabla$-constant polynomial on $\overline{M}$, then $\Pi_f$ (see n. 4.5.10) is a $\nabla^{\mathrm{S}^l}$ linear function on $\overline{M}^{\mathrm{S}^l}$.

This proposition combined with Proposition 4.5.17 prove the following assertion, which is key in the geometrical approach to invariance.

**Corollary.** $\nabla$-constant polynomials on a fat manifold $\overline{M}$ are invariant with respect to $\nabla$.

## 4.5.21

A geometric proof of the invariance of $\mathrm{Det}_{r,l}$ and similar functions is based on a due specialization of Corollary 4.5.20. The only fact we need to this end is formulated in the following analogue of Exercise 4.5.19.

**Exercise.** Let $\mathbf{T} \stackrel{\text{def}}{=} \mathbf{T}^{\gamma}_{t_0, t_1}$ be a parallel translation operator via a linear connection $\nabla$ on a fat manifold $\overline{M}$. Prove that the parallel translation via $\nabla^{\mathrm{End}}$ is

$$\varphi \mapsto \mathbf{T} \circ \varphi \circ \mathbf{T}^{-1},$$

$\varphi$ being an endomorphism of the fiber $\overline{m_0}$ over $m_0 = \gamma(t_0)$.

Recall that general fiber $F$ of $\overline{M}$ is a finite-dimensional vector space and the type of $\overline{M}$ is $\dim F$. Let $\nu$ be an ad-invariant polynomial on $\mathrm{End}(F)$, i.e.,

$$\nu(\varphi) = \nu\left(R \circ \varphi \circ R^{-1}\right), \qquad \forall \varphi \in \mathrm{End}(F), R \in \mathrm{Aut}(F).$$

We define a polynomial $f^{\nu}$ on $\overline{M}^{\mathrm{End}}$ whose restriction $f^{\nu}_m$ to the fiber $\overline{m}$ is defined by

$$f^{\nu}_m(\varphi) = \nu\left(I \circ \varphi \circ I^{-1}\right),$$

where $I : \overline{m} \to F$ is an isomorphism. Correctness of this definition, i.e., independence of a choice of $I$, is due to ad-invariance of $\nu$.

**Proposition.** The polynomial $f^{\nu}$ is invariant with respect to $\nabla^{\mathrm{End}}$ for every linear connections $\nabla$ on $\overline{M}$.

**Proof.** In view of Corollary 4.5.20 it suffices to prove that $f^{\nu}$ is $\nabla^{\mathrm{End}}$-constant. Let $\mathbf{T} \stackrel{\text{def}}{=} \mathbf{T}^{\gamma}_{t_0, t_1}$, $m_0 = \gamma(t_0)$, $m_1 = \gamma(t_1)$. Denote by $\mathbf{T}^{\mathrm{End}}$ the parallel translation from $t_0$ to $t_1$ along $\gamma$ in $\overline{M}^{\mathrm{End}}$ via $\nabla^{\mathrm{End}}$. Then, according to Exercise 4.5.21, $\mathbf{T}^{\mathrm{End}}(\varphi) = \varphi \mapsto \mathbf{T} \circ \varphi \circ \mathbf{T}^{-1}$. If $I_0 : \overline{m_0} \to F$ is an isomorphism, then $I_1 \stackrel{\text{def}}{=} I_0 \circ \mathbf{T}^{-1} : \overline{m_1} \to F$ is an isomorphism too, and

$$\left(f^{\nu}_{m_1} \circ \mathbf{T}^{\mathrm{End}}\right)(\varphi) = f^{\nu}_{m_1}\left(\mathbf{T} \circ \varphi, \circ \mathbf{T}^{-1}\right) = \nu\left(I_1 \circ \mathbf{T} \circ \varphi \circ \mathbf{T}^{-1} \circ I_1^{-1}\right)$$
$$= \nu\left(I_0 \circ \varphi \circ I_0^{-1}\right) = f^{\nu}_{m_0}(\varphi).$$

Thus $f^{\nu}_{m_0} = f^{\nu}_{m_1} \circ \mathbf{T}^{\mathrm{End}}$, i.e., $f^{\nu}$ is $\nabla^{\mathrm{End}}$-constant. $\square$

To conclude it remains to observe that for a fixed $t \in \mathbb{R}$, $\Phi \mapsto \det(\Phi - t\,\mathrm{id}_F)$, $\Phi \in \mathrm{End}(F)$, is an ad-invariant function on $\mathrm{End}(F)$. By this reason coefficients of the characteristic polynomial, which up to signs coincide with $C_l : \mathrm{End}(F) \to A$, are ad-invariant as well. Hence polynomials $f^{C_l}$, $l = 1, \ldots, r$, on $\overline{M}^{\mathrm{End}}$ are invariant with respect to every connection of the form $\nabla^{\mathrm{End}}$. If $P = \Gamma(\overline{M})$, then functions $C_{r,l}$ considered in n. 4.5.12 coincide with $f^{C_l}$. Hence $\nabla^{\mathrm{End}}$-invariance of $C_{r,l}$'s is a consequence of Proposition 4.5.21.

### 4.5.22 Dependence on $\nabla$

Let $\overline{M}$ and $\nabla$ be as before, and $P = \Gamma(\overline{M})$. Put $(R^\nabla)^s = R^\nabla \cdots R^\nabla$ ($s$ times; see n. 4.5.9). Then $R^{\nabla^s} \in \Lambda^{2s}(S^s(\mathrm{End}(P)))$. Since

$$\mathrm{d}_{(\nabla^{End})^{S^s}}\left((R^\nabla)^s\right) = \sum_{i=1}^{s} R^\nabla \cdots \mathrm{d}_{\nabla^{End}} R^\nabla \cdots R^\nabla = 0,$$

$\mathrm{Det}_{r,s}\left((R^\nabla)^s\right) \in \Lambda^{2s}(M)$ is a cocycle. Denote its cohomology class by $\zeta_s^\nabla \in \mathrm{H}^{2s}(M)$.

**Proposition.** The class $\zeta_s^\nabla$ does not depend on $\nabla$.

**Proof.** Let $\square$ be another linear connection in $\overline{M}$. Then $\nabla_t \stackrel{\mathrm{def}}{=} (1-t)\nabla + t\square$, $t \in \mathbb{R}$ is a family of connections such that $\nabla_0 = \nabla$, $\nabla_1 = \square$, and $\mathrm{d}_{\nabla_t} = \mathrm{d}_\nabla + th$, with $h = \mathrm{d}_\square - \mathrm{d}_\nabla \in \Lambda^1(\mathrm{End}(P))$. Recall that $\mathrm{d}_{\nabla_t^{End}}(h) = [\mathrm{d}_{\nabla_t}, h]^{\mathrm{gr}} = \mathrm{d}_{\nabla_t} \circ h + h \circ \mathrm{d}_{\nabla_t}$. So, $(\mathrm{d}/\mathrm{d}t)(\mathrm{d}_{\nabla_t}) = h$ and

$$\frac{\mathrm{d}}{\mathrm{d}t}\left(R^{\nabla_t}\right) = \frac{\mathrm{d}}{\mathrm{d}t}(\mathrm{d}_{\nabla_t} \circ \mathrm{d}_{\nabla_t}) = h \circ \mathrm{d}_{\nabla_t} + \mathrm{d}_{\nabla_t} \circ h = \mathrm{d}_{\nabla_t^{End}}(h).$$

Moreover, in view of the Bianchi identity, we have

$$\mathrm{d}_{(\nabla^{End})^{S^s}}\left(R^\nabla \cdots h \cdots R^\nabla\right) = R^\nabla \cdots \mathrm{d}_{\nabla^{End}}(h) \cdots R^\nabla$$

and hence

$$\frac{\mathrm{d}}{\mathrm{d}t}\left((R^{\nabla_t})^s\right) = \sum_{i=1}^{s} R^{\nabla_t} \cdots \frac{\mathrm{d}}{\mathrm{d}t}(R^{\nabla_t}) \cdots R^{\nabla_t}$$

$$= \sum_{i=1}^{s} R^{\nabla_t} \cdots \mathrm{d}_{\nabla^{End}}(h) \cdots R^{\nabla_t}$$

$$= \mathrm{d}_{(\nabla^{End})^{S^s}}\left(\sum_{i=1}^{s} R^{\nabla_t} \cdots h \cdots R^{\nabla_t}\right).$$

In other words,
$$\frac{d}{dt}\left((R^{\nabla_t})^s\right) = d_{(\nabla^{End})^{S^s}}(\Theta_t),$$
with $\Theta_t = \sum_{i=1}^{s} R^{\nabla_t} \cdots h \cdots R^{\nabla_t}$.

Finally, let $I : \Lambda^{\bullet}(S^s(\text{End}(P))) \mapsto \Lambda^{\bullet}(M)$ be a homomorphism of cd-modules. Then, with the Newton-Leibnitz formula we obtain:

$$I\left((R^{\nabla_1})^s\right) - I\left((R^{\nabla_0})^s\right) = \int_0^1 \frac{d}{dt}\left(I\left(R^{\nabla_t}\right)^s\right) dt = \int_0^1 \frac{d}{dt} I\left(\frac{d}{dt}(R^{\nabla_t})^s\right)$$

$$\overset{det}{=} \int_0^1 I\left(d_{(\nabla^{End})^{S^s}}(\Theta_t)\right) dt = \int_0^1 d\left(I\left(\Theta_t\right)\right) dt$$

$$= d\left(\int_0^1 I\left(\Theta_t\right) dt\right).$$

This proves that cocycles $I\left((R^{\nabla_0})^s\right)$ and $I\left((R^{\nabla_1})^s\right)$ are cohomologous. Thus with $I = \text{Det}_{r,s}$ we see that $\zeta_s^{\nabla} = \zeta_s^{\square}$. $\square$

### 4.5.23 Characteristic Classes

Proposition 4.5.22 allows to associate some cohomology classes with a fat manifold $\overline{M}$ by putting $\zeta_s(\overline{M}) = \zeta_s^{\nabla}(\overline{M}) \in H^{2s}(M)$. We shall also write $\zeta_s^{\nabla}(\xi)$ with $\xi : \overline{M} \to M$ being the projection. According to traditional terminology these classes are called *characteristic classes of* the vector bundle $\xi$. More exactly, classes $\zeta_s^{\nabla}(\xi)$ are trivial for odd $s$, as it will be shown later, while classes $\zeta_{2s}^{\nabla}(\xi)$ are, generally, nontrivial and $p_s(\xi) \overset{def}{=} (2\pi)^{-2s} \zeta_{2s}^{\nabla}(\xi)$ are known as *Pontrjagin classes of* $\xi$. Characteristic classes of the tangent bundle of a manifold $M$ are called its characteristic classes. In Riemannian geometry the Levi-Civita connection $\nabla$ is, as a rule, used in constructions of classes $\zeta_{2s}^{\nabla}$.

### 4.5.24 Naturalness of Characteristic Classes

An important property of characteristic classes is their naturalness with respect to fat maps. This means that for a fat map $\overline{f} : \overline{N} \to \overline{M}$

$$H^{\bullet} f^* (\zeta_s(\overline{M})) = \zeta_s(\overline{N}), \qquad (4.23)$$

where $H^{\bullet} f^*$ is the induced by $\Lambda^{\bullet} f^*$ map in de Rham cohomology. Indeed, let $\nabla$ and $\square$ be $\overline{f}$-compatible linear connections in $\overline{M}$ and $\overline{N}$, respectively.

Then, by definition, $\Lambda^\bullet \overline{f}^* : \Lambda^\bullet\left(\overline{M}\right) \mapsto \Lambda^\bullet\left(\overline{N}\right)$ is a homomorphism of cd-modules. So,

$$\Lambda^\bullet \overline{f}^* \circ R^\nabla = \Lambda^\bullet \overline{f}^* \circ d_\nabla^2 = d_\square \circ \Lambda^\bullet \overline{f}^* \circ d_\nabla$$
$$= d_\square^2 \circ \Lambda^\bullet \overline{f}^* = R^\square \circ \Lambda^\bullet \overline{f}^*.$$

The established relation $\Lambda^\bullet \overline{f}^* \circ R^\nabla = R^\square \circ \Lambda^\bullet \overline{f}^*$ is interpreted as

$$\Lambda^\bullet \overline{f}^*_{\text{End}}\left(R^\nabla\right) = R^\square,$$

where $\overline{f}_{\text{End}} : \overline{N}^{\text{End}} \to \overline{M}^{\text{End}}$ is a fat map naturally induced by $\overline{f}$. Similarly, it is not difficult to see that connections $\nabla^{\text{End}}$ and $\square^{\text{End}}$ are $\overline{f}_{\text{End}}$-compatible.

**Exercise.** Prove that connections $\left(\nabla^{\text{End}}\right)^{S^s}$ and $\left(\square^{\text{End}}\right)^{S^s}$ are $\left(\overline{f}_{\text{End}}\right)_{S^s}$-compatible if $\nabla$ and $\square$ are $\overline{f}$-compatible and use this fact in order to prove (4.23).

### 4.5.25

A refinement of the previously discussed construction of characteristic classes allows us to introduce more delicate ones for fat manifolds supplied with a gauge structure. In the rest of this section we shall show how to do it for fat bilinear forms and then present a general scheme. In fact, the construction we are going to describe can be reproduced in a much wider algebraic context. However, this requires a rather substantial algebraic preparation which goes beyond the scope of this book. Nevertheless, in order to stress an algebraic nature of the construction we shall use as before an algebraic flavor notation.

Namely, we fix a fat manifold $\overline{M}$ and put $A = C^\infty(M)$, $P = \Gamma\left(\overline{M}\right)$.

Let $b$ be a bilinear form on $P$, recall that $F$ denotes a general fiber of $\overline{M}$ and let $\beta$ a corresponding to $b$ general bilinear form on $F$. Denote by $\mathfrak{g}$ the Lie algebra of infinitesimal symmetries of $b$ (see n. 1.7.4). The Lie algebra $\mathfrak{g}$ is both an $A$-submodule and a Lie subalgebra of $\text{End}(P)$. Then a general fiber of $\overline{M}^\mathfrak{g}$ (see n. 1.1.2) is the Lie algebra $o(\beta)$ of infinitesimal symmetries of $\beta$.

Recall that $b^\Lambda$ is a natural extension of $b$ to $\Lambda^\bullet(P)$:

$$b^\Lambda\left(\omega \otimes p, \varkappa \otimes q\right) = \left(\omega \wedge \varkappa\right) b\left(p, q\right), \qquad \omega, \varkappa \in \Lambda^\bullet(A), p, q \in P$$

(see n. 3.6.10).

Then $\Lambda^\bullet(\mathfrak{g})$ is identified with graded infinitesimal symmetries of $\beta^\Lambda$, i.e.,

$$\beta^\Lambda\left(\Phi\left(\theta_1\right),\theta_2\right) + (-1)^{\deg\Phi\cdot\deg\theta_1}\beta^\Lambda\left(\theta_1,\Phi\left(\theta_2\right)\right) = 0 \qquad (4.24)$$

with $\theta_1,\theta_2 \in \Lambda^\bullet(P)$, $\Phi \in \Lambda^\bullet(\mathfrak{g})$. If $\theta = \omega \otimes p \in \Lambda^\bullet(P)$, $\Phi = \varkappa \otimes \chi$, with $\omega, \varkappa \in \Lambda^\bullet(A)$, $p \in P$ and $\chi \in \mathfrak{g}$, then by definition (see n. 3.6.10)

$$\Phi(\theta) = (\omega \wedge \varkappa) \otimes \chi(p).$$

Recall that (see n. 3.2.2) a connection $\nabla$ in $P$ preserves $b$ if $\mathrm{d}_{\nabla^{\mathrm{Bil}}}(b) = 0$, i.e.,

$$\mathrm{d}\left(b^\Lambda\left(\theta_1,\theta_2\right)\right) = b^\Lambda\left(\mathrm{d}_\nabla\left(\theta_1\right),\theta_2\right) + (-1)^{\deg\theta_1}b^\Lambda\left(\theta_1,\mathrm{d}_\nabla\left(\theta_2\right)\right),$$

$$\theta_1,\theta_2 \in \Lambda^\bullet(P).$$

The following two facts added to the previous general scheme give what we need in order to construct *special* characteristic classes, or, more exactly, $\beta$-*characteristic classes*. They are a graded version of Exercise 2.5.2, (2) and (3).

**Proposition.** Let $\nabla$ be a linear connection, preserving a fat bilinear form $\beta$. Then

(1) $\mathrm{d}_{\nabla^{\mathrm{End}}}\left(\Lambda^\bullet\left(\mathfrak{g}\right)\right) \subseteq \Lambda^\bullet\left(\mathfrak{g}\right)$;
(2) $R^\nabla \in \Lambda^2\left(\mathfrak{g}\right) \subseteq \Lambda^2\left(\mathrm{End}(P)\right)$.

**Proof.** (1). Let $p,q \in P$, $\chi \in \mathfrak{g}$. Then

$$b^\Lambda\left(\mathrm{d}_{\nabla^{\mathrm{End}}}(\chi)(p),q\right) + b^\Lambda\left(p,\mathrm{d}_{\nabla^{\mathrm{End}}}(\chi)(q)\right)$$
$$= b^\Lambda\left(\mathrm{d}_\nabla\left(\chi(p)\right),q\right) - b^\Lambda\left(\chi\left(\mathrm{d}_\nabla(p)\right),q\right)$$
$$+ b^\Lambda\left(p,\mathrm{d}_\nabla\left(\chi(q)\right)\right) - b^\Lambda\left(p,\chi\left(\mathrm{d}_\nabla(q)\right)\right).$$

But, since $\nabla$ preserves $b$,

$$b^\Lambda\left(\mathrm{d}_\nabla\left(\chi(p)\right),q\right) = \mathrm{d}\left(b^\Lambda\left(\chi(p),q\right)\right) - b^\Lambda\left(\chi(p),\mathrm{d}_\nabla(q)\right)$$

and (see (4.24))

$$b^\Lambda\left(\chi\left(\mathrm{d}_\nabla(p)\right),q\right) = -b^\Lambda\left(\mathrm{d}_\nabla(p),\chi(q)\right)$$

and similarly for remaining two terms in the last expression. Taking this into account one immediately gets:

$$b^\Lambda\left(\mathrm{d}_{\nabla^{\mathrm{End}}}(\chi)(p),q\right) + b^\Lambda\left(p,\mathrm{d}_{\nabla^{\mathrm{End}}}(\chi)(q)\right) = \mathrm{d}\left(b^\Lambda\left(\chi(p),q\right) + b^\Lambda\left(p,\chi(q)\right)\right) = 0.$$

This shows that $d_{\nabla\text{End}}(\chi) \in \Lambda^1(\mathfrak{g})$. Therefore $d_{\nabla\text{End}}(\mathfrak{g}) \subseteq \Lambda^1(\mathfrak{g})$ and hence $d_{\nabla\text{End}}(\Lambda^\bullet(\mathfrak{g})) \subseteq \Lambda^\bullet(\mathfrak{g})$.

(2). Similarly,

$$b^\Lambda\left(R^\nabla(p), q\right) + b^\Lambda\left(p, R^\nabla(q)\right) = b^\Lambda\left(d_\nabla^2(p), q\right) + b^\Lambda\left(p, d_\nabla^2(q)\right).$$

By observing that $b^\Lambda\left(d_\nabla^2(p), q\right) = b^\Lambda(d_\nabla(p), d_\nabla(q))$ and $b^\Lambda\left(p, d_\nabla^2(q)\right) = -b^\Lambda(d_\nabla(p), d_\nabla(q))$ we see that

$$b^\Lambda\left(R^\nabla(p), q\right) + b^\Lambda\left(p, R^\nabla(q)\right) = 0,$$

i.e., $R^\nabla \in \Lambda^\bullet(\mathfrak{g})$. □

**Corollary.** Set $d_{\nabla\mathfrak{g}} = d_{\nabla\text{End}}|_{\Lambda^\bullet(\mathfrak{g})}$. We have

(1) $(\Lambda^\bullet(\mathfrak{g}), d_{\nabla\mathfrak{g}})$ is a cd-submodule of $\Lambda^\bullet(\text{End}(P), d_{\nabla\text{End}})$;
(2) $d_{\nabla\mathfrak{g}}\left(R^\nabla\right) = 0$ (Bianchi identity).

Consider now the $s$-th symmetric power $S^s(\mathfrak{g})$ of $\mathfrak{g}$ and the corresponding cd-module $\left(\Lambda^\bullet(S^s(\mathfrak{g})), d_{(\nabla\mathfrak{g})^{S^s}}\right)$. Then by (2) of the above Corollary we see that

$$d_{(\nabla\mathfrak{g})^{S^s}}\left(\left(R^\nabla\right)^s\right) = 0. \tag{4.25}$$

**4.5.26**

At this point it remains to find a cd-module homomorphism $\left(\Lambda^\bullet(S^s(\mathfrak{g})), d_{(\nabla\text{End})^{S^s}}\right) \to (\Lambda^\bullet(A), d)$. To do that we just mimic n. 4.5.17.

Namely, let $I$ be an $s$-th degree Ad-invariant polynomial on $O(F, \beta)$ (see n. 4.5.21). Denote by $b_m$ (respectively, $\mathfrak{g}_m$) the restriction to the fiber $\overline{m}$ of the form $b$ (respectively, of the algebra $\mathfrak{g}$). By definition,

$$b_m(s_1(m), s_2(m)) = b(s_1, s_2)(m), \quad s_1, s_2 \in \Gamma(\overline{M}) = P.$$

Similarly, if $\chi \in \mathfrak{g} \subseteq \text{End}(\Gamma(\overline{M}))$ and $s \in \Gamma(\overline{M})$, then

$$\chi_m(s(m)) \stackrel{\text{def}}{=} \chi(s)(m) \quad \text{and} \quad \mathfrak{g}_m = \{\chi_m : \chi \in \mathfrak{g}\} \subseteq \text{End}(\overline{m}).$$

Obviously, $\mathfrak{g}_m$ is a Lie subalgebra of $\text{End}(\overline{m})$. Moreover, denote by $\text{Iso}_m$ the totality of isomorphisms $u : F \xrightarrow{\sim} \overline{m}$ that send $\beta$ to $b_m$, i.e., $b_m(u(e_1), u(e_2)) = \beta(e_1, e_2)$, with $e_1, e_2 \in F$. Then for all $u \in \text{Iso}_m$ the map $\chi \mapsto u \circ \chi \circ u^{-1}$, with $\chi \in O(F, \beta)$, identifies $O(F, \beta)$ and $\mathfrak{g}_m$. By Ad-invariance the polynomial $I \circ u^{-1}$ on $\mathfrak{g}_m$ does not depend on $u \in \text{Iso}_m$

and will be denoted by $I_m$. This allows to define a polynomial $I_{\overline{M}}$ on $\text{End}\left(\Gamma\left(\overline{M}\right)\right) = \text{End}(P)$ by putting

$$\left(I_{\overline{M}}(\chi)\right)(m) = I_m\left(\chi_m\right), \qquad \chi \in \mathfrak{g}.$$

In other words, if $I_{\overline{M}}$ is thought to be a function on $\overline{M}^{\mathfrak{g}}$, then $I_{\overline{M}}\big|_{\mathfrak{g}_m} = I_m$. Finally, denote by $L_I$ the $A$-linear function on $S^s(\mathfrak{g})$ associated with $I_{\overline{M}}$. Then the following assertion is literally proved as Proposition 4.5.20.

**Proposition.** The function $L_I : S^s(\mathfrak{g}) \to A$ is $\nabla$-invariant with respect to every linear connection $\nabla$ on $\overline{M}$ that preserves the gauge bilinear form $b$ on $\overline{M}$.

Recall that this means $L_I \circ d\nabla^{\mathfrak{g}} = d \circ L_I$ where we slightly abuse the notation by denoting by $L_I$ the $\Lambda(A)$-extension of $L_I$ to $\Lambda^\bullet(S^s(\mathfrak{g}))$. So, by formula (4.25), the differential form $L_I\left((R^\nabla)^s\right)$ on $M$ is closed.

### 4.5.27

**Proposition.** In the above notation, the cohomology class of $L_I\left((R^\nabla)^s\right)$ does not depend on the choice of a preserving $b$ linear connection.

**Proof.** Let $\square$ be another preserving $b$ linear connection. Then $h \stackrel{\text{def}}{=} d_\square - d_\nabla \in \Lambda^1(\mathfrak{g})$ and connections $\nabla_t \stackrel{\text{def}}{=} (t-1)\nabla + t\square = \nabla + th$ preserve $b$ as well, and $\nabla_0 = \nabla$, $\nabla_1 = \square$. Then arguments proving Proposition 4.5.22 show that

$$L_I\left(\left(R^\square\right)^s\right) - L_I\left((R^\nabla)^s\right) = L_I\left((R^{\nabla_1})^s\right) - L_I\left((R^{\nabla_0})^s\right)$$

$$= d\left(\int_0^1 I\left(\Theta_t^{\mathfrak{g}}\right) dt\right),$$

where $\Theta_t^{\mathfrak{g}}$ stands for the restriction of $\Theta_t \in \Lambda^{2s-1}(S^s(\text{End}(P)))$ to $\Lambda^{2s-1}(S^s(\mathfrak{g}))$. $\square$

### 4.5.28 $\beta$-Characteristic Classes

Proposition 4.5.27 assures correctness of the following definition.

**Definition.** The de Rham cohomology class $p_I\left(\overline{M}, b\right) \in H^{2s}(M)$ of the closed form $L_I\left((R^\nabla)^s\right)$ is called a $\beta$-*characteristic class* of a gauge bilinear form $b$ on $\overline{M}$.

This definition presupposes existence of linear connections preserving $b$. It can be proved by standard methods in Algebraic Topology, since such a connection may be interpreted as a section of a fiber bundle whose fibers are homotopy trivial (see, for instance, [Husemoller (1994), Chap. 2, Theorem 7.1]). Denote by $\mathcal{I}(\mathfrak{g})$ the algebra of ad-invariant polynomials on $\mathfrak{g}$. The map $w_{\mathfrak{g},\beta} : \mathcal{I}(\mathfrak{g}) \to H^\bullet(M)$ is called the *Weil homomorphism*.

One of the principal properties of $\beta$-characteristic classes is *naturalness*. Namely, let $\overline{f} : \overline{M}_1 \to \overline{M}_2$ be a fat map that sends a type $\beta$ gauge bilinear form $b_2$ on $\overline{M}_2$ to a similar from $b_1$ on $\overline{M}_1$, i.e.,

$$b_1\left(\overline{f}^*(p), \overline{f}^*(q)\right) = f^*(b_2(p,q)), \qquad p, q \in \Gamma\left(\overline{M}_2\right).$$

Then

$$\Lambda^\bullet \overline{f}^*\left(p_I\left(\overline{M}_2, b_2\right)\right) = p_I\left(\overline{M}_1, b_1\right).$$

For a proof it is just sufficient to repeat arguments in n. 4.5.24.

### 4.5.29

By varying the type of $\beta$ one obtains various types of characteristic classes, say, *orthogonal, symplectic*, etc. For instance, the orthogonal ones correspond to positive definite forms $\beta$, while symplectic to nondegenerate skew-symmetric.

Note that $\mathcal{I}(\mathfrak{g})$ contains polynomials $C_l(\mathrm{ad}_\chi)$, $\chi \in \mathfrak{g}$, that are, up to the sign, coefficients of the characteristic polynomial of the operator $\mathrm{ad}_\chi :$ $\mathfrak{g} \to \mathfrak{g}$, $\chi' \mapsto [\chi, \chi']$. The linear function $L_I$ for $I = C_l$ was denoted by $\mathrm{Det}_{r,l}$. Hence, the Pontrjagin classes are among $\beta$-characteristic classes. This obvious fact has various nontrivial consequences. One of them was already mentioned in nn. 4.5.8 and 4.5.23.

**Proposition.** Characteristic classes $\zeta_s$ are trivial for odd $s$.

**Proof.** A linear combination $\sum_i \lambda_i \beta_i$ of positive definite bilinear forms $\beta_i$ on a vector space $V$ with positive coefficients $\lambda_i$ is, obviously, positive definite too. By this reason, positive definite forms constitute a convex domain in the vector space of all symmetric bilinear forms. By standard topological arguments this fact implies existence of gauge metrics for any fat manifold. On the other hand, the algebra $o(V, \beta)$ for a positive definite $\beta$ is isomorphic to $\mathrm{so}(r)$ and operators $\mathrm{ad}_\chi, \chi \in \mathrm{so}(r)$ are skew-symmetric. But coefficients $C_s$ for skew-symmetric operators are trivial for odd $s$. □

In particular, the class tr $(\nabla)$ we started with (see n. 4.5.7) is trivial.

The following simple exercise shows that not all vector bundles admit gauge pseudo-metrics of a prescribed signature $(r, s)$.

**Exercise.** Prove that the tangent bundle of a 2-sphere does not admit gauge pseudo-metrics of signature $(1, 1)$.

### 4.5.30

Generally, existence of $\beta$-type bilinear forms on a given vector bundle reflects its topological complexity which can be captured by means of characteristic classes. This observation extends to all *special* characteristic classes we are going to discuss.

**Definition.** $\beta$-characteristic classes of the tangent bundle of a manifold $M$ are called *$\beta$-characteristic classes of $M$*.

**Exercise.** Show that $\beta$-characteristic classes of a torus are trivial.

### 4.5.31 Special Characteristic Classes

Now we shall describe the most general scheme of constructing characteristic classes via connections. Proof of all necessary for that facts are essentially the same as before. They are based on a rather elementary algebraic formalism concerning differential calculus in the tensor algebra associated with a given $A$-module $P$.

However, its description requires more 'space-time' than we have at disposal here. By this reason proofs are omitted, but they can be restored by analogy with $\beta$-characteristic classes.

First, we must introduce the necessary terminology. Let $F$ be a finite dimensional vector space. Recall that (see n. 3.6.2) a tensor type $\tau$ is understood an equivalence class in the tensor algebra $\mathrm{T}(F) = \oplus \mathrm{T}_q^p(F)$ of $F$ ($\mathrm{T}_q^p(F) = F \otimes \cdots \otimes F \otimes F^{\vee} \otimes \cdots \otimes F^{\vee}$; $p$ copies of $F$ and $q$ of $F^{\vee}$) under a natural action of $\mathrm{GL}(F)$ in $\mathrm{T}(F)$. In other words, $\tau$ is one of the orbits of this action. We shall write $\Psi \in \tau$ if a tensor $\Psi \in \mathrm{T}(F)$ belongs to the class $\tau$.

Put $\boldsymbol{\Psi} = (\Psi_1, \ldots, \Psi_l)$, with $\Psi_1, \ldots, \Psi_l \in \mathrm{T}(F)$, and $\boldsymbol{\tau} = (\tau_1, \ldots, \tau_l)$ with $\tau_i$'s being tensor types and write $\boldsymbol{\tau} \in \boldsymbol{\Psi}$ if $\Psi_i \in \tau_i$, $i = 1, \ldots, l$.

Fix a fat manifold $\overline{M}$ with a general fiber $F$ and $\boldsymbol{\Psi}$, and assume that $\overline{M}$ is supplied a $\boldsymbol{\Psi}$-type gauge structure, i.e., $\mathrm{C}^\infty(M)$-module $\Gamma\left(\overline{M}\right)$ is supplied with gauge structures $\Xi_1, \ldots, \Xi_l$ with gauge types $\Psi_1, \ldots, \Psi_l$, respectively.

Let $\operatorname{Sym} \Psi_i \subseteq \operatorname{End}(F)$ be the Lie algebra of infinitesimal symmetries of the tensor $\Psi_i$ and $\mathfrak{g}_i \subseteq \operatorname{End}(P)$ the $C^\infty(M)$–Lie algebra of infinitesimal symmetries of $\Xi_i$. Then $\mathfrak{g}_i$ is a gauge Lie algebra of gauge type $\operatorname{Sym} \Psi_i$. Put $\mathfrak{g} \stackrel{\text{def}}{=} \mathfrak{g}_1 \cap \cdots \cap \mathfrak{g}_l$. As before, for $I \in \mathcal{I}(\mathfrak{g})$, $\deg I = s$, denote b $L_I$ the associated with $I$ $C^\infty(M)$-linear function on $S^s$. Finally, assume $\nabla$ to be a linear connection on $\overline{M}$ that preserves all $\Xi$'s, i.e., such that $d_{\nabla^{\tau_i}}(\Xi) = 0$. Then $\Psi$-characteristic classes are defined according to Scheme 4.5.31.

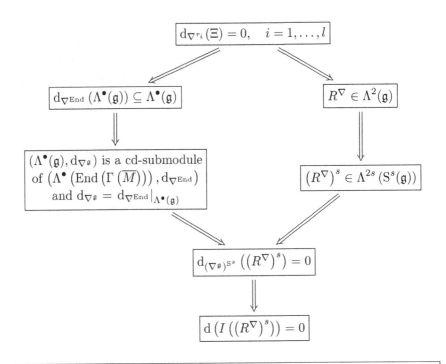

Scheme 4.5.31

In order to be in conformity with the standard terminology, when one deals with vector bundles, one may use the notation $p_I^\Psi(\pi)$ instead of $p_I^\Psi(\overline{M})$, with $\pi : \overline{M} \to M$ being the projection. $\Psi$-characteristic classes of a $\Psi$-structure on the tangent bundle of a manifold $M$ are called $\Psi$-characteristic classes of this $\Psi$-structure *on $M$*.

**Example.** Let $J$ be a complex structure on $F$. Then the tensor type of $J$ is also called 'complex structure', and 'complex characteristic classes' are nothing but famous Chern classes. The Lie algebra $\operatorname{Sym} J$ in this case is isomorphic to $\operatorname{gl}(l, \mathbb{C})$.

The algebra generated by $J$ in $\operatorname{End}(F)$ is isomorphic to $\mathbb{C}$. Scheme 4.5.31 allows to construct characteristic classes associated with a subalgebra of $\operatorname{End}(F)$. This illustrates the dimension of the obtained generalization of the classical theory.

The 'connection calculus' developed in this book allows to establish various types of characteristic classes and, in particular, various operations with them. This new 'characteristic' universe is not sufficiently explored, except its classical part (Pontrjagin and Chern classes), and promises interesting applications.

# Bibliography

Arnold, V. I. (2006). *Ordinary differential equations*, Universitext (Springer-Verlag, Berlin), translated from the Russian by Roger Cooke, Second printing of the 1992 edition.

Atiyah, M. F. and Macdonald, I. G. (1969). *Introduction to commutative algebra* (Addison-Wesley Publishing Co., Reading, Mass.-London-Don Mills, Ont.).

Berger, M. and Gostiaux, B. (1988). *Differential geometry: manifolds, curves, and surfaces*, Graduate Texts in Mathematics, Vol. 115 (Springer-Verlag, New York), translated from the French by Silvio Levy.

Bourbaki, N. (1989). *Algebra. I. Chapters 1–3*, Elements of Mathematics (Berlin) (Springer-Verlag, Berlin), translated from the French, Reprint of the 1974 edition.

Hartshorne, R. (1977). *Algebraic geometry*, Graduate Texts in Mathematics, Vol. 52 (Springer-Verlag, New York).

Hilton, P. J. and Stammbach, U. (1971). *A course in homological algebra*, Graduate Texts in Mathematics, Vol. 4 (Springer-Verlag, New York).

Husemoller, D. (1994). *Fibre bundles*, Graduate Texts in Mathematics, Vol. 20, 3rd edn. (Springer-Verlag, New York).

Lee, J. M. (2003). *Introduction to smooth manifolds*, Graduate Texts in Mathematics, Vol. 218 (Springer-Verlag, New York).

Mac Lane, S. (1971). *Categories for the working mathematician*, Graduate Texts in Mathematics, Vol. 5 (Springer-Verlag, New York).

Mac Lane, S. and Birkhoff, G. (1967). *Algebra* (The Macmillan Co., New York).

Moreno, G., Vinogradov, A. M. and Vitagliano, L. (in preparation). *Differential Cohomology: Theory of Integral and Leray-Serre Spectral Sequence*.

Nestruev, J. (2003). *Smooth manifolds and observables*, Graduate Texts in Mathematics, Vol. 220 (Springer-Verlag, New York), joint work of A. M. Astashov, A. B. Bocharov, S. V. Duzhin, A. B. Sossinsky, A. M. Vinogradov and M. M. Vinogradov, Translated from the 2000 Russian edition by Sossinsky, I. S. Krasil'schik and Duzhin.

Rubtsov, V. N. (1980). Cohomology of the Der-complex, *Uspekhi Mat. Nauk* **35**, 4(214), pp. 209–210.

Singer, I. M. and Thorpe, J. A. (1976). *Lecture notes on elementary topology* and geometry (Springer-Verlag, New York), reprint of the 1967 edition, Undergraduate Texts in Mathematics.

Vinogradov, A. M. (2001). *Cohomological analysis of partial differential equations and secondary calculus*, Translations of Mathematical Monographs, Vol. 204 (American Mathematical Society, Providence, RI), translated from the Russian manuscript by Joseph Krasil'shchik.

Vinogradov, A. M. and Vinogradov, M. M. (in preparation). Differential calculus in the category of dioles.

Vinogradov, A. M. and Vitagliano, L. (2006). Iterated differential forms: Tensors, *Dokl. Math.* **73**, 2, pp. 169–171.

# List of Symbols

In the following list, similar symbols are grouped. By this reason, they *do not* merely follow page number order. For each item it is indicated the subsection number, followed by the page number. Each group, as well as all first items of each group, are ordered according to page numbers.

| | | | |
|---|---|---|---|
| | | $\text{Diff}^+(P,Q)$ | 0.1.2, 3 |
| $P^\vee$ | 0.1.1, 1 | $\text{Diff}^{(+)}(P,Q)$ | 0.1.2, 3 |
| $\overline{X}^\vee$ | 1.4.1, 110 | | |
| $\nabla^\vee$ | 2.4.4, 160 | $D(P)$ | 0.1.3, 4 |
| | | $D_k(P)$ | 0.1.3, 4 |
| $[a,b]$ | 0.1.1, 3 | $D(A)_\varphi$ | 0.1.3, 4 |
| | 2.2.5, 150 | $D(M)$ | 0.4.1, 41 |
| $[a_r, a'_s]^{(\text{gr})}$ | 0.1.1, 3 | $D(M)_f$ | 0.4.12, 46 |
| $[\square]$ | 1.2.2, 99 | $D(M)_N$ | 0.4.18, 49 |
| | | $\overline{D}(\overline{M})$ | 1.3.1, 105 |
| $\text{Diff}_n(P,Q)$ | 0.1.2, 3 | $\overline{D}(\overline{M})_{\overline{f}}$ | 1.3.13, 110 |
| $\text{Diff}(P,Q)$ | 0.1.2, 3 | | |

| | | | |
|---|---|---|---|
| $\Lambda^\bullet$ | 0.1.5 ($^3$), 5 | $\overline{X}\|_N$ | 1.3.10, 108 |
| $\Lambda$ | 0.5.13, 58 | $\nabla\|_{\overline{N}}$ | 2.2.4, 149 |
| | 3.1.7, 176 | | 2.4.10, 164 |
| | 3.2.9, 184 | $\overline{\omega}\|_N$ | 4.2.1, 238 |
| | 3.6.2, 226 | | |
| | | $f^*$ | 0.2.11, 13 |
| $\|A\|$ | 0.2.1, 8 | | 0.3.6, 27 |
| $\|\varphi\|$ | 0.2.4, 9 | $\overline{f}^*$ | 0.3.13, 31 |
| | 0.3.5, 26 | $f_*$ | 0.4.2, 41 |
| | | $\alpha'$ | 0.5.9, 56 |
| $\widetilde{a}$ | 0.2.2, 8 | | |
| $\widetilde{A}$ | 0.2.2, 8 | $\mathfrak{SM}_a$ | 0.2.12, 13 |
| | | $\mathfrak{SM}_c$ | 0.2.13, 14 |
| $C^\infty$ | 0.2.3 ($^{10}$), 9 | $\mathfrak{VB}$ | 0.3.11, 31 |
| | 0.2.9, 12 | $\mathfrak{VB}_g$ | 0.3.19, 36 |
| $A\|_N$ | 0.2.6, 10 | $A\overline{\otimes}B$ | 0.2.24, 20 |
| $f\|_{N,N'}$ | 0.2.21, 18 | $\square_1 \boxtimes \square_2$ | 1.5.2, 115 |
| $X\|_N$ | 0.4.5, 43 | $\nabla \boxtimes \Delta$ | 2.4.5, 161 |
| $s\|_{N,\overline{N}}$ | 1.1.8, 95 | | |
| $s\|_N$ | 1.1.8, 95 | $\times$ | 0.2.25, 21 |

# List of Symbols

|  |  |  |  |
|---|---|---|---|
|  |  | $\Gamma(E_\pi)$ | 0.3.4, 25 |
| $T_mM$ | 0.2.27, 22 |  | 0.3.20, 37 |
| $\overline{T}_m\overline{M}$ | 1.3.4, 105 | $\Gamma(\overline{M})$ | 1.1.1, 92 |
| $\overline{T_M}$ | 1.6.13, 128 |  |  |
| $\mathbf{T}^\gamma_{t_0,t_1}$ | 3.3.17, 197 | $\varphi_m$ | 0.3.5, 26 |
|  |  | $\overline{f}_m$ | 0.3.5, 26 |
| $\mathrm{d}_n$ | 0.2.28, 22 |  | 0.3.8, 29 |
| $\mathrm{d}$ | 0.5.5, 54 | $X_m$ | 0.4.1, 41 |
|  | 0.5.11, 57 | $\omega_m$ | 0.5.4, 53 |
| $\frac{\mathrm{d}\omega_t}{\mathrm{d}t}\big|_{t=t_0}$, $\frac{\mathrm{d}}{\mathrm{d}t}\big|_{t=t_0}\omega_t$ | 0.6.18, 85 | $\overline{X}_n$ | 1.3.5, 106 |
| $\overline{\mathrm{d}}_n$ | 1.3.7, 107 |  | 1.3.13, 110 |
| $\overline{\frac{\mathrm{d}}{\mathrm{d}t}}$ | 1.6.1, 118 |  |  |
| $\overline{\mathrm{d}}$ | 3.1.8, 176 | S | 0.3.24 $(^{25})$, 40 |
| $\mathrm{d}_\nabla$ | 3.2.1, 178 | Bil | 2.4.7, 161 |
|  | 3.2.6, 183 |  |  |
| $\overline{\frac{\partial}{\partial t}}$ | 3.5.4, 211 | $\Lambda^s(P), \Lambda^\bullet(P)$ | 0.5.1, 52 |
| $\frac{\mathrm{d}\overline{\omega_t}}{\mathrm{d}t}\big|_{t=t_0}$, $\frac{\mathrm{d}}{\mathrm{d}t}\big|_{t=t_0}\overline{\omega_t}$ | 3.5.11, 216 | $\Lambda^s(M), \Lambda^\bullet(M)$ | 0.5.1, 52 |
|  |  | $\Lambda^s(M)_f, \Lambda^\bullet(M)_f$ | 0.5.22, 64 |
| $\Gamma(P)$ | 0.3.1, 24 | $\Lambda^s(M)_N, \Lambda^\bullet(M)_N$ | 0.5.35, 72 |
| $\Gamma(\pi)$ | 0.3.4, 25 | $\overline{\Lambda}^s(P), \overline{\Lambda}^\bullet(P)$ | 3.1.1, 172 |
|  | 0.3.20, 37 | $\overline{\Lambda}^s(\overline{M}), \overline{\Lambda}^\bullet(\overline{M})$ | 3.1.1, 172 |

| | | | |
|---|---|---|---|
| $\overline{\Lambda}^s(P;A)$ | 3.1.2, 172 | $\mathcal{L}_X$ | 0.6.19, 87 |
| $\overline{\Lambda}^s(M)$ | 3.1.2, 172 | $\mathcal{L}_{\overline{X}}$ | 3.5.12, 217 |
| $\Lambda^s(\overline{M}), \Lambda^\bullet(\overline{M})$ | 3.1.4, 173 | $\mathcal{L}_X^\nabla$ | 3.5.15, 219 |
| $\Lambda^s \overline{f}^*, \Lambda^\bullet \overline{f}^*$ | 3.3.8, 191 | | |
| $\Lambda^\bullet(\overline{M})_{\overline{f}}, \Lambda^s(\overline{M})_{\overline{f}}$ | 3.4.1, 201 | $F$ | 1.1, 92 |
| $\Lambda^\bullet(M,N)$ | 4.2.2, 238 | $\overline{M}$ | 1.1.1, 92 |
| $\Lambda^\bullet(\overline{M},\overline{N})$ | 4.2.2, 238 | $\overline{f}$ | 1.1.4, 93 |
| | | $\mathbf{K}$ | 2.2.10, 154 |
| $\omega_X$ | 0.5.18, 62 | | |
| | 0.5.28, 67 | $\overline{M}^\vee$ | 1.1.2, 92 |
| $\overline{\omega}_Y$ | 3.4.2, 202 | $\overline{M}^{\mathrm{End}}$ | 1.1.2, 92 |
| | | $\overline{M}^{S^r}$ | 1.1.2, 93 |
| $i_X$ | 0.5.18, 62 | $\nabla^{\mathrm{End}}$ | 2.4.3, 160 |
| | 0.5.28, 67 | $\nabla^\vee$ | 2.4.4, 160 |
| $\bar{i}_{X,P}$ | 0.5.18, 62 | $\nabla^{\mathrm{Bil}}$ | 2.4.7, 161 |
| $\bar{i}_X$ | 0.5.18, 62 | | |
| | 3.4.2, 202 | $\mathrm{Der}(P)$ | 1.2.3, 100 |
| | | $\mathrm{Der}(P)_{\overline{\varphi}}$ | 1.2.11, 104 |
| $X \lrcorner\, \omega$ | 0.5.18, 62 | | |
| | 0.5.28, 67 | $\mathrm{Hom}(\Box_1, \Box_2)$ | 1.5.1, 114 |
| $Y \lrcorner\, \overline{\omega}$ | 3.4.2, 202 | $\mathrm{Hom}(\nabla, \Delta)$ | 2.4.1, 159 |
| | | $\mathrm{Hol}(\nabla, m)$ | 3.3.18, 198 |

## List of Symbols

| | | | |
|---|---|---|---|
| | | $\nabla^{\text{End}}$ | 2.4.3, 160 |
| $\mathrm{O}(P,b)$ | 1.7.4, 133 | $\nabla^{\vee}$ | 2.4.4, 160 |
| $\mathrm{O}\left(\overline{M},g\right)$ | 1.7.4, 133 | $\nabla^{\text{Bil}}$ | 2.4.7, 161 |
| $\mathrm{o}\left(\overline{M},g\right)$ | 1.7.4, 134 | | |
| $\mathrm{o}(P,b)$ | 1.7.4, 134 | $\mathcal{J}^1(P)$ | 3.2.4, 179 |
| | | $j_1$ | 3.2.4, 179 |
| $\mathrm{GL}(P,\psi)$ | 1.7.4, 134 | $\mathcal{J}^1(\overline{M})$ | 3.2.4, 179 |
| $\mathrm{GL}\left(\overline{M},\psi\right)$ | 1.7.4, 134 | | |
| $\mathrm{gl}\left(\overline{M},\psi\right)$ | 1.7.4, 134 | $H_\nabla^\bullet(P), H_\nabla^s(P)$ | 4.2.3, 240 |
| $\mathrm{gl}(P,\psi)$ | 1.7.4, 134 | $H_\nabla^\bullet\left(\overline{M}\right), H_\nabla^s\left(\overline{M}\right)$ | 4.2.3, 240 |
| | | $H_\nabla^\bullet\left(\overline{M},\overline{N}\right)$ | 4.2.3, 240 |
| $\nabla_X$ | 2.1.1, 142 | $H_\nabla^s\left(\overline{M},\overline{N}\right)$ | 4.2.3, 240 |
| | 3.4.3, 203 | $H^\bullet(M)$ | 4.2.3, 240 |
| $\nabla_\xi$ | 2.1.10, 147 | $H^\bullet(A)$ | 4.2.3, 240 |
| $\nabla_{\overline{f}}$ | 3.4.6, 205 | | |
| | | $\displaystyle\rlap{$\int$}\int_a^b \omega$ | 4.2.8, 244 |
| $R^\nabla$ | 2.3.1, 157 | $\boxed{\nabla\!\!\int}_a^b \overline{\omega}$ | 4.2.11, 246 |
| | | $\int_a^b \overline{\omega_t}\,\mathrm{d}t$ | 4.4.3, 255 |

# Index

$A$-Lie algebra: 1.7.9, 138
adjoint equivalence: 0.1.6, 7
algebra: 0.1.1, 1
   cd-: 3.3.1, 188
   complete: 0.2.6, 11
   geometric: 0.2.2, 8
   graded: 0.1.1, 2
   graded commutative: 0.1.1, 2
   Lie a. over a commutative algebra: 1.7.9, 138
   of smooth functions: 0.2.9, 12
   restriction: 0.2.6, 10
   smooth: 0.2.7, 11
      with boundary: 0.2.7, 11
associated atlas: 0.2.14, 14
associated 1-form: 4.1.2, 234
associated homomorphism
   with a regular morphism of vector bundles: 0.3.13, 32
   with a smooth map: 0.2.11, 13
associated linear connection
   along a fat map: 3.4.6, 205
   in $\operatorname{End}(P)$: 2.4.3, 160
   in $\operatorname{Hom}(P,Q)$: 2.4.1, 159
   in $P \otimes Q$: 2.4.5, 161
associated map, with an isomorphism of modules of sections: 0.3.23, 39
associated vector field
   with a fat field: 1.4.5, 112
   with a one-parameter family: 0.6.11, 81
atlas, associated: 0.2.14, 14

base
   of a fat field: 1.3.1, 105
   of a fat point: 1.1.1, 92
   of a morphism in $\mathfrak{VB}_g$: 0.3.19, 36
   of a morphism of vector bundles: 0.3.8, 29
   of a vector bundle: 0.3.4, 25
basic differential
   covariant, see covariant differential
   fat: 3.1.8, 177
   semi-fat: 3.1.8, 177
$\beta$-characteristic class: 4.5.28, 276
   of a manifold: 4.5.29, 278
Bianchi identity: 3.2.14, 187

$C^\infty$-closed algebra: 0.2.8, 11
canonical morphism: 0.3.6, 28
Cartan formula: 0.6.20, 88
   covariant: 3.5.15, 219
cd-algebra: 3.3.1, 188
cd-module: 3.3.2, 188
   flat: 3.3.2, 188
characteristic class: 4.5.23, 272
   $\beta$-c. c.: 4.5.28, 276
   of a manifold: 4.5.29, 278
Christoffel symbols: 2.1.4, 144
closed submanifold: 0.2.20, 18
   fat: 1.1.8, 95
cochain complex: 0.1.1, 3
   der-: 3.1.8, 176
   semi-fat: 3.1.8, 177
cochain homomorphism: 0.1.1, 3

cohomological definite integral: 4.2.8, 244
   c.d. $\nabla$-i.: 4.2.11, 246
cohomology
   de Rham c. of a manifold: 4.2.3, 240
   de Rham c. of an algebra: 4.2.3, 240
   of a linear connection: 4.2.3, 240
      $s$-th: 4.2.3, 240
   relative, see relative cohomology
commutator: 0.1.1, 3
   graded: 0.1.1, 3
compatible fat fields: 1.3.8, 107
compatible linear connections: 3.3.12, 193
compatible vector fields: 0.4.3, 42
complete algebra: 0.2.6, 11
complex, cochain: 0.1.1, 3
   der-: 3.1.8, 176
   semi-fat: 3.1.8, 177
complex structure
   in a module: 1.7.2, 132
   inner: 1.7.2, 132
connection, see linear connection
constant function wrt a connection: 4.5.17, 268
constant rank: 0.1.1, 2
constant section wrt a connection: 2.2.7, 152
corresponding form: 0.6.17, 83
corresponding map of a smooth family: 0.6.8, 78
corresponding vector field of a smooth family: 0.6.10, 80
cosymbol map: 3.2.4, 180
cotangent bundle: 0.5.2, 52
counit of an adjunction: 0.1.6, 7
covariant derivative: 2.1.1, 142
   Lie: 3.5.15, 219
covariant differential: 3.2.1, 178; 3.2.6, 183
   along a fat map: 3.4.5, 204
   basic: 3.2.1, 178; 3.2.6, 183
curvature tensor: 2.3.1, 157
curve: 0.4.21, 50

fat, see fat curve
integral, see integral curve
cylinder, fat: 1.6.13, 128
de Rham
   cochain homomorphism induced by a smooth map: 0.5.16, 61
   cohomology of a manifold: 4.2.3, 240
   cohomology of an algebra: 4.2.3, 240
definite integral, cohomological: 4.2.8, 244
   c.d. $\nabla$-i.: 4.2.11, 246
degree, of a graded homomorphism: 0.1.1, 2
der-complex: 3.1.8, 176
der-operator: 1.2.3, 100
   along a homomorphism: 1.2.11, 104
   dual: 1.4.1, 110
   graded: 3.6.9, 231
   in a diole: 3.2.5, 182
   induced on a module of homomorphisms: 1.5.1, 114
   induced on a tensor product: 1.5.2, 115
   over a derivation: 1.2.5, 101
derivation
   along an algebra homomorphism: 0.1.3, 4
   graded: 0.1.3, 4
   of a $k$-algebra into a module: 0.1.3, 4
derivative
   covariant, see covariant derivative
   Lie, see Lie derivative
   of a smooth curve in a vector space: 0.5.9, 56
   of a time-dependent differential form: 0.6.18, 85
   of a time-dependent thickened form: 3.5.11, 216
diffeomorphism: 0.2.10, 12
differential
   basic, see basic differential
   covariant, see covariant differential

# Index

exterior, *see* exterior differential
fat, *see* fat differential
of a complex: 0.1.1, 3
of a smooth map: 0.2.28, 22
ordinary
    on a commutative algebra:
      0.5.5, 54
    on a smooth manifold: 0.5.5,
      54
semi-fat, *see* semi-fat differential
differential form
  along a map: 0.5.22, 64
  associated 1-form: 4.1.2, 234
  fat, *see* fat form
  local, *see* local differential form
  on a manifold: 0.5.1, 52
  on an algebra: 0.5.1, 52
  ordinary: 0.5.1, 52
  $P$-valued: 0.5.1, 52
  semi-fat, *see* semi-fat form
  thickened, *see* thickened form
  time-dependent: 0.6.17, 84
  with values in a module: 0.5.1, 52
differential operator
  graded: 3.6.9, 230
  linear: 0.1.2, 3
dimension of a smooth algebra: 0.2.7, 11
diole: 3.2.5, 182
direct sum: 0.1.1, 2
dual der-operator: 1.4.1, 110
dual linear connection: 2.4.4, 160
dual map: 0.2.4, 9
dual space: 0.2.1, 8

embedding: 0.2.28, 22
  fat, *see* fat embedding
  into a product: 0.2.26, 21
  into a relative interval: 0.6.3, 75
  of a submanifold: 0.2.20, 17
envelope
  homomorphism: 0.2.22, 19
  smooth e. algebra: 0.2.22, 19
equidimensional pseudobundle: 0.3.1, 24
equivalence of categories: 0.1.6, 7

extension of scalars: 0.1.1, 2
exterior differential
  fat: 3.1.8, 176
  on a commutative algebra: 0.5.11, 57
  on a smooth manifold: 0.5.11, 57
  semi-fat: 3.1.8, 177
exterior product of differential forms: 0.5.13, 59

$\overline{f}$-compatible fat fields: 1.3.8, 107
$\overline{f}$-compatible linear connections: 3.3.12, 193
$\overline{f}$-compatible vector fields: 0.4.3, 42
$\overline{f}$-related linear connections: 2.2.1, 147
faithful functor: 0.1.6, 7
fat curve: 1.1.9, 95
  integral: 1.6.8, 124
fat cylinder: 1.6.13, 128
fat diffeomorphism: 1.1.4, 93
fat differential
  basic: 3.1.8, 177
  exterior: 3.1.8, 176
  of a fat map: 1.3.7, 107
fat differential form, *see* fat form
fat embedding
  into a fat cylinder: 1.6.13, 128
  into a relative fat interval: 1.6.14, 128
  of a fat submanifold: 1.1.8, 95
fat field: 1.3.1, 105
  along a fat map: 1.3.13, 110
  induced by a fat map: 1.5.3, 116
  standard: 1.6.1, 118
fat flow: 1.6.15, 130
fat form
  on a fat manifold: 3.1.1, 172
  on a module: 3.1.1, 171
fat homotopy: 4.4.1, 253
fat identity map: 1.1.7, 95
fat inclusion: 1.1.11, 96
fat interval: 1.1.9, 95
  relative: 1.6.14, 128
fat manifold: 1.1.1, 92
fat orientable: 4.3.2, 248

fat map: 1.1.4, 93
  homotopic: 4.4.1, 253
    $\nabla$-homotopic: 4.4.2, 254
fat orientable fat manifold: 4.3.2, 248
fat point: 1.1.1, 92
fat projection
  of a fat cylinder: 1.6.13, 128
  of a relative fat interval: 1.6.14, 128
fat restriction: 1.1.11, 96
fat $s$-form, see fat form
fat submanifold: 1.1.8, 95
  closed: 1.1.8, 95
  open: 1.1.8, 95
fat tangent space: 1.3.4, 105
fat tangent vector: 1.3.2, 105
fat trajectory: 1.6.8, 124
  maximal: 1.6.8, 124
fat translation: 1.1.12, 97
fat vector field: 1.3.1, 105
  along a fat map: 1.3.13, 110
  induced by a fat map: 1.5.3, 116
  standard: 1.6.1, 118
fiber
  general: 0.3.4, 25
  of a uniform morphism: 0.3.17, 35
  of a vector bundle: 0.3.4, 25
  standard: 0.3.15, 33
field, see vector field
flat cd-module: 3.3.2, 188
flat linear connection: 2.3.3, 158
flow: 0.6.1, 73
  fat: 1.6.15, 130
form
  along a map: 0.5.22, 64
  associated 1-form: 4.1.2, 234
  fat, see fat form
  inner, bilinear: 1.7.3, 132
  local: 0.5.33, 71
  on a manifold: 0.5.1, 52
  on an algebra: 0.5.1, 52
  ordinary: 0.5.1, 52
  $P$-valued: 0.5.1, 52
  semi-fat, see semi-fat form
  thickened, see thickened form
  time-dependent: 0.6.17, 84
  with values in a module: 0.5.1, 52

full functor: 0.1.6, 7

gauge equivalent linear connections:
  4.1.1, 234
$g$-g.e.l.c.: 4.3.5, 251
gauge inner structure: 3.6.5, 228
gauge transformation: 1.1.7, 95
gauge trivial connection: 4.1.6, 236
$g$-g.t.c.: 4.3.5, 251
gauge type: 3.6.7, 229
general fiber: 0.3.4, 25
geometric algebra: 0.2.2, 8
geometric module: 0.3.2, 24
geometrization homomorphism: 0.3.2,
  24
geometrization module: 0.3.2, 24
gluing
  of fat fields: 1.3.11, 109
  of vector fields: 0.4.7, 44
graded algebra: 0.1.1, 2
graded commutative algebra: 0.1.1, 2
graded commutator: 0.1.1, 3
graded der-operator: 3.6.9, 231
graded derivation: 0.1.3, 4
graded differential operator: 3.6.9,
  230
graded homomorphism
  induced by a fat smooth map:
    3.3.8, 191
  of graded algebras: 0.1.1, 2
  of graded modules: 0.1.1, 2
    over a graded algebra: 0.1.1, 2
graded Jacobi identity: 3.6.9, 231
graded Leibnitz rule: 0.1.3, 4
graded Lie algebra structure: 3.6.9,
  231
graded module: 0.1.1, 2

holonomy group: 3.3.18, 198
homomorphism
  cochain: 0.1.1, 3
  of cd-modules: 3.3.3, 188
homotopic fat maps: 4.4.1, 253
  $\nabla$-h. f. m.: 4.4.2, 254
homotopy
  fat: 4.4.1, 253

∇-h.: 4.4.2, 254
homotopy operator: 4.4.4, 256

identification isomorphism: 0.3.16, 34
image of a vector field through a
    diffeomorphism: 0.4.2, 41
immersion: 0.2.28, 22
inclusion, fat: 1.1.11, 96
induced bundle: 0.3.6, 28
induced der-operator
    on a module of homomorphisms:
        1.5.1, 114
    on a tensor product: 1.5.2, 115
induced fat field: 1.5.3, 116
induced graded homomorphism:
    3.3.8, 191
induced homomorphism
    de Rham: 0.5.16, 61
    graded: 3.3.8, 191
induced linear connection
    by a fat map: 3.3.16, 197
        provisional definition: 2.2.4,
            149
    on the cross: 2.2.10, 155
induced map
    into a product: 0.2.25, 21
    of an induced bundle: 0.3.6, 28
induced vector bundle: 0.3.6, 28
infinitesimal symmetry
    of a bilinear form: 1.7.4, 134
    of an endomorphism: 1.7.4, 134
    of an inner form: 1.7.4, 134
inner complex structure: 1.7.2, 132
inner metric: 1.7.3, 133
inner pseudo-metric: 1.7.3, 133
inner structure: 3.6.3, 227
    complex: 1.7.2, 132
    gauge: 3.6.5, 228
insertion operator
    along fat maps: 3.4.2, 202
    along smooth maps: 0.5.28, 67
    of a derivation into a module of
        $P$-valued forms: 0.5.18, 62
    of a vector field: 0.5.18, 62
integral
    cohomological definite: 4.2.8, 244

c.d. ∇-i.: 4.2.11, 246
    of a time-dependent form: 4.4.3,
        255
integral curve: 0.4.22, 51
    fat: 1.6.8, 124
interval: 0.4.21, 50
    fat, see fat interval
    relative, see relative interval
invariant linear function
    with respect to a fat field: 4.5.9,
        263
    with respect to a linear connection:
        4.5.9, 263
invariant polynomial: 4.5.10, 264

Jacobi identity: 1.7.9, 138
    graded: 3.6.9, 231

$k$-algebra: 0.1.1, 1
$k$-point: 0.2.1, 8

Leibnitz rule: 0.1.3, 4
    graded: 0.1.3, 4
Levi-Civita connection: 2.5.4, 169
Lie algebra over a commutative
    algebra: 1.7.9, 138
Lie connection: 2.5.3, 168
Lie derivative
    along a fat field: 3.5.12, 217
    along a vector field: 0.6.19, 87
    covariant: 3.5.15, 219
lift
    ∇-l., see ∇-lift
    of a curve: 1.1.9, 96
    of a curve, by a linear connection:
        3.3.17, 197
        provisional definition: 2.2.5,
            150
    of a one-parameter family: 3.5.2,
        210
    of a tangent vector
        by a linear connection: 2.1.10,
            147
        by a linear connection along a
            fat map: 3.4.4, 203
linear connection: 2.1.1, 142

along a fat map: 3.4.3, 202
along a homomorphism: 3.4.3, 202
associated l.c. along a fat map:
   3.4.6, 205
associated l.c. in End $(P)$: 2.4.3,
   160
associated l.c. in Hom $(P,Q)$: 2.4.1,
   159
associated l.c. in $P \otimes Q$: 2.4.5, 161
compatible: 3.3.12, 193
dual: 2.4.4, 160
flat: 2.3.3, 158
gauge equivalent, see gauge
   equivalent linear
   connections
gauge trivial, see gauge trivial
   connection
in a module: 2.1.1, 142
induced by a fat map: 3.3.16, 197
   provisional definition: 2.2.4,
      149
induced on the cross: 2.2.10, 155
Levi-Civita: 2.5.4, 169
Lie: 2.5.3, 168
localization: 2.4.10, 164
on a fat manifold: 2.1.1, 142
on the cross: 2.2.10, 155
preserving a bilinear form: 2.5.2,
   166
preserving an endomorphism:
   2.5.1, 165
preserving an inner structure:
   3.6.4, 227
related: 2.2.1, 147
torsion-free: 2.5.4, 169
linear differential operator: 0.1.2, 3
linear function on a fat manifold:
   4.5.9, 262
   invariant
      with respect to a fat field:
         4.5.9, 263
      with respect to a linear
         connection: 4.5.9, 263
local differential form: 0.5.33, 71
   thickened: 3.5.9, 215
local fat map: 3.5.1, 209

local smooth map: 0.6.8, 79
local thickened form: 3.5.9, 215
local triviality, vector property of:
   0.3.18, 35
local vector field: 0.4.16, 48
localization of a linear connection:
   2.4.10, 164
loop: 3.3.18, 198

manifold: 0.2.9, 12
   fat, see fat manifold
   pseudo-Riemannian: 1.7.3, 133
   Riemannian: 1.7.3, 133
   with boundary: 0.2.9, 12
maximal trajectory: 0.4.22, 51
   fat: 1.6.8, 124
metric
   inner: 1.7.3, 133
   pseudo-Riemannian: 1.7.3, 133
   Riemannian: 1.7.3, 133
Minkowski spacetime: 4.3.1, 248
model structure: 3.6.7, 229
module: 0.1.1, 1
   cd-: 3.3.2, 188
      flat: 3.3.2, 188
   geometric: 0.3.2, 24
   geometrization: 0.3.2, 24
   graded: 0.1.1, 2
   of smooth sections: 0.3.4, 25
      in $\mathfrak{VB}_g$: 0.3.20, 37
morphism
   in $\mathfrak{VB}_g$: 0.3.19, 36
   of vector bundles
      over a map: 0.3.8, 29
      over the same base: 0.3.5, 26
   regular, see regular morphism
   uniform: 0.3.17, 35

$\nabla$-constant function: 4.5.17, 268
$\nabla$-homotopy: 4.4.2, 254
$\nabla$-integral, cohomological definite:
   4.2.11, 246
$\nabla$-lift
   of a curve: 3.3.17, 197
      provisional definition: 2.2.5,
         150

of a one-parameter family: 3.5.2, 210
of a tangent vector: 2.1.10, 147
  along a fat map: 3.4.4, 203
natural homomorphism into a smooth tensor product: 0.2.24, 20
natural topology on a dual space: 0.2.3, 9

one-parameter family
  of fat maps: 3.5.1, 209
  of smooth maps: 0.6.8, 79
one-parameter group
  generated by a fat field: 3.5.12, 216
  generated by a vector field: 0.6.6, 77
open submanifold: 0.2.20, 18
  fat: 1.1.8, 95
ordinary differential
  on a commutative algebra: 0.5.5, 54
  on a smooth manifold: 0.5.5, 54
ordinary differential form: 0.5.1, 52
orthogonal group
  of a bilinear form: 1.7.4, 133
  of an inner form: 1.7.4, 133

$P$-valued form: 0.5.1, 52
parallel translation: 3.3.17, 197
  provisional definition: 2.2.6, 151
point, see $k$-point
Pontrjagin class: 4.5.23, 272
positive bilinear form: 1.7.3, 132
preserving linear connection
  of a bilinear form: 2.5.2, 166
  of an endomorphism: 2.5.1, 165
  of an inner structure: 3.6.4, 227
product
  exterior, see exterior product
  of manifolds: 0.2.25, 20
  wedge, see wedge product
projectable vector field: 0.4.14, 47
projecting vector field: 0.4.14, 47
projection map
  fat, see fat projection
  of a product: 0.2.25, 21

of a vector bundle: 0.3.4, 25
pseudo-metric, inner: 1.7.3, 133
pseudo-Riemannian manifold: 1.7.3, 133
pseudo-Riemannian metric: 1.7.3, 133
pseudobundle: 0.3.1, 23
  equidimensional: 0.3.1, 24
pull-back vector bundle: 0.3.6, 28

$\mathbb{R}$-algebra: 0.1.1, 1
rank, constant: 0.1.1, 2
regular morphism: 0.3.8, 29
  in $\mathfrak{VB}_g$: 0.3.19, 36
regular section: 0.3.1, 23
related linear connections: 2.2.1, 147
relative cohomology: 4.2.3, 240
  $s$-th: 4.2.3, 240
relative form: 4.2.2, 238
relative interval: 0.6.3, 74
  fat: 1.6.14, 128
  open: 0.6.3, 74
relative thickened form: 4.2.2, 238
restriction
  algebra: 0.2.6, 10
  fat: 1.1.11, 96
  homomorphism: 0.2.6, 10
  of a fat field: 1.3.10, 108
  of a form to a closed submanifold: 4.2.1, 238
  of a linear connection
    to a closed fat submanifold: 2.2.4, 149
    to an open fat submanifold: 2.4.10, 164
  of a smooth section to a fat submanifold: 1.1.8, 95
  of a thickened form to a closed fat submanifold: 4.2.1, 238
  of a vector field to an open submanifold: 0.4.5, 43
  of vector fields along maps: 0.4.15, 47
Riemannian manifold: 1.7.3, 133
Riemannian metric: 1.7.3, 133
ring: 0.1.1, 1

$s$-form
  along a map: 0.5.22, 64
  associated 1-form: 4.1.2, 234
  fat, *see* fat form
  local: 0.5.33, 71
  on a manifold: 0.5.1, 52
  ordinary: 0.5.1, 52
  $P$-valued: 0.5.1, 52
  semi-fat, *see* semi-fat form
  thickened, *see* thickened form
  time-dependent: 0.6.17, 84
$s$-th cohomology of a linear
    connection: 4.2.3, 240
  relative: 4.2.3, 240
section
  constant wrt a connection: 2.2.7, 152
  of an object in $\mathfrak{VB}_g$: 0.3.20, 36
  regular s. of a pseudobundle: 0.3.1, 23
  smooth, *see* smooth section
semi-fat complex: 3.1.8, 177
semi-fat differential
  basic: 3.1.8, 177
  exterior: 3.1.8, 177
semi-fat form
  on a fat manifold: 3.1.2, 172
  on a module: 3.1.2, 172
smooth algebra: 0.2.7, 11
  with boundary: 0.2.7, 11
smooth envelope
  algebra: 0.2.22, 19
  homomorphism: 0.2.22, 19
smooth family
  of diffeomorphisms: 0.6.8, 79
  of differential forms: 0.6.17, 83
  of fat diffeomorphisms: 3.5.1, 209
  of fat maps: 3.5.1, 209
  of local differential forms: 0.6.17, 84
  of local fat maps: 3.5.1, 209
  of local smooth maps: 0.6.8, 79
  of local thickened forms: 3.5.10, 216
  of local vector fields: 0.6.10, 80
  of smooth maps: 0.6.8, 78
  of thickened forms: 3.5.10, 216
  of vector fields: 0.6.10, 80
smooth fat map: 1.1.4, 93
smooth fat vector field, *see* fat field
smooth function: 0.2.9, 12
  on the cross: 2.2.10, 154
  wrt an atlas: 0.2.13, 13
smooth manifold: 0.2.9, 12
  with boundary: 0.2.9, 12
smooth map: 0.2.10, 12
  fat: 1.1.4, 93
  wrt atlases: 0.2.13, 13
smooth section
  module of: 0.3.4, 25
    in $\mathfrak{VB}_g$: 0.3.20, 36
  of a vector bundle: 0.3.4, 25
    in $\mathfrak{VB}_g$: 0.3.20, 36
smooth set: 0.2.20, 18
smooth tensor product: 0.2.24, 20
standard fat field
  on a relative fat interval: 3.5.4, 211
  on a standard trivial fat manifold
      over an interval: 1.6.1, 118
standard fiber: 0.3.15, 33
standard trivial bundle: 0.3.15, 33
standard vector field
  fat, *see* standard fat field
  on a relative interval: 0.6.3, 75
  on an interval: 0.4.21, 50
structure, inner: 3.6.3, 227
submanifold: 0.2.20, 17
  closed, *see* closed submanifold
  fat, *see* fat submanifold
  open, *see* open submanifold
  with boundary: 0.2.20, 17
submersion: 0.2.28, 22
symbol
  Christoffel: 2.1.4, 144
  of a differential operator: 1.2.2, 99
symmetry
  infinitesimal, *see* infinitesimal symmetry
  of a bilinear form: 1.7.4, 133
  of an endomorphism: 1.7.4, 134
  of an inner form: 1.7.4, 133
symmetry group: 1.7.4, 134

tangent bundle: 0.4.9, 44
tangent space: 0.2.27, 22
  fat: 1.3.4, 105
tangent vector: 0.2.27, 21
  fat: 1.3.2, 105
tensor, curvature: 2.3.1, 157
tensor product: 0.1.1, 2
  smooth: 0.2.24, 20
tensor type: 3.6.2, 226
thickened form
  along a fat map: 3.4.1, 201
  local: 3.5.9, 215
  on a fat manifold: 3.1.4, 173
  on a module: 3.1.4, 173
  relative: 4.2.2, 238
  time-dependent: 3.5.10, 216
time-dependent differential form: 0.6.17, 84
  thickened: 3.5.10, 216
time-dependent vector field: 0.6.10, 80
  associated with a one-parameter family of diffeomorphisms: 0.6.11, 81
torsion-free linear connection: 2.5.4, 169
total space: 0.3.4, 25
trajectory: 0.4.22, 51
  fat, see fat trajectory
  maximal: 0.4.22, 51
translation
  fat: 1.1.12, 97
  parallel: 3.3.17, 197
    provisional definition: 2.2.6, 151
triangular identities: 0.1.6, 7
trivial bundle: 0.3.15, 33
  standard: 0.3.15, 33
trivial connection
  gauge t.c., see gauge trivial connection
  in a free module: 2.1.3, 143
  on a standard trivial fat manifold: 2.1.3, 143

triviality, local, vector property of: 0.3.18, 35
trivializing morphism: 0.3.15, 33
type
  of a fat manifold: 1.1.1, 92
  tensor t.: 3.6.2, 226

uniform morphism: 0.3.17, 35
unit of an adjunction: 0.1.6, 7

valued form, $P$-: 0.5.1, 52
vector bundle: 0.3.4, 25
  induced by a smooth map: 0.3.6, 28
  pull-back: 0.3.6, 28
  trivial: 0.3.15, 33
    standard: 0.3.15, 33
vector field: 0.4.1, 41
  along a map: 0.4.12, 45
  associated with a fat field: 1.4.5, 112
  associated with a one-parameter family: 0.6.11, 81
  fat, see fat field
  local: 0.4.16, 48
  on the cross: 2.2.10, 154
  standard, see standard vector field
  time-dependent: 0.6.10, 80
vector property of local triviality: 0.3.18, 35

wedge product
  for fat, semi-fat and thickened forms: 3.1.7, 175
  of an $\mathrm{End}(P)$–valued form and a $P$-valued form: 3.2.9, 184
  of differential forms along a map: 0.5.27, 66
  of $\mathrm{End}(P)$–valued forms: 3.2.9, 184
  of ordinary differential forms: 0.5.13, 59